Randomness and Recurrence in Dynamical Systems

A Real Analysis Approach

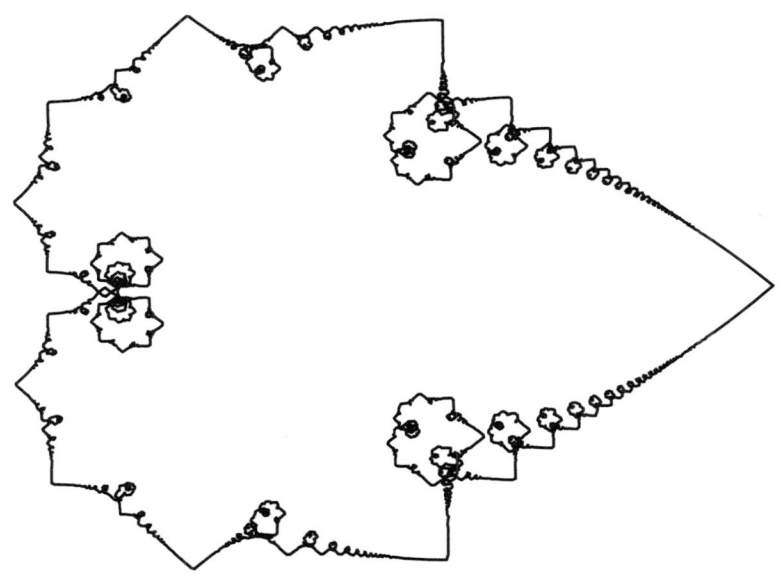

The illustration on the half title page is of the parametric curve $(x(t), y(t))$ in 2-dimensional space given by

$$x(t) = \sum_{n=1}^{500} \frac{\cos(n^2 \pi t)}{n^2}$$

and

$$y(t) = \sum_{n=1}^{500} \frac{\sin(n^2 \pi t)}{n^2}$$

© 2010 by the Mathematical Association of America, Inc.

Library of Congress Catalog Card Number 2010924993

ISBN 978-0-88385-043-5

Printed in the United States of America

Current Printing (last digit):
10 9 8 7 6 5 4 3 2 1

The Carus Mathematical Monographs

Number Thirty-One

Randomness and Recurrence in Dynamical Systems

A Real Analysis Approach

Rodney Nillsen
University of Wollongong, Australia

Published and Distributed by
THE MATHEMATICAL ASSOCIATION OF AMERICA

THE
CARUS MATHEMATICAL MONOGRAPHS

Published by

THE MATHEMATICAL ASSOCIATION OF AMERICA

Council on Publications and Communications
Frank Farris, *Chair*

Carus Mathematical Monographs Editorial Board

Robert L. Devaney, *Editor*

Lida K. Barrett
Donna L. Beers
Michael C. Berg
David M. Bressoud
Satyan L. Devadoss
Andrew Granville
Jerrold W. Grossman

The Carus Mathematical Monographs are an expression of the desire of Mrs. Mary Hegeler Carus and of her son, Dr. Edward H. Carus, to contribute to the dissemination of mathematical knowledge by making accessible a series of expository presentations of the best thoughts and keenest research in pure and applied mathematics. The publication of the first four of these monographs was made possible by Mrs. Carus as sole trustee of the Edward C. Hegeler Trust Fund. The sales from these have resulted in the Carus Monograph Fund, and the Mathematical Association of America has used this as a revolving book fund to publish the succeeding monographs.

The expositions of mathematical subjects that the monographs contain are set forth in a manner comprehensible not only to teachers and students specializing in mathematics, but also to scientific workers in other fields. More generally, the monographs are intended for the wide circle of thoughtful people familiar with basic graduate or advanced undergraduate mathematics encountered in the study of mathematics itself or in the context of related disciplines who wish to extend their knowledge without prolonged and critical study of the mathematical journals and treatises.

The following Monographs have been published:

1. *Calculus of Variations,* by G. A. Bliss (out of print)
2. *Analytic Functions of a Complex Variable,* by D. R. Curtiss (out of print)
3. *Mathematical Statistics,* by H. L. Rietz (out of print)
4. *Projective Geometry,* by J. W. Young (out of print)
5. *A History of Mathematics in America before 1900,* by D. E. Smith and Jekuthiel Ginsburg (out of print)
6. *Fourier Series and Orthogonal Polynomials,* by Dunham Jackson (out of print)
7. *Vectors and Matrices,* by C. C. MacDuffee (out of print)
8. *Rings and Ideals,* by N. H. McCoy (out of print)
9. *The Theory of Algebraic Numbers,* second edition, by Harry Pollard and Harold G. Diamond
10. *The Arithmetic Theory of Quadratic Forms,* by B. W. Jones (out of print)
11. *Irrational Numbers,* by Ivan Niven
12. *Statistical Independence in Probability, Analysis and Number Theory,* by Mark Kac
13. *A Primer of Real Functions,* third edition, by Ralph P. Boas, Jr.

14. *Combinatorial Mathematics,* by Herbert J. Ryser
15. *Noncommutative Rings,* by I. N. Herstein (out of print)
16. *Dedekind Sums,* by Hans Rademacher and Emil Grosswald
17. *The Schwarz Function and its Applications,* by Philip J. Davis
18. *Celestial Mechanics,* by Harry Pollard
19. *Field Theory and its Classical Problems,* by Charles Robert Hadlock
20. *The Generalized Riemann Integral,* by Robert M. McLeod
21. *From Error-Correcting Codes through Sphere Packings to Simple Groups,* by Thomas M. Thompson
22. *Random Walks and Electric Networks,* by Peter G. Doyle and J. Laurie Snell
23. *Complex Analysis: The Geometric Viewpoint,* by Steven G. Krantz
24. *Knot Theory,* by Charles Livingston
25. *Algebra and Tiling: Homomorphisms in the Service of Geometry,* by Sherman Stein and Sándor Szabó
26. *The Sensual (Quadratic) Form,* by John H. Conway assisted by Francis Y. C. Fung
27. *A Panorama of Harmonic Analysis,* by Steven G. Krantz
28. *Inequalities from Complex Analysis,* John P. D'Angelo
29. *Ergodic Theory of Numbers,* Karma Dajani and Cor Kraaikamp
30. *A Tour through Mathematical Logic,* Robert S. Wolf
31. *Randomness and Recurrence in Dynamical Systems: A Real Analysis Approach,* Rodney Nillsen

MAA Service Center
P.O. Box 91112
Washington, DC 20090-1112
1-800-331-1MAA   FAX: 1-301-206-9789

SYLVIA, ANN
and
CHRISTOPHER

Contents

Foreword   xiii

Preface   xv

1 Background Ideas and Knowledge   1
    1.1 Dynamical systems, iteration and orbits 1
    1.2 Information loss and randomness in dynamical systems . 4
    1.3 Assumed knowledge and notation 10
    Appendix: Mathematical reasoning and proof 12
    Exercises 17
    Investigations 19
    Notes 19
    Bibliography 22

2 Irrational Numbers and Dynamical Systems   25
    2.1 Introduction: irrational numbers and the infinite 25
    2.2 Fractional parts and points on the unit circle 28
    2.3 Partitions and the Pigeon-hole Principle 33
    2.4 Kronecker's Theorem 34
    2.5 The dynamical systems approach to Kronecker's Theorem 41
    2.6 Kronecker and chaos in the music of Steve Reich 46
    2.7 The ideas in Weyl's Theorem on irrational numbers . . . 52
    2.8 The proof of Weyl's Theorem 57
    2.9 Chaos in Kronecker systems 70
    Exercises 75

	Investigations	80
	Notes	83
	Bibliography	85

3 Probability and Randomness — 87

- 3.1 Introduction: probability, coin tossing and randomness . 87
- 3.2 Expansions to a base 92
- 3.3 Rational numbers and periodic expansions 101
- 3.4 Sets, events, length and probability 106
- 3.5 Sets of measure zero 117
- 3.6 Independent sets and events 122
- 3.7 Typewriters, recurrence, and the Prince of Denmark . . . 125
- 3.8 The Rademacher functions 136
- 3.9 Randomness, binary expansions and a law of averages . 142
- 3.10 The dynamical systems approach 148
- 3.11 The Walsh functions 158
- 3.12 Normal numbers and randomness 163
- 3.13 Notions of probability and randomness 173
- 3.14 The curious phenomenon of the leading significant digit . 180
- 3.15 Leading digits and geometric sequences 186
- 3.16 Multiple digits and a result of Diaconis 193
- 3.17 Dynamical systems and changes of scale 198
- 3.18 The equivalence of Kronecker and Benford systems . . . 205
- 3.19 Scale invariance and the necessity of Benford's law . . . 208
- Exercises 220
- Investigations 225
- Notes 227
- Bibliography 230

4 Recurrence — 235

- 4.1 Introduction: random systems and recurrence 235
- 4.2 Transformations that preserve length 239
- 4.3 Poincaré recurrence 248
- 4.4 Recurrent points 253
- 4.5 Kac's result on average recurrence times 256

CONTENTS

4.6	Applications to the Kronecker and Borel systems	266
4.7	The standard deviation of recurrence times	270
	Exercises	283
	Investigations	284
	Notes	286
	Bibliography	287

5 Averaging in Time and Space — 289

5.1	Introduction: averaging in time and space	289
5.2	Outer measure	293
5.3	Invariant sets	300
5.4	Measurable sets	303
5.5	Measure-preserving transformations	307
5.6	Poincaré recurrence ... again	310
5.7	Ergodic systems	312
5.8	Birkhoff's Theorem on time and space averages	318
5.9	Weyl's Theorem from the ergodic viewpoint	320
5.10	The Ergodic Theorem and expansions to an arbitrary base	323
5.11	Kac's recurrence formula: the general case	327
5.12	Mixing transformations and an example of Kakutani	328
5.13	Lüroth transformations and continued fractions	333
	Exercises	342
	Investigations	345
	Notes	346
	Bibliography	347

Index of Subjects — 351

Index of Symbols — 355

About the Author — 357

Foreword

Mathematics is an ancient discipline, and it persists, and will continue to persist, in a changing world. But as a discipline it is also facing many challenges. At one time, mathematics was a natural choice of university and college study for many serious students. Now there is a vast range of studies and disciplines whose more immediate relationship to society, and the perceived rewards that accompany that immediacy, make mathematics a less natural choice. At the same time, mathematical techniques are permeating more and more disciplines and becoming more essential to those disciplines. This often occurs at the expense of a deeper understanding of mathematics, and at the expense of the precision, clarity of thought, and recognition of complexity that mathematics provides.

I had the pleasure of reading an early draft of this fine book. It reflects the author's desire to make mathematics accessible to a wide audience, without sacrificing the beauty and precision that characterizes the field—an excellent addition to the Carus Monograph Series. At its most exciting, mathematics reveals an order, and surprising connections and relationships, often where one may be least expecting them. And yet, it expresses that order, and the unexpected connections, in a very precise and unambiguous way, supported by arguments that carry conviction and are, with effort, universally recognizable. That is possibly what Max Dehn meant when he said that mathematics was the only subject that could be presented in an entirely undogmatic way.

Randomness and dynamical systems are the underlying and unifying themes. The main questions are easily stated: if we consider the orbit of a point in a dynamical system, how do we express notions of randomness and order for the orbit, and what phenomena of randomness and order occur? This book considers these questions in relation to various dynamical systems. They range from one associated with the famous 19th century result of Kronecker on the density of the sequence of fractional parts of the multiples of an irrational number, to the recent characterization of the logarithmic distribution as the only single probability distribution that sequences of data can satisfy at all possible scales. Along the way, Weyl's theorem on the uniform distribution of the sequence of fractional parts of the multiples of an irrational number is set in the context of the uniform distribution of the points in an orbit of an associated dynamical system. This viewpoint is taken elsewhere, including Kronecker's Theorem and Borel's theorem on normal numbers.

A notable feature of this book is that important and interesting results of mathematical analysis are made accessible at the undergraduate level, whereas traditionally this material has belonged to the graduate level and the world of research. The elimination of traditional prerequisites helps bring the world of undergraduate analysis closer to the world of current mathematical endeavor and research. Accordingly, it should be of interest to both students and researchers, and there are references where possible to the current research literature. Students and researchers from other disciplines will find this an accessible introduction to an important and interesting area of mathematics whose ideas are of increasing relevance in our complex world.

This is made possible by not assuming a knowledge of measure theory and integration, or of functional analysis. What is needed is gently introduced as needed. In fact, the only integration theory that is used in most of the book is integration of step functions on intervals. Use is also made of finitely additive set functions, or measures, but the countably additivity properties of measures are avoided. So, the book presents a certain area of mathematical analysis from the point of view of finitely additive measures on intervals, and shows that a good deal more can be done in this limited set-up than might be expected. A final part of the work presents Birkhoff's individual ergodic theorem and applications to further dynamical systems, at an expository level with measure theory.

Kenneth A. Ross
Eugene, Oregon
March 15, 2010

Preface

This work had its origins some years ago, when I found myself assigned to teach a new subject, to be called *Topology and Chaos*. At the time I knew quite a lot of topology, but very little of the area known as chaos theory. Conditions in Australian universities were changing, with the "commercialization" of learning, scholarship and research, and changing attitudes in society were influencing universities, making the teaching environment difficult in an area like real analysis, especially as traditionally taught at the undergraduate level. This project has been a result of these changing circumstances.

Over a long period I have received help from many people, and I cannot mention everyone who has had some beneficial effect upon the ideas in this work. But I would like to mention some of those who made a very specific contribution. Peter Siminski prepared drafts of some early illustrations and helped me with advice on some early parts of the work. Martin Bunder read large parts of an early draft of the work and made helpful comments. Graham Williams also read a draft of the manuscript and made many suggestions, as well as pointing out some better proofs. I am very grateful for the time and effort they spent in doing this, as well as providing me with discussions and responses on the general issues involved. Keith Tognetti pointed out some improvements, especially on the ideas arising in the discussion on Benford's Law. Pamela Davy was able to help me with suggestions concerning Birkhoff's Theorem. I have also profited from conversations with Philip Broadbridge, Michael Cwikel, Mark Nelson, Peter Nickolas, Susumu Okada, Geoff Wood,

Annette Worthy and others. I have had a great deal of feedback and response from mathematics students at the University of Wollongong who have been exposed to some parts of this work. However, I am most of all indebted to Kenneth A. Ross, who read in great detail later drafts of the work, making many comments and suggestions. His interest, thoroughness and insights have influenced many details of the work, as well as its final overall structure and presentation. I am also indebted to him for suggesting that the work could be suitable for publication by the *Mathematical Association of America* (MAA). I appreciate the consideration of the manuscript by the Carus Monographs Editorial Board of the MAA, by Barbara L. Osofsky and, later, by David M. Bressoud, members of the Editorial Board. I also received detailed and helpful comments from mathematicians who provided input on the manuscript to the Editorial Board. I express my gratitude for all this interest, time and help from so many. I also wish to thank Rebecca Elmo, Elaine Pedreira and Beverly Ruedi in the MAA office, who have been of such very great help in the preparation of the manuscript for publication.

In a book review of Bourbaki's *Algèbre* in 1953, Emil Artin wrote:

> We all believe that mathematics is an art. The author in a book, the lecturer in a classroom, tries to convey the structural beauty of mathematics to his readers, to his listeners. In this attempt, he must always fail. Mathematics is logical to be sure; each conclusion is drawn from previously derived statements. Yet the whole of it, the real piece of art, is not linear; worse than that its perception should be instantaneous. We have all experienced on some rare occasions the feeling of elation in realizing that we have enabled our listeners to see at a moment's glance the whole architecture and its ramifications.

This work will be no exception to Artin's comment that every attempt must have an accompanying failure. But despite all the help that I have received, the responsibility for failures, errors and infelicities remains my own.

The writing of this work has been spread out over a long period, but it has had an impact on my family and I wish to acknowledge their

Preface

patience and tolerance of the many hours spent at home working on and thinking about the ideas presented here. I also wish to acknowledge the long-lasting, loving and unvarying encouragement and support I have received from my father Victor Nillsen and from my late mother Audrey Nillsen.

The approach in this work is to take a number of specific topics, and to discuss them in a way that minimizes the need for prior knowledge. Most of these topics lie outside the traditional undergraduate curriculum, and it is an aim to make these topics more accessible. The approach also endeavors to connect the material with some recent research, with the aim of giving a more "contemporary" feel to the material and contributing in a small way towards "bridging the gap" between undergraduate teaching and the world of current mathematical ideas and research.

In some ways the work should be regarded as "lecture notes", in that questions of motivation of the ideas, and thinking about the results from different viewpoints, are considered perhaps more than is usual. There is a certain emphasis on how we might think "metaphorically" about the possible *meanings* of concepts and results, and not only about the *formal* aspects of the results. Occasionally, there is mention of connections to other areas of enquiry.

The various topics are unified by the concept of an abstract dynamical system, and so this work has close connections with many recent books on chaos theory—the main difference is that the emphasis here is on what might be termed "probabilistic chaos theory" and "randomness", rather than "deterministic" or "topological" chaos theory, which is the theme of most of these books.

The topics have been chosen to be compatible with the pedagogical and research objectives above, and it might be said that they have a strong feeling of probability theory rather than real analysis. While this may be true, the spirit of the treatment is that of classical real analysis. However, the work should not be regarded as a standard book on real analysis. For example, there is no discussion of the integration of continuous functions. But this should not be too surprising because, in the theory of probability and measure, the continuous functions play no special role. Also, the work can be regarded as showing the extent to which some topics generally associated with probability and statistics can be treated entirely within an elementary real analysis framework, without any need of measure theory.

The work has been organized to make the various topics as independent of each other as possible. The flow diagrams at the beginning of each chapter indicate the logical relationships between the various chapters and sections.

Rodney Nillsen
July 2010

Chapter 1

Background Ideas and Knowledge

> When we have pushed the analysis to the end ... and finally have come to considering a few natures understood only by themselves without prerequisites and needing nothing outside themselves to be conceived, then we have arrived at *perfect knowledge* of the proposed thing. When this thing merits it, we must try and have this perfect knowledge present in our minds all at once, and that is done by repeating the analysis several times until it seems to us that we see it as a complete whole in a single act of the mind.
>
> *Gottfried Wilhelm von Leibniz* (1646–1716)

1.1 Dynamical systems, iteration and orbits

Let A and B be sets and let f be a function with domain A and codomain B. That is, for each $x \in A$ there corresponds $f(x) \in B$. We may indicate this situation symbolically by writing $f : A \longrightarrow B$. In the case when the domain and codomain are equal to a set S say, and $f : S \longrightarrow S$, we say that f is a *transformation on* S. In such a case, we call the pair (S, f) a *dynamical system*. Thus, a dynamical system consists of a set together with a transformation on the set. Given a dynamical system (S, f), S may be called the *phase space* and the elements of S may be called *states*. If $f : A \longrightarrow B$ and $g : B \longrightarrow C$, the *composition of* g with f is the function $g \circ f : A \longrightarrow C$ given by $(g \circ f)(x) = g(f(x))$ for all $x \in A$.

Suppose we have a dynamical system (S, f). The process of taking successive compositions of f with itself is called *iteration*. Each of these compositions is a transformation on S. The *first iterate* of f is f itself, the *second iterate* of f is $f \circ f$, and so on. In general, the nth *iterate of* f is denoted by f^n and is given by

$$f^n = f \circ f \circ \cdots \circ f,$$

where f appears n times and the composition symbol $\circ$ appears $n - 1$ times. It is convenient to define the 0th iterate of f to be f^0, where $f^0(x) = x$, for all $x \in S$. That is, f^0 is the identity function on S. It is straightforward to verify that $f^{m+n} = f^m \circ f^n$ for all $m, n = 0, 1, 2, 3, \ldots$.

If we think of each application of f as a description of how the system is going to change in one time unit, we see that iteration describes the evolution of the system over time. Specifically, let $x \in S$, which we may call the *initial state* of the system. Then the sequence of points

$$x, f(x), f^2(x), f^3(x), \ldots$$

is called the *orbit* of x. If x is the initial state of the system, then $f(x)$ may be thought of as the state of the system after one time unit, $f^2(x)$ as the state of the system after 2 time units, and so on. Thus, the orbit of x describes the behavior of the system over time, given that it started off in the state x. A knowledge of the behaviors of the different orbits of the system would give us a good understanding of the overall behavior of the system over time, giving insight into such questions as stability, unchanging behavior, repetitive behavior, expansion, contraction and uncertainty of behavior. In the case of a transformation on an interval, the orbit of a point may be interpreted graphically, see Figure 1.1.

We conclude this section with some further definitions and terminology concerning functions. If a function f has domain A and codomain B, this may be indicated by writing $f : A \longrightarrow B$, and we may say that f *maps* A *into* B. If $f : A \longrightarrow B$, the subset $\{f(x) : x \in A\}$ of B is called the *range* of f, and f is called *onto* if its range equals B, in which case we may say that f maps A *onto* B. If $f : A \longrightarrow B$,

1.1. Dynamical systems, iteration and orbits

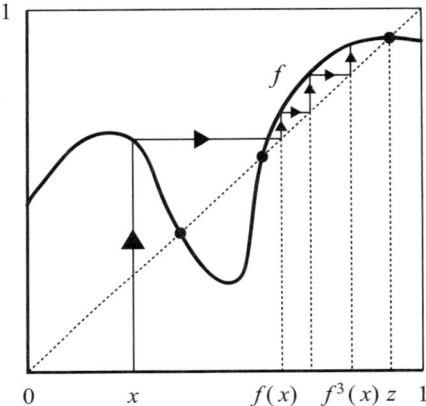

Figure 1.1. The graph indicates a transformation f on $[0, 1]$. The points where the graph of f intersects the graph of the line $y = x$ are the points x such that $f(x) = x$—these are called *fixed points* of the transformation. The arrows depict the successive points in the orbit x, $f(x)$, $f^2(x), \ldots$ of x. In terms of 2-dimensional coordinates, we start at the point $(x, 0)$, and then proceed consecutively to the points at the end of each vertical arrow. Thus, we obtain the sequence of points
$$(x, f(x)), (f(x), f^2(x)), \ldots, (f^{n-1}(x), f^n(x)), \ldots$$
in the Cartesian plane. The corresponding sequence of X-coordinates, namely
$$x, f(x), f^2(x), \ldots, f^{n-1}(x), \ldots$$
gives the points of the orbit of x. Note that the points $(x, f(x))$, $(f(x), f^2(x)), \ldots$ appear to be getting closer to the point (z, z), where $f(z) = z$, so that z is a fixed point of f.

f is called *one-to-one* if $f(x) = f(y) \implies x = y$, in which case there is a function f^{-1}, called the *inverse* of f. The domain of the inverse is the range of f, and f^{-1} is given by $f^{-1}(y) = x$, where x is the (unique) element of A such that $f(x) = y$. If $f : A \longrightarrow B$ and $C \subseteq A$, we write $f(C) = \{f(x) : x \in C\}$; while if $D \subseteq B$ we write $f^{-1}(D) = \{x : f(x) \in D\}$. If $f : A \longrightarrow B$ and $C \subseteq A$, the function $f|C : C \longrightarrow B$ given by $(f|C)(x) = f(x)$ for all $x \in C$ is called the *restriction* of f to C. In some situations, we may not wish to emphasize the domain or codomain of the function so much, and we may indicate the function simply by writing $x \longmapsto f(x)$. The *identity function* on a set is the function $x \longmapsto x$, and may be denoted by ι.

1.2 Information loss and randomness in dynamical systems

This is optional, and is not technically essential for any subsequent chapter of this work. Its aim is to set the scene for some of the recurring ideas. Since the time of Charles Darwin, a common and widely accepted idea in biology has been that species gradually evolve. In his book *The Selfish Gene*, Richard Dawkins popularizes the idea that the underlying force in this evolutionary process is the "wish" of DNA molecules to make *replicas of themselves*. He writes in [7, pp. 16 and 24]:

> At some point a particularly remarkable molecule was formed by accident. We will call it the *Replicator*. It may not necessarily have been the biggest or most complex molecule around, but it has the extraordinary property of being able to create copies of itself. This may seem a very unlikely accident to happen. And so it was. It was exceedingly improbable ... DNA molecules do two important things. Firstly, they replicate, that is to say they make copies of themselves. This has gone on non-stop ever since the beginning of life, and the DNA molecules are now very good at it indeed ... when you were first conceived, you were just a single cell, endowed with one master copy of the architect's plans. This cell divided into two, and each of the two cells received its own copy of the plans. Successive divisions took the number of cells up to 4,8,16,32, and so on into the billions. At every division the DNA plans were faithfully copied with scarcely any mistakes.

The idea of self replication, producing "self similar" objects, also occurs in other contexts, including mathematics, where it has been especially prominent in discussions of chaos and the complex geometrical objects known as fractals. There seems to have been less discussion in

1.2. Information loss and randomness in dynamical systems

mathematics around the notion of self similarity in relation to functions. As this idea is relevant to some later considerations, it is considered here as part of the discussion on functions and iteration. However, the discussion is not treated in the most general possible way; rather it is merely indicative of an idea. It is intended that the discussion also gives some insight into what can happen when functions are successively composed.

Definition. Let A, B and C be given sets, and let

$$f : A \longrightarrow C \quad \text{and} \quad g : B \longrightarrow C$$

be functions with the indicated domains and codomain. Then, f is a *copy*, or a *replica*, of g if there is a one-to-one and onto function $h : A \longrightarrow B$ such that

$$f = g \circ h.$$

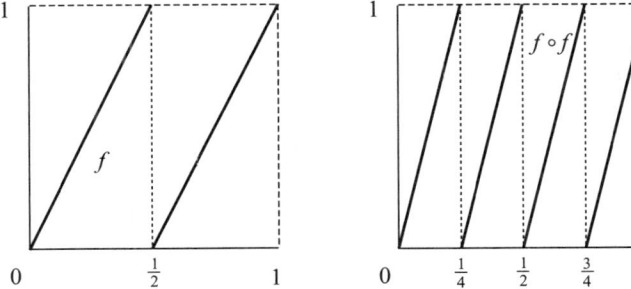

Figure 1.2. The picture on the left is of the graph of the transformation f on $[0, 1)$ given by

$$f(x) = \begin{cases} 2x, & \text{if } 0 \leq x < 1/2, \\ 2x - 1, & \text{if } 1/2 \leq x < 1. \end{cases}$$

The picture on the right is of the graph of $f \circ f$. Note that the restriction $(f \circ f)|[0, 1/2)$ of $f \circ f$ to the subinterval $[0, 1/2)$ is a copy of f for, if h is the one-to-one mapping from $[0, 1/2)$ onto $[0, 1)$ given by $x \longmapsto 2x$, then $f \circ h = f \circ f$. Similarly, the restriction $(f \circ f)|[1/2, 1)$ of $f \circ f$ to $[1/2, 1)$ is a copy of f. So, in this example, composing f with itself has produced *two* copies of f. Further iterations produce 2, 4, 8, ... copies of f, indefinitely. Note that f itself consists of two copies of the identity function $x \longmapsto x$ on $[0, 1)$.

Note that as h is one-to-one and onto, it has an inverse h^{-1}. Consequently, if $f = g \circ h$ then $g = f \circ h^{-1}$ and it follows that *if f is a copy of g, then g is a copy of f*. It is easy to see that if f, g are copies of each other, then they have the same range, and that if one of them is one-to-one, then so is the other. The idea of one function being a copy of the other is that each function mimics the behavior of the other function, but possibly on a different domain and at a different scale.

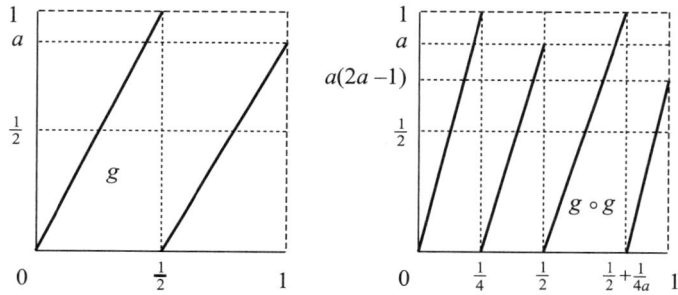

Figure 1.3. Let $1/2 < a < 1$. The picture on the left depicts the transformation g of $[0, 1)$ given by

$$g(x) = \begin{cases} 2x, & \text{if } 0 \leq x < 1/2, \\ 2ax - a, & \text{if } 1/2 \leq x < 1. \end{cases}$$

The function depicted on the right is $g \circ g$. Note that the restriction of $g \circ g$ to $[0, 1/2)$ is a copy of g, but the restriction of $g \circ g$ to $[1/2, 1)$ is not a copy of g. The reason that $g \circ g$ on $[1/2, 1)$ is not a copy of g on $[0, 1)$ is that if $y \in (a(2a-1), a)$, the equation $g \circ g(x) = y$ has a unique solution for $x \in [1/2, 1)$, but the equation $g(x) = y$ has two solutions for $x \in [0, 1)$. This behavior should be compared with that of the function in Figure 1.2. When g is iterated to obtain $g \circ g$, some of the full range of the behavior of g is lost, and this continues with further iterations (see the Investigations at the end of this chapter). Thus, $g \circ g$ produces a copy of g, but also a "mutant" of g. Note that the figure illustrates the case where $(1 + \sqrt{5})/4 < a < 1$, which ensures that $a(2a - 1) > 1/2$.

As an example, if $f : [0, 1] \longrightarrow [0, 1]$ is given by $f(x) = x$, and if $g : [0, 1/2] \longrightarrow [0, 1]$ is given by $g(x) = 2x$, then g is a copy of f because the function $h : [0, 1/2] \longrightarrow [0, 1]$ given by $h(x) = 2x$ is one-to-one and onto and $f \circ h = g$. Note that in this example, $h = g$.

1.2. Information loss and randomness in dynamical systems

In fact, and more generally, if g is a one-to-one function mapping a set X onto itself, then g is a copy of the identity function on X.

Taking composition may produce copies of a function, as we see in the following proposition. Note that if $f : A \longrightarrow B$ is a given function and $C \subseteq A$, then the restriction $f|C$ of f to C is one-to-one if f is one-to-one but, if f is onto, $f|C$ is not necessarily onto.

Proposition 1.1. *Let $f : A \longrightarrow A$ be a function that is onto and assume that there is a subset U of A such that the restriction $f|U$ of f to U is one-to-one on U and has range A. Then $(f \circ f)|U$ is a copy of f.*

Proof. The function $f|U$ is one-to-one on U and maps U onto A. Thus, $f|U$ has an inverse function $h : A \longrightarrow U$, which is also one-to-one and onto. Then, it is sufficient to check that $\big((f \circ f)|U\big) \circ h = f$. But for $x \in A$ we have

$$\big((f \circ f)|U\big) \circ h(x) = (f \circ f)(h(x)) = f(f(h(x))) = f(x),$$

as the inverse of $f|U$ is h. □

In the situation of Proposition 1.1, the procedure applied to the function f may now be repeated indefinitely. For, if we put $U_1 = U$ and

$$U_2 = \{x : x \in U \text{ and } f(x) \in U\},$$

we can check that $f \circ f$ is one-to-one on U_2, has range A on U_2 and that $(f \circ f \circ f)|U_2$ is a copy of f. In general, if $U_n = \{x : x \in U_{n-1} \text{ and } f(x) \in U_{n-1}\}$, then $f^{n+1}|U_n$ is a copy of f (see Exercise 4).

This concept of "self similarity" is especially important in studying successive composition of functions, and in chaos theory. The above discussion is preliminary only, and avoids some of the complexities that may occur in some cases. An important function showing self similarity properties appears on the left in Figure 1.2. This function is used extensively in Chapter 3.

The result of Proposition 1.1, and the subsequent discussion, show that if the function f satisfies the given conditions, then the process of forming $f, f^2, f^3, \ldots$ produces successively more and more copies of

8 Chapter 1. Background Ideas and Knowledge

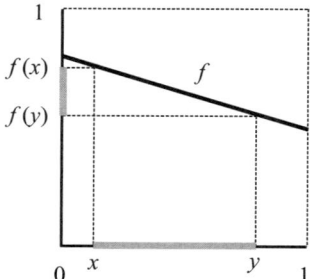

Figure 1.4. The figure depicts the graph of a transformation on $[0, 1]$, whose constant slope s is such that $|s| < 1$. Let us imagine that the true initial state of the system $([0, 1], f)$ is x, but owing to our imperfect knowledge of the state of the system, we take the initial state to be y. Then at the next stage, the true state of the system will be $f(x)$—but, because we take the initial state to be y, we take the next state of the system to be $f(y)$. However, as $|s| < 1$, the error in taking $f(y)$ in place of the true value $f(x)$ is *less* than the error in taking y to be the initial state of the system instead of taking the true state x. That is, the distance between $f(x)$ and $f(y)$ is less than the distance between x and y. In this sense, we have *gained* information about the true state of the system at the next stage. In a case like this, iteration starting from an initial point leads to an orbit convergent to a fixed point.

the original function—that is, over time, the system is self replicating and produces more and more copies of itself. Bearing in mind this biological metaphor, we might ask: if the system as described produces exact replicas of the original system, what type of system would produce "errors" in the attempt of the system to replicate itself by means of iteration and successive composition? Here is a partial answer.

Assume that a function f has domain A and codomain B, and that U is a subset of A on which the function is one-to-one, but where the the range of f restricted to U is "not quite" equal to B. Then $f \circ f$ restricted to U will not generally produce a copy of f, although if $f(U)$ is nearly equal to the range of f, $f \circ f$ restricted to U may be very close to a copy of f. See Figure 1.3, and for further ideas see the Investigations at the end of this chapter.

1.2. Information loss and randomness in dynamical systems

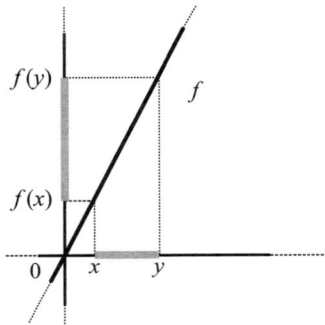

Figure 1.5. Here the graph is of a transformation f of $\mathbb{R}$, given by $f(x) = sx$ where $s > 1$. In this case, where y is an approximation to the true initial state x of the system $(\mathbb{R}, f)$, we see that $f(y)$ is a *worse* approximation to the true state $f(x)$ of the system at the next stage. So, we have *lost* information about the true state of the system as the system evolves through one stage. As pictured, the phase space is unbounded and orbits, with one exception, will go off to ∞ or $-\infty$—but were the phase space bounded, the orbits would be prevented from going off to ∞ and $-\infty$ and would be "forced back" into the phase space, mixing up points and producing recurring phenomena—that is, "randomness".

Questions also arise concerning how information is lost or gained in a dynamical system over time. For example: if the initial state of the system is x, when can we expect to gain information about the long-term behavior of the system simply by knowing x, or perhaps by approximately knowing x? Another question is : if we observe after the elapse of n time units that the system is in state y, can we infer or gain information about the initial state of the system? Such questions can be put in a different way, indicating an analogy with historical inquiry: (i) Given an (approximate) knowledge of the present state of affairs, what can we say about the future? (ii) Given an (approximate) knowledge of the present state of affairs, what can we say about the past? That is, what can we say about the origins of the present state of affairs? Aspects of such questions are discussed in Figures 1.4 and 1.5.

Now it seems that a "random" dynamical system should have at least two aspects to it. On the one hand, it should "mix things up" and be "unpredictable" in some sense. On the other hand, such a system should be "mindless", which could mean, for example, that it does not favor certain parts of the phase space over others. Or, "mindlessness" could mean that certain patterns may not be precisely predictable but that, in an averaging sense, patterns will be repeated because we don't expect the system to deliberately avoid those patterns. Or again, "mindlessness" could mean that the system has no "memory", which could be expressed by the future of the system being independent of the past. We might expect "randomness" to arise in a system where information is lost over time because the system is expanding, but where the space is bounded, so that the expansion of the system is restrained by the boundedness of the phase space, leading to a "mixing up" of the system, and to recurring states (see the comments in Figure 1.5).

1.3 Assumed knowledge and notation

We assume a knowledge of set theory, the notion of a function and the associated terminology, complex numbers, and elementary calculus and real analysis. Suitable background references for these topics are [2, 3, 4, 6, 8, 9, 14, 17, 18, 20, 21], and there are many others. When a result from these areas is called upon in this work a reference, as needed, will be given to a source for the result. Given sets A and B, $A \cup B$ denotes their union, $A \cap B$ denotes their intersection, $A \times B$ denotes their Cartesian product, and A^c denotes the complement of A. The situation where A is a subset of B is written as $A \subseteq B$. If A is a finite set, $|A|$ denotes the number of elements in the set. The empty set is denoted by $\emptyset$. Sets A and B are *disjoint* if $A \cap B = \emptyset$. A family of sets is called a *partition* of a given set S if the distinct sets in the family are disjoint, and the union of all sets in the family is S. We denote the set of natural numbers $\{1, 2, 3, \ldots\}$ by $\mathbb{N}$. The set of integers is denoted by $\mathbb{Z}$, the set of non-negative integers by $\mathbb{Z}_+$, the set of rational numbers by $\mathbb{Q}$, the set of real numbers by $\mathbb{R}$ and the set of complex numbers by $\mathbb{C}$. Occasionally, the product of two numbers a, b may be denoted by $a \cdot b$. If J is an interval of real numbers, $\mu(J)$ denotes its length. A subset of $\mathbb{R}$ is *open* if it is

1.3. Assumed knowledge and notation

a union of open intervals, and it is *closed* if it is the complement of an open set. If A is a closed and bounded subset of $\mathbb{R}$, and if $(A_\alpha)_{\alpha \in I}$ is a family of open sets whose union contains A, the *Heine-Borel Theorem* asserts that there are $\alpha_1, \alpha_2, \ldots, \alpha_n \in I$ such that A is contained in the (finite) union of $A_{\alpha_1}, A_{\alpha_2}, \ldots, A_{\alpha_n}$.

Definitions and terminology concerning functions have been discussed in Section 1.1.

A *sequence* $u_1, u_2, u_3, \ldots$ of points in a given set is generally denoted by (u_n) and u_n is the nth *term* of the sequence. A sequence (A_n) of sets is *non-decreasing* if $A_n \subseteq A_{n+1}$ for all n, and (A_n) is *non-increasing* if $A_{n+1} \subseteq A_n$ for all n. A sequence (u_n) of real numbers is *non-decreasing* if $u_n \le u_{n+1}$ for all n, and (u_n) is *non-increasing* if $u_{n+1} \le u_n$ for all n. A sequence (u_n) of numbers is *convergent with limit ℓ* if, for every $\varepsilon > 0$, for all sufficiently large n we have $|u_n - \ell| < \varepsilon$. In this case, we write $\lim_{n \to \infty} u_n = \ell$. Following the terminology of Igor Kluvánek[1], a sequence (u_n) of numbers is *summable* if there is a number M such that $\sum_{j=1}^n |u_j| \le M$ for all $n \in \mathbb{N}$. In this case, $\lim_{n \to \infty} \sum_{j=1}^n u_j$, exists, this limit is independent of the order of the terms of the sequence, and it is called the *sum* of the sequence (u_n) and is denoted by $\sum_{n=1}^\infty u_n$. In particular, if $p > 1$, the sequence $(1/n^p)$ is summable. If (u_n) is a sequence of numbers that is not summable, but for which $\lim_{n \to \infty} \sum_{j=1}^n u_j$ exists, then (u_n) is *conditionally summable* and its sum is $\lim_{n \to \infty} \sum_{j=1}^n u_j$. If (u_n) is a sequence of non-negative terms, we may write $\sum_{n=1}^\infty u_n < \infty$ if it is summable, but if the sequence is not summable we may write $\sum_{n=1}^\infty u_n = \infty$. A double sequence (u_{mn}) is *summable* if there is a number M such that $\sum_{j=1}^m \sum_{k=1}^n |u_{mn}| \le M$ for all m, n. In this case the sum $\sum_{m=1}^\infty \sum_{n=1}^\infty u_{mn}$ exists and has the same value independently of how the summation is carried out. A set A is *countable* if its elements can be made to appear as the terms of a sequence—that is, we can write $A = \{u_1, u_2, u_3, \ldots\}$, for suitable $u_1, u_2, \ldots$. In particular, $\mathbb{Q}$ is countable.

The work as a whole requires only a limited use of integration. In Chapters 2, 3 and 4, the only place where integration is used is in Chapter 3 where step functions are integrated over intervals, and we assume the usual properties of integration in this context (see [2, pp. 50–55] and [3, p. 141], for example, for the definition and properties of the integrals of step functions). Chapter 5 is, in part, an exposition of the general theory of measure, with some reference to integration in the measure theory context. The end of a proof is denoted by □.

Appendix
Mathematical reasoning and proof

Mathematics starts with basic concepts and assumptions, and then builds upon these by a process of logical reasoning and argument to make assertions and arrive at definite conclusions. The process of logical reasoning and argument called upon to establish the claimed truth of an assertion is called a *proof*. Mathematicians generally regard the proof as establishing an absolute truth, but one that is relative to the given assumptions. That is, the mathematician aims to say something like: "if it is the case that A, B, C hold, then X, Y, Z are (necessarily) true".

A consequence of proving statements under carefully stated assumptions is that the conclusions of mathematical enquiry are not generally subject to dispute. But an individual who reads a mathematical proof has to make an individual judgment about the validity of the proof, and not take it upon authority. It is a remarkable feature of mathematics that the necessity to make individual judgments nevertheless does not generally produce variations in judgment as to the validity of a proof. This markedly distinguishes mathematics from most other areas of intellectual enquiry. There is a clarity in mathematics as to what constitutes valid argument and what constitutes error so that, despite the reader of a proof making an individual judgment, the possibility of dispute between individuals about the validity of a proof generally vanishes upon repeated inspection of the proof. The validity of a proof is a question of technical verification or, in the case of an invalid attempt at proof, a question of identifying errors. There is a superficial similarity to the game of chess, where there is no dispute as to what constitutes a valid move, or who is

Appendix: Mathematical reasoning and proof

the winner of the game if there is one, even though the players make individual judgments about the validity of the moves and the final state of the game. We therefore observe a paradox: although it is the individual person who makes judgments as to which mathematical statements are true, which are not proven, or which are false, nevertheless mathematical statements tend to *transcend* individual judgment.[2] For this reason, many mathematicians believe that in mathematics we *discover* knowledge, rather than *construct* it. Mathematics achieves a special status of accuracy and objectivity, making it unique among fields of human enquiry. The mathematician Hermann Weyl (1885–1955) put it like this:

> Besides language and music, it [mathematics] is one of the primary manifestations of the free creative power of the human mind, and it is the universal organ for world-understanding through theoretical construction.

There are several factors that are important in understanding how mathematical argument achieves this special status. One is the *clarity* and *lack of ambiguity* in the basic concepts, a clarity which is achieved by giving precise definitions of concepts so that it is clear what is needed to establish true statements involving the concepts.

A second is the *logical argument* that is a characteristic of mathematics as presented in its final form. In logical argument the underlying assumptions are stated clearly, or are at least understood and agreed upon, as a basis for the argument. Once we are clear about the precise assumptions we are making, and we have clear definitions of the objects of the enquiry, we can then apply logical argument to see what conclusions can be drawn. We must remain aware that any conclusions we draw are dependent upon the assumptions we make.[3] The assumptions in a situation may be indicated by the wording "let such-and-such" be the case. For example, we might say "let x be a number with $x > 1$" to indicate that we are assuming that x is a number and that x is greater than 1. As the argument continues, the assumptions may change or new assumptions introduced, but when this occurs it will generally be indicated. It is a general obligation upon the good mathematical writer to indicate to the reader what assumptions are being made in any given situation. This is an aim of this work.

A third factor is the *use of symbols* in indicating the mathematical argument. The use of symbols gives a terseness to the argument, but this can have the effect of clarifying the argument, leaving one free to think about the logic of the argument rather than spending time describing the underlying objects the argument is about. The intention in the use of symbols is to provide intellectual clarity and efficiency. Mathematics is concerned with statements such as: "if P is true then Q is true". This may be abbreviated as: $P \implies Q$, which might also be read as: "P implies Q". We may also have a situation where: "if P is true then Q is true, and if Q is true then P is true". In this case we say: "P is true if and only if Q is true", and we may write this as $P \iff Q$, and we say that the statements P and Q are (logically) *equivalent*. Another way of expressing that P and Q are equivalent is to say "P *if and only if* Q".

A fourth factor is the *attention to detail* in the process as a whole. The clarification of the concepts by means of definitions, the explicit identification of assumptions, the close reasoning, and the systematic use of symbols, mean that the process of gaining mathematical knowledge requires discipline and an attention to detail at each stage.

The clarity of mathematical argument and its conclusions are, of course, achieved at a cost. The restriction of argument to unambiguous concepts, and reasoning with those concepts in a strict way, means that mathematics narrows its field of view to concepts and environments where these requirements can be satisfied. It means that vast areas of human experience seem far removed from mathematics. On the other hand, it is a remarkable fact that mathematics remains such an effective way of analyzing problems that may seem far removed from the objects studied by mathematicians. This is the subject of a famous paper by Eugene Wigner [22], "The unreasonable effectiveness of mathematics in the physical sciences". And the philosopher Bertrand Russell comments in his autobiography [19]:

> With equal passion I have sought knowledge. I have wished to understand the hearts of men. I have wished to know why

Appendix: Mathematical reasoning and proof

> the stars shine. And I have tried to apprehend the Pythagorean
> power by which number holds sway above the flux.

The power of mathematics is evident in providing the conceptual framework for so much of our understanding of the physical world, extending even to the whole universe as we perceive it. Nevertheless, the source of this power, and the reasons for it, remain as mysterious as ever.

If P and Q are statements, we may write $P \implies Q$ for "P implies Q". Also, we may write $P \iff Q$ for "P implies Q and Q implies P." If P is a statement, the negation of P is denoted by $\sim P$. We sometimes prove statements by *contradiction*. For example, if we are trying to prove that P implies Q, we might proceed by assuming that P is true and Q is not true. Then, we prove that

$$\sim Q \implies \sim P.$$

Thus, our assumption has led to the negation of P, which was assumed to be true. So, Q must be true and we deduce that

$$P \implies Q.$$

Sometimes, we may need to *disprove* statements. We assume some knowledge of elementary set theory and its standard notation. Let A be a set and let Z denote a property that an element of A may or may not have. Then, the statement

$$x \in A \implies x \text{ has property } Z, \tag{1.1}$$

means that every element of A has property Z. Alternatively, (1.1) means that for all $x \in A$, x has property Z. One way to *disprove* (1.1) is to show the following:

there is $x_0 \in A$ such that x_0 does not have property Z.

The point x_0 is then a *counterexample* to the statement or conjecture (1.1). A counterexample is a specific instance that shows that a general statement is not true, or that a conjecture is false. The production of counterexamples has a long history in mathematics, but its importance in science, where it appears in a slightly different guise as refuting or falsifying a scientific conjecture, only seems to have been recognized explicitly more recently.[4]

The preceding discussion may convey the idea that mathematical argument and proof are mere technical constructions that are found and verified in an automatic manner, but this is far from being the case. A proof aims to present an argument in a way that convinces through its abstract reasoning and logic, but the proof has to come *from* somewhere. The constructor of a proof may have a flash of inspiration or insight so that a proof comes effortlessly, like the goddess Athene who, as narrated by Robert Graves, gave us the science of numbers and was born fully armed, springing from the skull of her father, Zeus, with a mighty shout [11, pp. 46 and 96]. But it is more likely that a proof will require an examination and re-examination of ideas, sifting out what is useful and what isn't. The most important single quality here is *imagination*, for it is this that will give us the needed ideas.[5] It is the *interplay* between logic and imagination that gives mathematics its distinctive qualities as an intellectual activity. To the reader disposed to see it, a proof may have a function that lies beyond simply showing a certain statement is true—for example when a proof strikes one with the beauty of its ideas and its overall structure. This was expressed by Paul Erdős in his references to "The Book", in which God keeps the perfect proofs of mathematical theorems (see the introduction to "Proofs from THE BOOK", by M. Aigner and G. M. Ziegler [1]). As well, a proof has a type of moral function, for it requires both the reader and the constructor of a proof to make a judgment on its validity, adequacy and truth. For, despite the claims to intrinsic truth implied by a proof, as R. W. Hamming says in [12]:

> In science and mathematics we do not appeal to authority but rather you are responsible for what you believe.

This statement has even more force in mathematics, where there is more of an obligation upon the student and the mathematical practitioner to verify, wherever possible, the totality of the reasoning leading to a conclusion. In science, one often has to take observations or the results of experiments conducted by others on trust, for it is not practicable for each individual scientist to satisfy himself or herself as to the correctness of observations, or the conclusions of an experiment, by repeating the observations of the experiment for themselves in every case. Similar difficulties arise in mathematics, and mathematicians very often accept the truth of a result on the basis that others have checked it, or perhaps that a computer calculation has verified it. Nevertheless, when presenting new research results of one's own, and although it may depend upon the circumstances, there is a strong obligation in obtaining conclusions based on the previous work of others to do so *not* by saying "X says this is true so it must be true", but rather by saying "I have examined X's work and have satisfied myself as to its correctness and the validity of its proofs".

Exercises

1. Let X be a set and let $f : X \longrightarrow X$ and $h : X \longrightarrow X$ be functions. Assume that h is one-to-one and onto and put $g = h^{-1} \circ f \circ h$. Prove by induction that
$$g^n = h^{-1} \circ f^n \circ h,$$
for all $n = 1, 2, 3, \ldots$. The point about this result is that, under the stated assumptions, if we know the iterates of f, we can calculate the iterates of g.

2. If X is a set, let $\Theta(X)$ denote the set of all functions $h : X \longrightarrow \mathbb{R}$. Now, let $\theta : A \longrightarrow B$ be a given function, and let $\theta^* : \Theta(B) \longrightarrow \Theta(A)$ be the function given by
$$\theta^*(h) = h \circ \theta, \text{ for all } h \in \Theta(B).$$
Prove the following.
   (i) If θ is onto, then θ^* is one-to-one.
   (ii) If θ is one-to-one, then θ^* is onto.

3. Let J be a bounded interval and let h be a transformation on J that is of the form $h(x) = sx + c$ for all $x \in J$, where s, c are constants. Thus, the graph of h is a straight line segment. Prove that $|s| \leq 1$.

4. Let $f : A \longrightarrow A$ be a function whose range is A. Let $U_1, U_2, \ldots$ be nonempty subsets of A such that, for each n, the restriction $f|U_n$ of f to U_n is one-to-one and has range A. Put $f^0 = \iota$, the identity function on A, put $V_0 = A$, $V_1 = U_1$ and for $n \geq 2$ put

$$V_n = \{x : x \in U_n \text{ and } f(x) \in V_{n-1}\}.$$

This defines $V_0, V_1, V_2, \ldots$ as subsets of A by induction.

   (i) Prove that for all $n \geq 1$ the restriction $f^{n-1}|V_{n-1}$ of f^{n-1} to V_{n-1} is one-to-one and maps V_{n-1} onto A.

   (ii) Prove that the restriction $f^n|V_{n-1}$ of f^n to V_{n-1} is a copy of f.

   (iii) Discuss these results in relation to each of the functions f and g mapping $[0, 1)$ into $[0, 1)$ given by

$$f(x) = \begin{cases} 2x, & \text{if } 0 \leq x < 1/2, \\ 2x - 1, & \text{if } 1/2 \leq x < 1/2; \end{cases}$$

and

$$g(x) = \begin{cases} 3x, & \text{if } 0 \leq x < 1/3, \\ 3x - 1, & \text{if } 1/3 \leq x < 2/3, \\ 3x - 2, & \text{if } 2/3 \leq x < 1. \end{cases}$$

5. Let A be a set and assume that there is a one-to-one function $f : A \longrightarrow \mathbb{N}$ that has infinite range. Prove that there is a one-to-one and onto function $g : A \longrightarrow \mathbb{N}$, and deduce that A is countable.

6. Let A be a set and let B be a countable set. If $A \cup B$ is not countable, prove that A is not countable.

7. Let Σ be the set of all sequences of 0s and 1s. Thus, $x \in \Sigma$ means that $x = (x_n)$ and $x_n \in \{0, 1\}$ for all n. Prove that Σ is not countable. [HINTS: assume that Σ is countable. Then, there is a sequence (z_n) in Σ such that $\Sigma = \{z_n : n \in \mathbb{N}\}$. Put $z_n = (z_{nr})_{r=1}^\infty$, for all $n \in \mathbb{N}$. Let $w \in \Sigma$ be given by

$$w_r = \begin{cases} 1, & \text{if } z_{rr} = 0; \\ 0, & \text{if } z_{rr} = 1, \end{cases}$$

for all $r \in \mathbb{N}$. Deduce that $w \notin \{z_1, z_2, \ldots\}$.]

Investigations

1. This investigation develops Exercise 4 in a special case. Let $0 < a \leq 1$ and let $g : [0, 1] \longrightarrow [0, 1]$ be the function given by

   $$g(x) = \begin{cases} 2x, & \text{for } 0 \leq x < 1/2; \\ a(2x - 1), & \text{for } 1/2 \leq x \leq 1. \end{cases}$$

   (a) Calculate $g^2(x)$ in the cases $a = 1$, $1/2 < a < 1$, $a = 1/2$ and $0 < a < 1/2$ (see Figure 2.2 for a special case).
   (b) Sketch the graphs of g and g^2 for the cases: $a = 1$, $1/2 < a < 1$, $a = 1/2$ and $0 < a < 1/2$. Describe in words the qualitative difference between the cases where $0 < a < 1/2$ and $1/2 < a < 1$.
   (c) Does g^2 contain copies of g, obtained by restricting g^2 to an appropriate subset of $[0, 1]$? Does the situation depend upon the parameter a? What happens for the iterates $g^3, g^4, \ldots$? Do the iterates produce some copies of g, do they they produce some "partial" copies of g, and is there a part of g^n that "degenerates" as $n \to \infty$?

2. Let X be a set, and let U, V, W be subsets of X such that $U \cap V = \emptyset$ and $U \cup V = X$. Let $f : X \longrightarrow X$ be a function such that f is one-to-one on U and the range of f restricted to U is X, and f is one-to-one on V and the range of f restricted to V is W. Note that in the case when $W = X$, we know from Proposition 1.1 that $f \circ f$ restricted to either to U or to V is a copy of f.

   Under the stated assumptions, the restriction $(f \circ f)|U$ of $f \circ f$ to U is a copy of f. Investigate under what conditions $(f \circ f)|V$ is not a copy of f. If $(f \circ f)|V$ is not a copy of f describe, if possible, in what sense it may be some sort of "partial" copy of f. This situation may correspond to where a DNA molecule produces an imperfect copy of itself. Compare this with the preceding investigation, and try to determine whether this situation can be extended along the lines of Exercise 4.

Notes

1. [Page 11] Igor Kluvánek (1931–1993) was a Slovak-Australian mathematician who was noted especially for his work on vector measures and integration, and associated applications. His view was that the term "summable sequence" should correspond to the term "integrable function", and that the sum of a summable sequence corresponds to the integral of an integrable

function. He regarded the term "series" as more-or-less superfluous, pointing out that the general theory of integration simply does without any equivalent. Note that in [20, p. 439], Michael Spivak uses the term "summable sequence" for a sequence that has a sum. This means that Spivak's definition of a summable sequence includes the sequences that are summable in the sense used here, as well as those here that are called conditionally summable. Spivak's approach is entirely reasonable, in that it makes sense to say that a sequence that has a sum is going to be called summable. On the other hand, in integration theory, an integrable function f is a function such that $\int |f| < \infty$—there is no unique way, in general, in which one can speak of $\int f$ for a function that is a mixture of positive and negative values. So, if one wants the term "summable" to correspond to the term "integrable" in integration theory, one is forced to adopt the Kluvánek terminology.

2. [Page 13] This view of mathematics and the status of mathematical proof is not accepted by all. For example, R. W. Hamming wrote in [12, p. 645]:

> To simple people, who believe whatever they read and do not question things for themselves, a proof is a proof is a proof, but to others a proof merely provides a way of thinking about the theorem and it is up to the individual to form an opinion... For me theorems are true or false pretty much independent of their corresponding proofs; my internal beliefs must be the final arbiter of whether I accept or reject the mathematics I see. But the purists believe that the postulates, definitions, and the accepted logic determine such things!

One might reply to this that a valid proof is an argument that convinces because it has been subject to repeated individual scrutiny and critical thought, and not because it has been accepted on authority. Also, yes, a proof *does* provide a way of thinking about the problem, but there is no incompatibility between these two functions of proof. The following comments by G. Gonthier [10, p. 1382], in relation to *strictly* formal proofs, bring out the point that such a proof may provide ways of thinking about the problem and may lead to new insights:

> ... to produce a formal proof one must make explicit every single logical step of a proof; this both provides new insight into the structure of the proof, and forces one to use this insight to discover every possible symmetry, simplification, and generalization, ... Perhaps this is the most promising aspect of formal

proof: it is not merely a method to make absolutely sure we have not made a mistake in our proof, but also a tool that shows us and compels us to understand why a proof works.

Although Gonthier is talking about "strictly formal" proofs, rather than "ordinary" mathematical proofs, his comment remains valid for the latter type of proof—for the purpose of rigour and formality in mathematical argument is not authoritarian, nor should it be, but rather it is to provide insight and a level of certainty that is not available through other means.

As well, while one can agree with Hamming that one's own belief must be the final arbiter of acceptance or rejection of a proof, this is not inconsistent with the idea that the logic of the argument is what determines the truth or falsity of the conclusion. It is our perception of the logic of the argument that convinces us, not reliance on our own, possibly imperfect, intuitions. One might admit that in fact proofs have been and may be presented that are incomplete, obscure, incomprehensible, difficult for others to read because they are long and involved, depend upon computer calculations that are hard to verify, or that standards of logical rigour have changed over time and vary in different contexts, and so on. None of this affects the substance of the situation, although it may affect our confidence in the truth of what is claimed. The remarkable thing about mathematics is not that some proofs have been or are incomplete or obscure and so on, and so should lead to contention; rather it is the relative *lack* of contention as to what constitutes valid knowledge that is remarkable in mathematics.

Hamming's view that a proof "merely" provides a way of thinking about a theorem has echos in the ideas of the philosopher Imre Lakatos. In [13] there is a wonderfully entertaining discussion between a teacher and the pupils, as they attempt to find a valid version of Euler's formula $V - E + F = 2$ (V is the number of vertices, E is the number of edges and F is the number of faces in a polyhedron), which seems to suggest that whatever mathematical knowledge is, it is "only" socially constructed and probably no convincing proof of anything is possible. However, one could look at Lakatos' discussion as simply reminding us that in presenting a convincing proof, it may be crucially important to have appropriate assumptions and clear definitions of the objects in the discussion. The issues raised by the status of mathematics and proof are complex, and in addition include: what activities one chooses to call mathematics, the imagination and the uncertainties involved in trying to find a proof, the status of proofs that depend upon computer calculations, and the practicalities of checking long and complex proofs. Further discussion of some of these issues may be found in the Preface of [5]. There are also the questions of the logical consistency of mathematics and the provability of mathematical statements arising from the work of Kurt Gödel, and the extent to which mathematicians should take these into account (see [15, pp.120-121] for a pointed discussion in relation to the work of Nicolas Bourbaki).

3. [Page 13] One way of getting broader insights into situations is to identify how the conclusions change in accordance with changes in the assumptions. This is true in situations involving mathematics itself, but it is equally true when we apply mathematics to situations outside mathematics (such as physics, economics or social policy) where the mathematical assumptions we should make to ensure a type of compliance with external reality may be less clear, and may require an individual judgment more than a purely analytical or technical verification.

4. [Page 16] The idea that science is distinguished from other intellectual activities by proposing conjectures and by falsifying or attempting to falsify existing theories is associated with Karl Popper (see [16] for example).

5. [Page 16] There is a story, which may be apocryphal, about the mathematician David Hilbert that reports a conversation between Hilbert and a colleague as follows:

Hilbert's colleague: Have you heard that X's research student is going to give up mathematics? He says that he is going to become a poet.

Hilbert: I'm not all that surprised. I don't think he has enough imagination to be a mathematician.

Bibliography

[1] M. Aigner and G. M. Ziegler, *Proofs from THE BOOK*. Springer-Verlag, Berlin and Heidelberg, 2004.

[2] T. M. Apostol, *Calculus, volume 1*, Blaisdell, NewYork, 1962.

[3] R. G. Bartle and D. R. Scherbert, *Introduction to Real Analysis*, third edition, Wiley, NewYork, 2000.

[4] S. K. Berberian, *A First Course in Real Analysis*, Springer-Verlag, Berlin, 1994.

[5] D. M. Bressoud, *Proofs and Confirmations: the story of the alternating sign matrix conjecture*, Mathematical Association of America and Cambridge University Press, Washington DC, 1999.

[6] D. M. Bressoud, *A Radical Approach to Real Analysis*, second edition, Mathematical Association of America, Washington DC, 2007.

Bibliography

[7] R. Dawkins, *The Selfish Gene*, Oxford University Press, Oxford, 1976.

[8] R. Devaney, *A First Course in Chaotic Dynamical Systems*, Addison Wesley, Massachusetts, 1992.

[9] E. Gaughan, *Introduction to Analysis*, Brooks/Cole, Belmont, 1968.

[10] G. Gonthier, *Formal proof—The Four Color Theorem*, Notices Amer. Math. Soc., **55**, (2008), 1382–1393.

[11] R. Graves, *The Greek Myths, volume 1*, Penguin Books, Harmondsworth, 1977.

[12] R. W. Hamming, *Mathematics on a distant planet*, Amer. Math. Monthly, **105** (1998), 640–650.

[13] I. Lakatos, *Proofs and Refutations; the logic of mathematical discovery (J. Worrall and E. Zahar eds.)*, Cambridge University Press, 1976.

[14] S. R. Lay, *Analysis*, Prentice-Hall, New Jersey, 1986.

[15] M. Mashaal, *Bourbaki: a secret society of mathematicians*, American Mathematical Society, 2006.

[16] K. R. Popper, *Conjectures and Refutations*, Routledge, London, 1963.

[17] K. A. Ross, *Elementary Analysis: The Theory of Calculus*, Springer-Verlag, New York, 1980.

[18] W. Rudin, *Principles of Mathematical Analysis*, McGraw-Hill, New York, 1964.

[19] B. Russell, *The Autobigraphy of Bertrand Russell, 1872–1914*, Allen and Unwin, London, 1967.

[20] M. Spivak, *Calculus*, second edition, Publish or Perish, Wilmington DE, 1980.

[21] K. R. Stromberg, *An Introduction to Classical Real Analysis*, Wadsworth, Belmont, 1981.

[22] E. Wigner, *The unreasonable effectiveness of mathematics in the physical sciences*, Communications in Pure and Applied Mathematics, **13** (1960), 1–14.

Chapter 2 Flow Chart

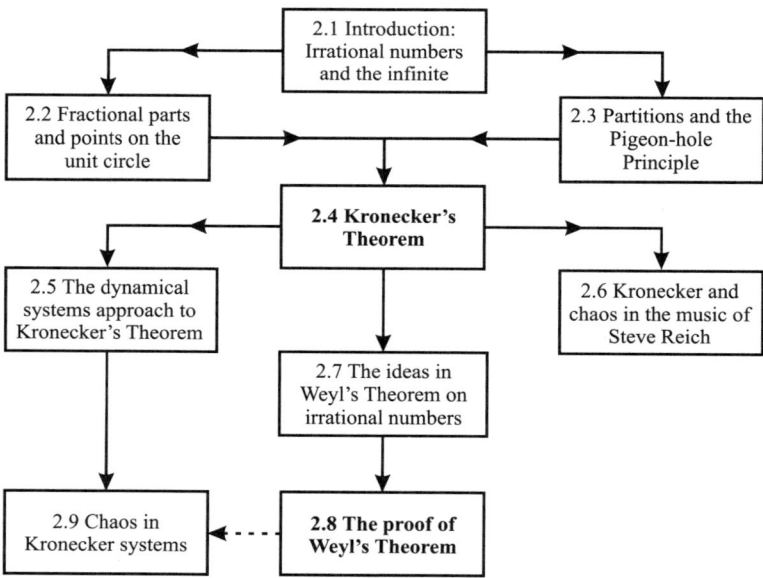

The solid arrows indicate the sections that are necessary for using subsequent sections. The dashed arrow indicates a preceding section that is complementary to the subsequent section. The sections with the main results are in bold type.

Chapter 2

Irrational Numbers and Dynamical Systems

> All our philosophy is a correction of the common usage of words
> *Georg Christoph Lichtenberg* (1742–1799)

2.1 Introduction: irrational numbers and the infinite

The discovery of irrational numbers in ancient Greece is attributed to the School of Pythagoras[1] in 500–400 BCE. The discovery seems to have caused a type of crisis in Greek science and mathematics, which had until then considered that measurement must consist in counting natural units and considering their ratios—that is, it was considered that measurement is a *discrete* process, or a process of *counting*. This was the view of Pythagoras, who regarded nature as an expression of whole numbers [5, pp. 154–155]. Sir Karl Popper regarded the discovery of irrational numbers as leading to major philosophical adjustments in Greek thought. He writes [12, p. 75]:

> My thesis here is that Plato's central philosophical doctrine, the so-called Theory of Forms or Ideas, cannot be properly understood except ... in the context of the critical *problem situation* in Greek science (mainly in the theory of matter) which developed as the result of *the discovery of the irrationality of the square root of 2*.[2]

Chapter 2. Irrational Numbers and Dynamical Systems

I think it is fair to say that the problems raised by the discovery of irrational numbers are essentially unresolved to this day. For, once we admit the notion of irrational numbers, we admit entities whose description essentially requires an infinite sequence of discrete and independent pieces of information. The crucial word here is *"infinite"*. The infinite cannot be fully and humanly grasped, short perhaps of some rare mystical experience, so that entities such as irrational numbers immediately take us beyond the realm of what can be directly experienced, even by the mind and the intellect at their most rarefied levels. The general cultural and human significance of irrational numbers and other such paradoxical entities is commented on by the Italian mathematician Ennio De Giorgi:

> In antiquity there there was a lively discussion over irrational numbers which was, at bottom, a philosophical discussion on foundations: are numbers only the integers, or can we conceive that there might be more? In this discussion, ... one sees a great affinity between modern problems and those of antiquity ... this interest, according to me, of the cultured person of the Greek world in the paradox of the liar or the problem of the existence of irrational numbers and analogous curiosities, could be—in fact should be—central elements of debate and reflection in modern culture.[3]

Reflection upon the irrational numbers may produce in us a sense of imaginative and intellectual limitation in that they cannot be fully grasped either by a metaphysical leap or by our innate reasoning ability; but on the other hand they present a host of special phenomena, many of a "random" or "chaotic" nature, which reveal concretely the hidden complexity implied by the underlying concept. They present us with the infinite in its "unknowability", but in a way which is also very precise and concrete.

A main aim of this chapter is to investigate a particular dynamical system that is associated with irrational numbers. The system arises from a result of Leopold Kronecker, which says that if α is an irrational number, then the sequence of fractional parts of the sequence $\alpha, 2\alpha, 3\alpha, \ldots$ is *dense* in $[0, 1)$. Precise definitions are below, but denseness of the se-

2.1. Introduction: irrational numbers and the infinite

quence simply means that each number in [0, 1) may be approximated to within any degree of accuracy by some element of the sequence. If we consider the sequence of fractional parts of the multiples of a given number α, this sequence is periodic if and only if α is rational; whereas we see from Kronecker's Theorem that this sequence is dense in [0, 1) if and only if α is irrational. In terms of the associated dynamical system, Kronecker's result shows in the irrational case that the points in each orbit are "spread out" through the system, whereas, in the rational case, each orbit is periodic and thus has only a finite number of distinct points. Kronecker's result is developed in a more refined form as Weyl's Theorem, which for an irrational number says that the corresponding sequence of fractional parts is "uniformly distributed" throughout the interval [0, 1), so that the sequence is distributed throughout [0, 1), but its distribution favors no part of the interval [0, 1) over any other part. This appears to be at least a requirement for a "random" system—further comments on this issue are in Section 2.5. Some chaotic properties of the associated dynamical system are discussed, and Weyl's Theorem says in this context that as the system evolves, the "chaotic" behavior of the system is distributed evenly throughout the system. The "chaotic" behavior of the system can be regarded as a form of "information loss" or "randomness" in the system.

There are two equivalent settings for discussing Kronecker's Theorem and Weyl's Theorem. One is the interval [0, 1), mentioned above, and the other is the unit circle $\mathbb{T}$, consisting of all complex numbers of absolute value 1. The equivalence of these settings arises from the fact that there is a one-to-one correspondence between [0, 1) and $\mathbb{T}$, given by the function $t \mapsto e^{2\pi i t}$ that maps [0, 1) onto $\mathbb{T}$. In this work, [0, 1) is the main setting, because this allows for the easier use of figures to illustrate graphs of functions, and ideas that are used in the course of the discussion. At the same time, the reader is encouraged to develop facility in seeing the ideas in both settings, and "switching" from one setting to the other, and various figures are included to assist in doing this by pointing out how the ideas in one setting may be interpreted in the other. However, note that the equivalence of the two settings is not complete— for the function $t \mapsto e^{2\pi i t}$ is continuous from [0, 1) onto $\mathbb{T}$, but the inverse of this function from $\mathbb{T}$ onto $[0, 1)$ is not continuous. This leads to the situation where a certain type of chaotic behavior on $[0, 1)$, which is associated with the dynamical systems interpretation of Kronecker's Theorem, is not reflected by a corresponding behavior on $\mathbb{T}$.

2.2 Fractional parts and points on the unit circle

If α is a real number, make the definition that $\text{int}(\alpha)$ is the largest integer that is less than or equal to α. Then make the definition that

$$\text{frac}(\alpha) = \alpha - \text{int}(\alpha).$$

Note that $\text{int}(\alpha)$ is called the *integer part* of α, $\text{frac}(\alpha)$ is called the *fractional part* of α, and that $\text{frac}(\alpha)$ is in $[0, 1)$. If α is an integer then $\text{frac}(\alpha) = 0$, and if $0 \leq \alpha < 1$ then $\text{frac}(\alpha) = \alpha$. Also, it follows from the definition of $\text{frac}(\alpha)$ that

$$\alpha = \text{int}(\alpha) + \text{frac}(\alpha). \tag{2.1}$$

Note that (2.1) implies that if α is irrational, then $\text{frac}(\alpha)$ is also irrational. Also, it is an important fact, proved in Proposition 2.1 below, that the equation $\alpha = \text{int}(\alpha) + \text{frac}(\alpha)$ expresses α *uniquely* as a sum of two numbers, one in $\mathbb{Z}$ and one in $[0, 1)$.

Given the number α, the numbers $\text{frac}(\alpha), \text{frac}(2\alpha), \text{frac}(3\alpha), \ldots$ lie in $[0, 1)$. The aim in this section is to understand something of how this sequence is distributed within $[0, 1)$. For example, we might ask whether the terms of the sequence cluster together in one part of the interval—convergence of the sequence would be an extreme form of this. Whatever the answer, we might ask how this can be expressed in precise mathematical terms. In what ways might the answers to such questions depend upon the number α? We shall see that if α is rational, then there are only a finite number of distinct values of the sequence and that these values occur in a strictly determined order. When α is irrational, all terms of the sequence are distinct, and they are spread throughout the interval in a

2.2. Fractional parts and points on the unit circle

"random" way, which we seek to make mathematically precise. Thus, if α is rational, the terms of the sequence have a "rational" or "predictable" behavior; while if α is irrational, the terms of the sequence may appear to have an "irrational" or "unpredictable" behavior. The fractional parts of the multiples of an irrational number provide an elementary setting for investigating ideas of randomness and order.

Note that in studying a sequence of the form (frac$(n\alpha)$), it suffices to assume that $\alpha \in [0, 1)$. For,

$$\begin{aligned}\text{frac}(n\alpha) &= n\alpha - \text{int}(n\alpha) \\ &= n\,\text{frac}(\alpha) + n\,\text{int}(\alpha) - \text{int}(n\alpha) \\ &= \text{frac}(n\,\text{frac}(\alpha)) + \text{int}(n\,\text{frac}(\alpha)) + n\,\text{int}(\alpha) - \text{int}(n\alpha),\end{aligned}$$

so that frac$(n\alpha) - $ frac$(n\,\text{frac}(\alpha))$ equals an integer belonging to $(-1, 1)$. We deduce that this integer is 0 and that frac$(n\alpha) = $ frac$(n\,\text{frac}(\alpha))$. So, the sequences (frac$(n\alpha)$) and (frac$(n\,\text{frac}(\alpha))$) are the same, so that as frac$(\alpha) \in [0, 1)$ we could assume that $\alpha \in [0, 1)$. (This argument is related to the uniqueness of the decomposition of a number into its integer and fractional parts—see Proposition 2.1 below.)

Another way of thinking of these questions arises as follows. Let $\mathbb{T} = \{z : z \in \mathbb{C} \text{ and } |z| = 1\}$. The points of $\mathbb{T}$ may be regarded as comprising a circle of center 0 and radius 1. Owing to this geometrical interpretation, $\mathbb{T}$ is called the *unit circle*. Then if $\alpha \in \mathbb{R}$, the sequence frac(α), frac(2α), frac$(3\alpha), \ldots$ in $[0, 1)$ may be interpreted in an equivalent way as a sequence $z, z^2, z^3, \ldots$ of complex numbers in the unit circle $\mathbb{T}$. To see this, observe that for $\alpha \in \mathbb{R}$,

$$e^{2\pi i \alpha} = e^{2\pi i (\text{int}(\alpha) + \text{frac}(\alpha))} = e^{2\pi i\,\text{frac}(\alpha)}. \tag{2.2}$$

This shows that $e^{2\pi i \alpha}$ depends only upon frac(α). Also from (2.2) we see that

$$2\pi\,\text{frac}(\alpha) = \text{the argument of } e^{2\pi i \alpha}.$$

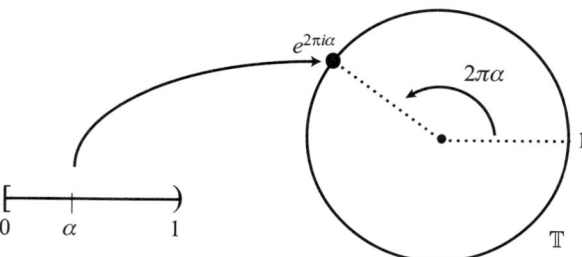

Figure 2.1. The function on $[0, 1)$ given by $\alpha \mapsto e^{2\pi i \alpha}$ is one-to-one and maps $[0, 1)$ onto the unit circle $\mathbb{T}$ which is the set $\{z : z \in \mathbb{C} \text{ and } |z| = 1\}$. Note that $e^{2\pi i \alpha} = \cos 2\pi\alpha + i \sin 2\pi\alpha$, and that $2\pi\alpha$ is the argument of $e^{2\pi i \alpha}$. As α increases from 0 to 1, $e^{2\pi i \alpha}$ goes completely around the circle counterclockwise starting from 1. The function $\alpha \mapsto e^{2\pi i \alpha}$ maps a subinterval of $[0, 1)$ onto an arc of $\mathbb{T}$. By identifying $\alpha \in [0, 1)$ with the point $e^{2\pi i \alpha} \in \mathbb{T}$ we can identify $[0, 1)$ with the unit circle $\mathbb{T}$. Under this identification, a rational number in $[0, 1)$ corresponds to a root of unity, and an irrational number corresponds to a complex number that is not a root of unity. If $z = e^{2\pi i \beta}$, then the function $\alpha \mapsto e^{2\pi i \alpha}$ maps frac$(n\beta)$ to z^n for all $n = 1, 2, \ldots$. In fact, for all $z \in \mathbb{C}$, if the argument of z is $2\pi\beta$, the argument of z^n equals 2π frac$(n\beta)$. See Exercise 17 for more on these properties of $\mathbb{T}$ and the map $\alpha \mapsto e^{2\pi i \alpha}$.

Putting $z = e^{2\pi i \alpha}$, we now have, for $n = 1, 2, 3, \ldots$,

$$2\pi \text{ frac}(n\alpha) = \text{the argument of } e^{2\pi i n\alpha}$$
$$= \text{the argument of } (e^{2\pi i \alpha})^n$$
$$= \text{the argument of } z^n.$$

Thus, bearing in mind the geometric meaning of the argument of a complex number, we see that the question of how the points frac(α), frac(2α), frac(3α), ... are distributed in $[0, 1)$ can be seen equivalently as the question of how the points $z, z^2, z^3, \ldots$ are distributed around the unit circle $\mathbb{T}$. This is illustrated in Figures 2.1 and 2.2.

The following proposition describes the uniqueness of the decomposition in (2.2), and also shows that the fractional part of a number is unaffected by adding an integer.

Proposition 2.1. (1) *Let* $\alpha \in \mathbb{R}$, $m \in \mathbb{Z}$ *and* $\gamma \in [0, 1)$ *be such that* $\alpha = \gamma + m$. *Then,* $\gamma = \text{frac}(\alpha)$ *and* $m = int(\alpha)$.

2.2. Fractional parts and points on the unit circle

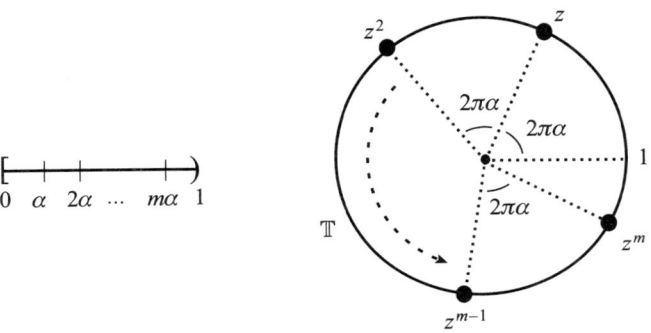

Figure 2.2. Let $\alpha \in (0, 1)$. Put $z = e^{2\pi i \alpha} \in \mathbb{T}$. Let m be the largest number in $\mathbb{N}$ such that $m\alpha < 1$. Now when we multiply two complex numbers, we add their arguments. Thus, under the mapping $\alpha \mapsto e^{2\pi i \alpha}$, α maps to z, 2α maps to z^2 and so on, and $m\alpha$ maps to z^m. When $n > m$ the pattern continues, and $\mathrm{frac}(n\alpha)$ maps into z^n, as in Figure 2.2. In this way the distribution of the points of the sequence $(\mathrm{frac}(n\alpha))$ in $[0, 1)$ becomes a question of the distribution of the points of the sequence (z^n) in $\mathbb{T}$.

(2) *Let* $\alpha, \beta \in \mathbb{R}$. *Then,* $\alpha - \beta \in \mathbb{Z} \iff \mathrm{frac}(\alpha) = \mathrm{frac}(\beta)$.
(3) *Let* $\alpha \in \mathbb{R}$ *and* $m \in \mathbb{Z}$. *Then,* $\mathrm{frac}(\alpha) = \mathrm{frac}(\alpha + m)$.

Proof. (1) Let $\alpha = \gamma + m$, where $m \in \mathbb{Z}$ and $\gamma \in [0, 1)$. Then,

$$\mathrm{frac}(\alpha) + \mathrm{int}(\alpha) = \alpha = \gamma + m \implies \mathrm{frac}(\alpha) - \gamma = m - \mathrm{int}(\alpha).$$

Here, $\mathrm{frac}(\alpha) - \gamma \in (-1, 1)$, but $m - \mathrm{int}(\alpha) \in \mathbb{Z}$. Hence,

$$\mathrm{frac}(\alpha) - \gamma = m - \mathrm{int}(\alpha) \in (-1, 1) \cap \mathbb{Z} = \{0\}.$$

Thus, $\gamma = \mathrm{frac}(\alpha)$ and $m = \mathrm{int}(\alpha)$, which proves (1).
(2) Observe that

$$\mathrm{frac}(\alpha) - \mathrm{frac}(\beta) = \alpha - \mathrm{int}(\alpha) - \beta + \mathrm{int}(\beta),$$

Then if $\alpha - \beta \in \mathbb{Z}$, we deduce that $\mathrm{frac}(\alpha) - \mathrm{frac}(\beta) \in (-1, 1) \cap \mathbb{Z} = \{0\}$, and $\mathrm{frac}(\alpha) = \mathrm{frac}(\beta)$. Conversely, if $\mathrm{frac}(\alpha) = \mathrm{frac}(\beta)$, we see that $\alpha - \mathrm{int}(\alpha) - \beta + \mathrm{int}(\beta) = 0$, which gives immediately that $\alpha - \beta \in \mathbb{Z}$.

(3) We have
$$\alpha + m = \big(\mathrm{frac}(\alpha) + \mathrm{int}(\alpha)\big) + m = \mathrm{frac}(\alpha) + \big(\mathrm{int}(\alpha) + m\big),$$
and note that $\mathrm{frac}(\alpha) \in [0, 1)$ and $\mathrm{int}(\alpha) + m \in \mathbb{Z}$. Then, by (1), $\mathrm{frac}(\alpha + m) = \mathrm{frac}(\alpha)$. □

Whether α is rational or irrational is expressed by different properties of the sequence
$$\mathrm{frac}(\alpha), \mathrm{frac}(2\alpha), \mathrm{frac}(3\alpha), \ldots.$$
One of these distinguishing properties is the following.

Definition A sequence $x_1, x_2, x_3, \ldots$ is called *periodic* if there is $r \in \mathbb{N}$ such that
$$x_{n+r} = x_n, \text{ for all } n \in \mathbb{N}.$$

Thus, a periodic sequence is one which, after some finite number r of terms, exactly repeats its behavior over the next r terms, and so on indefinitely.

Proposition 2.2. *The number α is rational if and only if the sequence $\big(\mathrm{frac}(n\alpha)\big)$ is periodic, in which case $\mathrm{frac}(n\alpha) = 0$ for some n. Alternatively, α is irrational if and only if the sequence $\big(\mathrm{frac}(n\alpha)\big)$ is not periodic. Also, when α is irrational, all terms of the sequence $\big(\mathrm{frac}(n\alpha)\big)$ are distinct.*

Proof. Let α be rational, and put $\alpha = p/q$ where $p \in \mathbb{Z}$ and $q \in \mathbb{N}$. Let $x_n = \mathrm{frac}(n\alpha)$, for $n = 0, 1, 2, 3, \ldots$. Then, using (3) of Proposition 2.1,
$$x_{n+q} = \mathrm{frac}\big((n+q)\alpha\big) = \mathrm{frac}\left((n+q)\frac{p}{q}\right) = \mathrm{frac}\left(\frac{np}{q} + p\right)$$
$$= \mathrm{frac}\left(\frac{np}{q}\right) = \mathrm{frac}(n\alpha) = x_n.$$

Thus, the sequence (x_n) is periodic, and we also see that $0 = x_q = \mathrm{frac}(q\alpha)$. Thus, the sequence $\big(\mathrm{frac}(n\alpha)\big)$ is periodic when α is rational.

Now, let α be irrational, and assume that there are two terms of the sequence $(\text{frac}(n\alpha))$ which are equal. Then there are $m, n \in \mathbb{N}$ with $m \neq n$ such that $\text{frac}(m\alpha) = \text{frac}(n\alpha)$. Using (2) of Proposition 2.1 shows that there is $s \in \mathbb{Z}$ such that $m\alpha - n\alpha = s$. But this implies that $\alpha = s/(m-n)$, which contradicts the assumption that α is irrational. Hence, if α is irrational, all terms of the sequence $(\text{frac}(n\alpha))$ are distinct, and this also implies that the sequence is not periodic. □

There is another way of thinking about Proposition 2.2. The terms of the sequence may be considered as a *set* rather than a sequence. In this case we refer to the *set* $\{\text{frac}(\alpha), \text{frac}(2\alpha), \text{frac}(3\alpha), \ldots\}$, instead of the sequence $(\text{frac}(n\alpha))$. Then, Proposition 2.2 implies the following for a real number α: *α is rational if and only if the set $\{\text{frac}(\alpha), \text{frac}(2\alpha), \text{frac}(3\alpha), \ldots\}$ is finite*; alternatively, *α is irrational if and only if the set $\{\text{frac}(\alpha), \text{frac}(2\alpha), \text{frac}(3\alpha), \ldots\}$ is infinite*. Complementing the remarks in Section 1, this observation shows there is another sense in which irrational numbers may be associated with notions of the infinite, while rational numbers are associated with the finite.

2.3 Partitions and the Pigeon-hole Principle

Suppose that we have a certain number of pigeons and a smaller number of pigeonholes in which they go to sleep each night. Then it is surely true that at least two pigeons must share a single pigeonhole. This fact, expressed in various forms, is known as *the Pigeon-hole Principle*.[4] In this section, we mathematically formalize this principle, with a view to future applications. Recall from Section 1.3 of Chapter 1 that a *partition* of a set X is an expression of X as a disjoint union of some family of subsets of X.

The Pigeon-Hole Principle 2.3. *Let X be a set and let $X_1, X_2, \ldots, X_n$ be a finite partition of X. Assume further that $x_1, x_2, \ldots, x_r$ are points in X with $r > n$. Then there are $i \in \{1, 2, \ldots, n\}$ and $j, k \in \{1, 2, \ldots, r\}$ with $j \neq k$ such that $x_j \in X_i$ and $x_k \in X_i$.*

Proof. Observe that if there are $j, k \in \{1, 2, \ldots, r\}$ such that $j \neq k$ and $x_j = x_k$, the result holds. So, we may assume that all the points $x_1, x_2, \ldots, x_r$ are distinct. Now, if each of the n sets $X_1, X_2, \ldots, X_n$ contains at most one point of the form x_j for $j = 1, 2, \ldots, r$, the total number of points x_j is at most n and so $r \leq n$. But this contradicts the given fact that $n < r$, and establishes the result. □

Proposition 2.4. *Let $n \in \mathbb{N}$ and let $x_1, x_2, \ldots, x_{n+1}$ be $n + 1$ points in $[0, 1)$. Then there are $i, j \in \{1, 2, \ldots, n+1\}$ such that $i \neq j$ and $|x_i - x_j| < 1/n$.*

Proof. The sets

$$\left[0, \frac{1}{n}\right), \left[\frac{1}{n}, \frac{2}{n}\right), \ldots, \left[\frac{j-1}{n}, \frac{j}{n}\right), \ldots, \left[\frac{n-2}{n-1}, \frac{n-1}{n}\right), \left[\frac{n-1}{n}, 1\right),$$

clearly form a partition of $[0, 1)$ into n intervals. So, by the Pigeon-hole Principle, there are $i, j \in \{1, 2, \ldots, n+1\}$ such that $i \neq j$ and both x_i and x_j belong to a single one of these intervals. But each of these intervals is half open and has length $1/n$. It follows that $|x_i - x_j| < 1/n$. □

Proposition 2.4 is very useful in the following section concerning the distribution in the interval $[0, 1)$ of the fractional parts of the multiples of an irrational number.

2.4 Kronecker's Theorem

In Section 2.2, we have seen that when α is irrational, all terms of the sequence $\mathrm{frac}(\alpha), \mathrm{frac}(2\alpha), \mathrm{frac}(3\alpha), \ldots$ are different and so comprise an infinite set of points in $[0, 1)$. It is natural, then, to ask the following questions: how are these points distributed throughout $[0, 1)$? Is it possible, for example, for the terms of such a sequence to "cluster" in one part of the interval rather than another, thus giving the appearance of a non-random process; or are the terms of the sequence distributed throughout the interval? The main result in this section, Kronecker's Theorem, shows that when α is irrational, the terms of the sequence $(\mathrm{frac}(n\alpha))$ are spread throughout the interval, in that every non-empty open subinterval of $[0, 1)$ contains terms of the sequence.

2.4. Kronecker's Theorem

Theorem 2.5 (Kronecker's Theorem[5]). *Let α be an irrational number. Then for each non-empty open subinterval U of $[0, 1]$, there is $m \in \mathbb{N}$ such that the fractional part of $m\alpha$ is in U.*

Ideas in the proof of Kronecker's Theorem   It suffices to show that for each number $x \in [0, 1]$, and for each $\varepsilon > 0$, there is $m \in \mathbb{N}$ such that $|x - \text{frac}(m\alpha)| < \varepsilon$. Observe that as α is irrational, the terms of the sequence $(\text{frac}(n\alpha))$ are all distinct. If $0 < \varepsilon < 1$ is given, using the Pigeon-hole Principle we deduce that there are distinct terms of the sequence $(\text{frac}(n\alpha))$ whose distance apart is less than ε. This enables us to show that there are $j, k \in \mathbb{N}$ with $j > k$ such that $|\text{frac}(j\alpha) - \text{frac}(k\alpha)| < \varepsilon$. We put $\ell = j - k$ and consider when $\text{frac}(j\alpha) > \text{frac}(k\alpha)$. The properties of the frac function mean that we must have $0 < \text{frac}(\ell\alpha) < \varepsilon$. We see from this that there is $r \in \mathbb{N}$ such that $|x - r\,\text{frac}(\ell\alpha)| < \varepsilon$. Putting $m = r\ell$, we deduce that $|x - \text{frac}(m\alpha)| < \varepsilon$. In the case when $\text{frac}(k\alpha) > \text{frac}(j\alpha)$, we apply a similar argument, but to $1 - x$ in place of x. In either case, we deduce that there is $m \in \mathbb{N}$ such that $|x - \text{frac}(m\alpha)| < \varepsilon$, which suffices for the result.

The analysis here is based upon the interval $[0, 1)$, but it is very natural to think of the above ideas and the lemmas geometrically in terms of complex numbers on the unit circle. Letting $\mathbb{T}$ denote the unit circle, Kronecker's Theorem in its circle interpretation says: *if $z \in \mathbb{T}$ and z is not a root of unity, and if U is a non-empty open arc of $\mathbb{T}$, then there is $n \in \mathbb{N}$ such that $z^n \in U$.* Kronecker's Theorem from this viewpoint is illustrated in Figures 2.4 and 2.5. We prepare for the proof of Kronecker's Theorem with two lemmas.

Lemma 2.6. *Let $0 < \beta < \varepsilon < 1$ and let $x \in [0, 1]$. Then there is $r \in \mathbb{N}$ with $r\beta < 1$ such that $|x - r\beta| < \varepsilon$.*

Proof. Let n be the largest number in $\mathbb{N}$ such that $n\beta < 1$. Then, $n\beta < 1 \le (n+1)\beta$ and we have

$$[0, 1] = [0, \beta) \cup [\beta, 2\beta) \cup [2\beta, 3\beta) \cup \cdots \cup [(n-1)\beta, n\beta) \cup [n\beta, 1].$$

36   Chapter 2. Irrational Numbers and Dynamical Systems

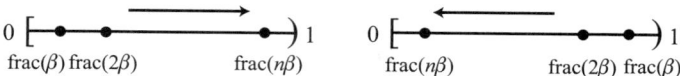

Figure 2.3. This illustrates the ideas in Lemma 2.6, and these ideas are used also in the proof of Kronecker's Theorem. We have $0 < \beta < \varepsilon < 1$. The points $\beta, 2\beta, \ldots, n\beta$ "mark a chain across the interval $(0, 1)$ whose mesh is less than ε", as it is expressed by Hardy and Wright [4, p. 376]. Thus, if x is a point in $[0, 1]$, it lies close to one of these points, and within a distance of ε. That is, there is $r \in \{1, 2, \ldots, n\}$ such that $|x - r\beta| < \varepsilon$.

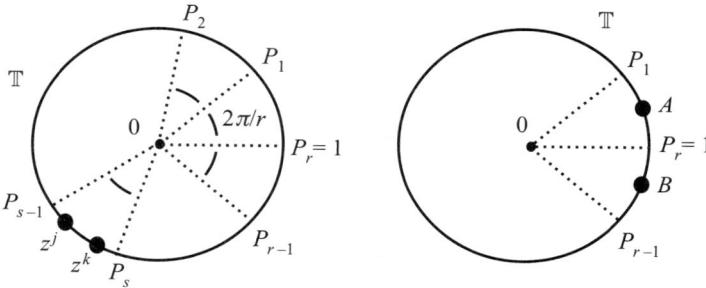

Figure 2.4. Figures 2.4 and 2.5 illustrate a proof of Kronecker's Theorem in its formulation on the unit circle $\mathbb{T}$. In the left picture, the unit circle is partitioned into r arcs of equal length, each subtended at the center by an angle $2\pi/r$. The end points of the arcs are $P_0 = 1, P_1, P_2, \ldots, P_{s-1}, P_s, \ldots, P_{r-1}, P_r = P_0 = 1$. Each point in $\mathbb{T}$ belongs to one of the r arcs going from P_{j-1} to P_j for some $j \in \{1, 2, \ldots, r\}$. Now let $z \in \mathbb{T}$, where z is not a root of unity, and consider the $r + 1$ (distinct) points $z, z^2, \ldots, z^r, z^{r+1}$. By the Pigeon-hole Principle, two of these points must belong to one of these arcs. Thus, there are distinct $j, k \in \{1, 2, \ldots, r + 1\}$ with $j > k$ and $s \in \{1, 2, \ldots, r\}$ such that z^j, z^k lie in the arc from P_{s-1} to P_s. Note that z^j is illustrated as being clockwise from z^k, but in fact it could be the other way around. Put $w = z^{j-k}$. As z^j, z^k both lie in the arc from P_{s-1} to P_s, w is at A in the arc from P_0 to P_1, or w is at B in the arc from P_{r-1} to P_r (of course, $P_0 = P_r = 1$). This is illustrated in the picture on the right.

2.4. Kronecker's Theorem

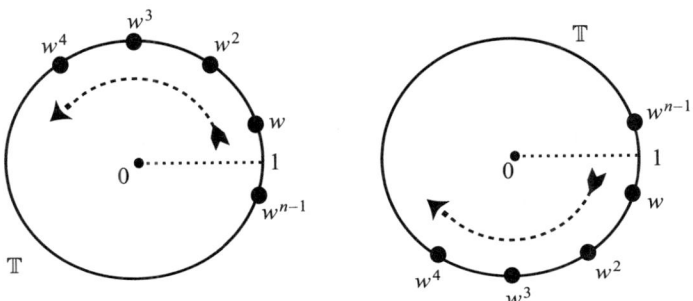

Figure 2.5. Following on from Figure 2.4, we have $w = z^{j-k}$ and w is in position A or B, as indicated in the right-hand picture in Figure 2.4. If w is in position A, the numbers $w, w^2, w^3, \ldots$ are spread equally around $\mathbb{T}$ in a counterclockwise direction, as indicated on the left of Figure 2.5, with n being the least value in $\mathbb{N}$ such that 1 is in the arc from w^{n-1} to w^n. Now let $u \in \mathbb{T}$. Then u must belong to one of the arcs going from w^{q-1} to w^q for some $q \in \{1, 2, \ldots, n-1\}$, or to the arc from w^{n-1} to 1. In any case, u belongs to an arc, one of whose endpoints is of the form w^q for some $q \in \mathbb{N}$. Every such arc has length less than $2\pi/r$, so if we put $m = (j-k)q \in \mathbb{N}$. we have

$$|u - z^m| = |u - z^{(j-k)q}| = |u - w^q| < \frac{2\pi}{r}. \qquad (*)$$

If w is in position B, the numbers $w, w^2, w^3, \ldots$ spread equally around $\mathbb{T}$ in a *clockwise* direction, as indicated on the right-hand side of Figure 2.5, but the substance of the argument is not affected, and again we obtain $(*)$. Now, for any given $\varepsilon > 0$, observe that in $(*)$, if r is sufficiently large, we get $|u - z^m| < \varepsilon$. Thus, every non-empty open arc in $\mathbb{T}$ contains a point z^m, for some $m \in \mathbb{N}$, which is Kronecker's Theorem for $\mathbb{T}$.

Thus, if $x \in [0, 1]$, x must belong to $[n\beta, 1]$ or to one of the intervals $[(j-1)\beta, j\beta)$ for some $j \in \{1, 2, \ldots, n\}$. Note that all of these intervals have length at most β, and each interval has an endpoint that is of the form $j\beta$ for some $j \in \{1, 2, \ldots, n\}$. Thus, there is $r \in \mathbb{N}$ such that $r\beta < 1$ and $|x - r\beta| \leq \beta < \varepsilon$. □

Lemma 2.7. *The following hold.*

(1) *Let $\gamma \in \mathbb{R}$ and let $n \in \mathbb{N}$ be such that $n \operatorname{frac}(\gamma) < 1$. Then, $n \operatorname{frac}(\gamma) = \operatorname{frac}(n\gamma)$.*

(2) Let $\alpha, \beta \in \mathbb{R}$. Then, if $\operatorname{frac}(\alpha) > \operatorname{frac}(\beta)$, $\operatorname{frac}(\alpha - \beta) = \operatorname{frac}(\alpha) - \operatorname{frac}(\beta)$.

(3) If $\alpha \in \mathbb{R}$ and $\alpha \notin \mathbb{Z}$, $\operatorname{frac}(-\alpha) = 1 - \operatorname{frac}(\alpha)$.

Proof. (1) As $0 \leq n \operatorname{frac}(\gamma) < 1$ we have, using (1) of Proposition 2.1,

$$n \operatorname{frac}(\gamma) = \operatorname{frac}(n \operatorname{frac}(\gamma)) = \operatorname{frac}(n\gamma - n\operatorname{int}(\gamma)) = \operatorname{frac}(n\gamma).$$

(2) Observe that

$$\begin{aligned}\operatorname{frac}(\alpha - \beta) &= \alpha - \beta - \operatorname{int}(\alpha - \beta) \quad (2.3)\\ &= \operatorname{frac}(\alpha) + \operatorname{int}(\alpha) - \operatorname{frac}(\beta) - \operatorname{int}(\beta) - \operatorname{int}(\alpha - \beta)\\ &= \bigl(\operatorname{frac}(\alpha) - \operatorname{frac}(\beta)\bigr) + \bigl(\operatorname{int}(\alpha) - \operatorname{int}(\beta) - \operatorname{int}(\alpha - \beta)\bigr).\end{aligned}$$

However, $\operatorname{frac}(\alpha) > \operatorname{frac}(\beta)$, so that $\operatorname{frac}(\alpha) - \operatorname{frac}(\beta) \in [0, 1)$. Hence, (2.4) expresses $\operatorname{frac}(\alpha) - \operatorname{frac}(\beta)$ as the sum of a number $\operatorname{frac}(\alpha) - \operatorname{frac}(\beta) \in [0, 1)$ and an integer. So, by (1) of Proposition 2.1, we have

$$\operatorname{frac}(\alpha - \beta) = \operatorname{frac}(\alpha) - \operatorname{frac}(\beta).$$

(3) Note that because $\alpha \notin \mathbb{Z}$, $\operatorname{frac}(\alpha) \in (0, 1)$. We have

$$\begin{aligned}\alpha &= \operatorname{frac}(\alpha) + \operatorname{int}(\alpha) \Longrightarrow -\alpha\\ &= -\operatorname{frac}(\alpha) - \operatorname{int}(\alpha) \Longrightarrow -\alpha\\ &= (1 - \operatorname{frac}(\alpha)) + (-\operatorname{int}(\alpha) - 1).\end{aligned}$$

$\square$

Note that the conclusion in (3) of Lemma 2.7 fails if $\alpha \in \mathbb{Z}$.

Kronecker's Theorem. We show that for every given irrational number α, for every number $x \in [0, 1]$, and for every $\varepsilon > 0$, there is $m \in \mathbb{N}$ such that $|x - \operatorname{frac}(m\alpha)| < \varepsilon$.

Let $x \in [0, 1]$, let ε be any number such that $0 < \varepsilon < 1$, and let $r \in \mathbb{N}$ be such that $1/r < \varepsilon$. Let α be an irrational number, and observe that if $j, k \in \mathbb{N}$ with $j \neq k$ then, by Proposition 2.2, $\operatorname{frac}(j\alpha) \neq \operatorname{frac}(k\alpha)$. So, the points $\operatorname{frac}(\alpha), \operatorname{frac}(2\alpha), \ldots, \operatorname{frac}((r + 1)\alpha)$ are $r + 1$ distinct points

2.4. Kronecker's Theorem

in $[0, 1)$. By Proposition 2.4 there are $j, k \in \{1, 2, \ldots, r, r+1\}$ such that $j \neq k$ and

$$|\operatorname{frac}(j\alpha) - \operatorname{frac}(k\alpha)| < \frac{1}{r} < \varepsilon. \qquad (2.4)$$

Now, (2.4) remains the same if we interchange j and k. So, as $j \neq k$, we may assume without losing any generality in our argument that $j > k$. Put $\ell = j - k$. Now, we consider two possible cases.

Case I. $\operatorname{frac}(j\alpha) - \operatorname{frac}(k\alpha) > 0$.

Using (2) of Lemma 2.7 and (2.4), we have

$$0 < \operatorname{frac}(\ell\alpha) = \operatorname{frac}(j\alpha - k\alpha)$$
$$= \operatorname{frac}(j\alpha) - \operatorname{frac}(k\alpha)$$
$$= |\operatorname{frac}(j\alpha) - \operatorname{frac}(k\alpha)| < \varepsilon.$$

Thus, $0 < \operatorname{frac}(\ell\alpha) < \varepsilon$, and we can apply Lemma 2.6 with $\beta = \operatorname{frac}(\ell\alpha)$ to deduce that there is $r \in \mathbb{N}$ with $r \operatorname{frac}(\ell\alpha) < 1$ and

$$|x - r \operatorname{frac}(\ell\alpha)| < \varepsilon. \qquad (2.5)$$

But, as $r \operatorname{frac}(\ell\alpha) < 1$, (1) of Lemma 2.7 gives

$$r \operatorname{frac}(\ell\alpha) = \operatorname{frac}(r\ell\alpha).$$

Thus, putting $m = r\ell$ and using (2.5) gives

$$|x - \operatorname{frac}(m\alpha)| = |x - \operatorname{frac}(r\ell\alpha)| = |x - r \operatorname{frac}(\ell\alpha)| < \varepsilon,$$

which establishes the result for Case I.

Case II. $\operatorname{frac}(j\alpha) - \operatorname{frac}(k\alpha) < 0$.

Using (2) of Lemma 2.7 and (2.4) we have

$$0 < \operatorname{frac}(-\ell\alpha)$$
$$= \operatorname{frac}(k\alpha - j\alpha)$$
$$= \operatorname{frac}(k\alpha) - \operatorname{frac}(j\alpha) \qquad (2.6)$$
$$= |\operatorname{frac}(k\alpha) - \operatorname{frac}(j\alpha)|$$
$$< \varepsilon.$$

Thus, (2.6) gives $0 < \text{frac}(-\ell\alpha) < \varepsilon$, and we can apply Lemma 2.6 with $\beta = \text{frac}(-\ell\alpha)$ and $1 - x$ in place of x to deduce that there is $r \in \mathbb{N}$ with $r \,\text{frac}(-\ell\alpha) < 1$ and

$$|1 - x - r\,\text{frac}(-\ell\alpha)| < \varepsilon. \tag{2.7}$$

But, as $r \,\text{frac}(-\ell\alpha) < 1$, (1) of Lemma 2.7 gives

$$r \,\text{frac}(-\ell\alpha) = \text{frac}(-r\ell\alpha).$$

Thus, using (3) of Lemma 2.7, putting $m = r\ell$ and using (2.7) gives

$$\begin{aligned}|x - \text{frac}(m\alpha)| &= |1 - x - 1 + \text{frac}(r\ell\alpha)| \\ &= |1 - x - \text{frac}(-r\ell\alpha)| \\ &= |1 - x - r\,\text{frac}(-\ell\alpha)| \\ &< \varepsilon,\end{aligned}$$

which establishes the result for Case II. □

Kronecker's Theorem may be stated in various ways. In the proof given, we showed that if α is an irrational number, then for every $x \in [0, 1]$ and every $\varepsilon > 0$, there is $m \in \mathbb{N}$ such that $|x - \text{frac}(m\alpha)| < \varepsilon$. Now, a sequence or set is called *dense* in $[0, 1]$ if it has the property that for every $x \in [0, 1]$ and every $\varepsilon > 0$, there is an element y in the sequence or set such that $|x - y| < \varepsilon$. Thus, when α is irrational, Kronecker's Theorem says that the sequence of terms $\text{frac}(\alpha), \text{frac}(2\alpha), \ldots$ is dense in $[0, 1]$. As mentioned at the beginning of this section, this is consistent with the idea that the sequence has some characteristics of "randomness", although more remains to be said on this. In view of the fact that when α is rational, the sequence $\text{frac}(\alpha), \text{frac}(2\alpha), \ldots$ has only a finite number of terms, we see from Kronecker's Theorem that *a real number α is irrational if and only if the sequence $\text{frac}(\alpha), \text{frac}(2\alpha), \ldots$ is dense* in $[0, 1]$.

2.5 The dynamical systems approach to Kronecker's Theorem

Recall that if we have a dynamical system (S, f), and if $x \in S$, the *orbit* of x is the sequence of points $x, f(x), f^2(x), \ldots$ in S. Kronecker's Theorem says that for an irrational number α, points of the sequence $\text{frac}(\alpha), \text{frac}(2\alpha), \text{frac}(3\alpha) \ldots$ occur in every open and non-empty subinterval of $[0, 1)$. We will show that for a suitable transformation τ_α of $[0, 1)$, the sequence $\text{frac}(\alpha), \text{frac}(2\alpha), \text{frac}(3\alpha) \ldots$ is the orbit of 0 under the transformation τ_α. Thus, Kronecker's Theorem can be regarded as saying that every open and non-void subinterval of $[0, 1)$ will contain points in the orbit of 0 under τ_α.

Now, let $\alpha \in \mathbb{R}$. We will see below that the transformation on $[0, 1)$ relevant to the dynamical systems formulation of Kronecker's Theorem for α is τ_α, as given by

$$\tau_\alpha(x) = \text{frac}(x + \alpha). \tag{2.8}$$

A transformation of the form τ_α is called a *translation* on $[0, 1)$, and when α is irrational it is called an *irrational translation*. Sometimes, τ_α may be called the *translation by α* on $[0, 1)$. If $\alpha > 0$ the translation "slides" or "translates" points of $[0, 1)$ to the right by a distance α. When $\alpha < 0$, we regard the translation as a negative translation to the right by α—geometrically, it translates to the left. If a point is thus translated "too far" to the right or left, so that it lies outside of $[0, 1)$, we simply "start again" at 0 or at 1, as the case may be, until the full translation by α has occurred, giving a point in $[0, 1)$. In the case when $\alpha \in [0, 1)$, (2.8) and the definition of the fractional part give

$$\tau_\alpha(x) = \begin{cases} x + \alpha, & 0 \leq x < 1 - \alpha; \\ x + \alpha - 1, & 1 - \alpha \leq x < 1. \end{cases} \tag{2.9}$$

Note that when $\alpha = 0$, $\tau_\alpha(x) = x$ for all $x \in [0, 1)$. That is, τ_0 is the identity transformation on $[0, 1)$. However, if $0 < \alpha < 1$, τ_α is a linear or "straight line function" on each of the intervals $[0, 1 - \alpha)$ and

$[1-\alpha, 1)$, with a discontinuity at $1-\alpha$. The transformation τ_α is one-to-one and maps $[0, 1)$ onto $[0, 1)$. If the interval $[0, 1)$ is identified with the unit circle $\mathbb{T}$, under the mapping $t \mapsto e^{2\pi i t}$, translation by α on $[0, 1)$ corresponds to a *rotation* on $\mathbb{T}$, through an angle of $2\pi\alpha$ in an counterclockwise direction. Equivalently, translation by α on $[0, 1)$ corresponds to the transformation $z \mapsto e^{2\pi i \alpha} z$ of $\mathbb{T}$. These ideas, and some more detailed mapping properties of τ_α, including aspects of information loss and randomness of the system, are illustrated in Figures 2.6 and 2.7

Definition A dynamical system of the form $\bigl([0, 1), \tau_\alpha\bigr)$ is called a *Kronecker system*.

Lemma 2.8. *Let $\alpha, \beta \in \mathbb{R}$. Then, $\tau_\alpha = \tau_\beta \iff \alpha - \beta \in \mathbb{Z}$. In particular, $\tau_\alpha = \tau_{\mathrm{frac}(\alpha)}$.*

Proof. We have from (2) of Proposition 2.1 that

$$\tau_\alpha = \tau_\beta \iff \mathrm{frac}(x+\alpha) = \mathrm{frac}(x+\beta) \text{ for all } x \in [0, 1) \iff \alpha - \beta \in \mathbb{Z}.$$

In particular, as $\alpha - \mathrm{frac}(\alpha) = \mathrm{int}(\alpha) \in \mathbb{Z}$, we have also that $\tau_\alpha = \tau_{\mathrm{frac}(\alpha)}$. □

We have indicated that a translation on $[0, 1)$ can be thought of as causing points to "slide" to the right by a certain amount. So if we have two translations, one by α and another by β, the composition of the two translations would cause points to "slide" through an amount α followed by an amount β. That is, the composition of the two would cause a "slide" through the amount $\alpha + \beta$. Alternatively, if we think of this in terms of rotations on $\mathbb{T}$, as described in Figures 2.6 and 2.7, it is clear that if we rotate counterclockwise through an angle $2\pi\alpha$, and rotate counterclockwise again through an angle $2\pi\beta$, then we would expect this to result overall in an counterclockwise rotation through an angle $2\pi(\alpha + \beta)$. These conclusions correspond respectively to the identities

$$\mathrm{frac}(\alpha + \mathrm{frac}(\beta)) = \mathrm{frac}(\alpha + \beta) \text{ and } e^{2\pi i \alpha} e^{2\pi i \beta} = e^{2\pi i (\alpha+\beta)},$$

which hold for all $\alpha, \beta \in \mathbb{R}$. The following is the formal result in the case of translations.

2.5. The dynamical systems approach to Kronecker's Theorem 43

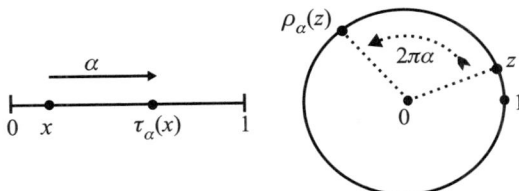

Figure 2.6. The picture on the left illustrates the transformation τ_α on $[0, 1)$ for $0 < \alpha < 1$, where x moves to the right through a distance α, giving the point $\tau_\alpha(x)$. The picture on the right illustrates the transformation ρ_α on $\mathbb{T}$ which rotates points in an counterclockwise direction around $\mathbb{T}$ through an angle of $2\pi\alpha$. Thus, ρ_α is the transformation $z \mapsto e^{2\pi i \alpha} z$ on $\mathbb{T}$. Under the identification of $[0, 1)$ with $\mathbb{T}$, described in Figure 2.2, τ_α corresponds to ρ_α. That is, "a translation on $[0, 1)$ is (equivalent to) a rotation on $\mathbb{T}$". If $x \in [0, 1)$ we have $e^{2\pi i \tau_\alpha(x)} = e^{2\pi i \, \text{frac}(x+\alpha)} = e^{2\pi i (x+\alpha)} = e^{2\pi i \alpha} e^{2\pi i x} = \rho_\alpha(e^{2\pi i x})$. Thus, the correspondence between τ_α and ρ_α is formally expressed by the equation
$$e^{2\pi i \tau_\alpha(x)} = \rho_\alpha(e^{2\pi i x}).$$

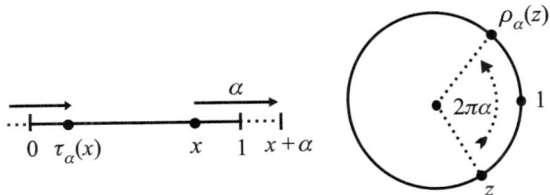

Figure 2.7. As in Figure 2.6, this illustrates the transformation τ_α. This time, when x is moved to the right through a distance α, x passes through 1 and the point $x + \alpha$ is outside the interval $[0, 1)$. In this case, to calculate $\tau_\alpha(x)$, we take the fractional part of $x + \alpha$, so $\tau_\alpha(x)$ reappears near the "beginning" of the interval $[0, 1)$, at $x + \alpha - 1$. In terms of the unit circle, there is no need to consider this as a separate case, because if we put $z = e^{2\pi i x}$, $\rho_\alpha(z)$ is simply found, as before, by rotating z counterclockwise through the angle $2\pi\alpha$.

Proposition 2.9. Let $\alpha, \beta \in \mathbb{R}$. Then,

$$\tau_\alpha \circ \tau_\beta = \tau_{\alpha+\beta}. \tag{2.10}$$

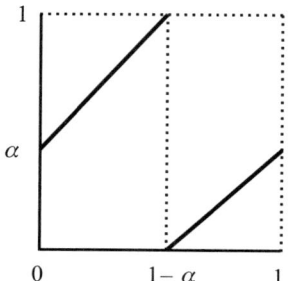 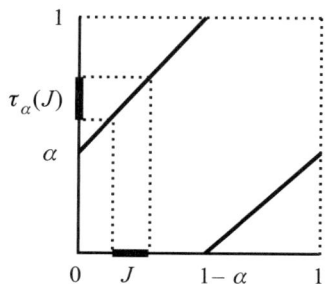

Figure 2.8. The picture on the left is of the graph of the transformation τ_α on $[0, 1)$, for $0 < \alpha < 1$. The discontinuity τ_α has at $1 - \alpha$ corresponds to when $\tau_\alpha(x)$ reaches 1 as x increases to $1 - \alpha$ and "reappears" at 0 as x passes through $1 - \alpha$. The graph consists of two line segments of slope 1, with a discontinuity at $1 - \alpha$. Within $[0, 1 - \alpha)$ and $[1 - \alpha, 1)$, information on the first iteration is neither lost nor gained, in the sense of Section 1.2 of Chapter 1, because the slopes are 1 in each case. However, at the discontinuity $1 - \alpha$, information is lost in the first iteration. As iteration continues, the system will lose information at the successive *single* points $1 - \alpha, 1 - \text{frac}(2\alpha), 1 - \text{frac}(3\alpha), \ldots$. When α is irrational, by Kronecker's Theorem these points of information loss spread throughout the system, leading to a form of "semi-random" behavior (see the comments in Section 2.9). But a more extreme form of randomness occurs with the system $([0, 1), f)$ where $f(x) = 2x$ for $0 \leq x < 1/2$ and $f(x) = 2x - 1$, for $1/2 \leq x < 1$. This is because at every point of the domain, except for a finite number of discontinuities, the slope of f^n is greater than 1, namely it is 2^n, so that information is going to be lost at every point on the next iteration. The system $[0, 1), f)$ is discussed extensively in Chapter 3.

On the right we see that if an interval J is contained in $[0, 1 - \alpha)$, then $\tau_\alpha(J)$ is an interval having the same length as J. In the circle interpretation, both J and $\tau_\alpha(J)$ correspond to arcs, having equal lengths.

Proof. Recall from (3) of Proposition 2.1 that if $y \in \mathbb{R}$ and $m \in \mathbb{N}$, then $\text{frac}(y + m) = \text{frac}(y)$. Using this, we see that for all $x \in [0, 1)$,

$$(\tau_\alpha \circ \tau_\beta)(x) = \text{frac}(\alpha + \text{frac}(x + \beta))$$
$$= \text{frac}(\alpha + x + \beta - \text{int}(x + \beta))$$
$$= \text{frac}(\alpha + \beta + x)$$
$$= \tau_{\alpha+\beta}(x).$$

2.5. The dynamical systems approach to Kronecker's Theorem

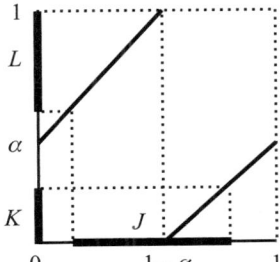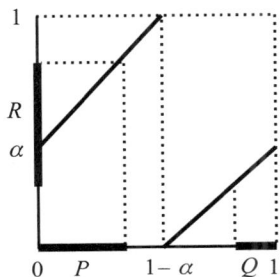

Figure 2.9. Each picture again shows the graph of the transformation τ_α. On the left we see that if J is an interval with $1-\alpha \in J$, then $\tau_\alpha(J)$ is a union of the two intervals K, L. In the circle interpretation, J and $K \cup L$ correspond to arcs. On the right, we see that if P, Q are two intervals as shown, then $\tau_\alpha(P \cup Q)$ is the interval R. Thus we see from Figures 2.8 and 2.9 that if a subset A of $[0, 1)$ is either an interval, or the union of two intervals respectively having 0 and 1 as endpoints, then $\tau_\alpha(A)$ is a set of the same type. In the circle interpretation, all this is saying is that when an arc is rotated counterclockwise around the circle, we obtain an arc. These ideas are used in Section 2.8 (see Figures 2.11 and 2.12).

□

If ρ_α denotes the rotation on $\mathbb{T}$ corresponding to the translation τ_α on $[0, 1)$, then the identity corresponding to (2.10) is $\rho_\alpha \circ \rho_\beta = \rho_{\alpha+\beta}$. The identity (2.10) is important for interpreting Kronecker's Theorem in terms of the dynamical system $([0, 1), \tau_\alpha)$, as we now see. For, if $\alpha \in \mathbb{R}$, we will have

$$\tau_\alpha^2 = \tau_\alpha \circ \tau_\alpha = \tau_{\alpha+\alpha} = \tau_{2\alpha}, \quad \tau_\alpha^3 = \tau_\alpha \circ \tau_\alpha \circ \tau_\alpha = \tau_{\alpha+\alpha+\alpha} = \tau_{3\alpha},$$

and so on.

In general we have that for all $n = 1, 2, 3, \ldots$, $\tau_\alpha^n = \tau_{n\alpha}$. So we now have that for all $n = 1, 2, 3, \ldots$,

$$\text{frac}(n\alpha) = \tau_{n\alpha}(0) = \tau_\alpha^n(0). \quad (2.11)$$

This has proved the following statement which establishes a connection between the sequence (frac($n\alpha$)) and an orbit in a Kronecker system.

Proposition 2.10. *Let α be a given a real number. Then the sequence* $\operatorname{frac}(\alpha)$, $\operatorname{frac}(2\alpha)$, $\operatorname{frac}(3\alpha)$, ... *is the sequence which is the orbit of 0 in the Kronecker system* $([0, 1), \tau_\alpha)$. *In this system, the orbit of 0 is finite if and only if α is rational and, when α is irrational, the orbit of 0 is dense in* $[0, 1)$.

Proof. As noted, the first statement follows from (2.11). The remaining statements follow from (2.11) in the light of Proposition 2.2 and Kronecker's Theorem. □

Now if Proposition 2.10 tells us about the orbit of 0 in the dynamical system $([0, 1), \tau_\alpha)$, what can be said about the orbit of points other than 0? This is the subject of Exercises 10 and 12, but in view of its importance, it is stated here.

Theorem 2.11. *Let α be an irrational number and let $x \in [0, 1)$. Then the orbit of x in the dynamical system $([0, 1), \tau_\alpha)$ is the sequence* $\operatorname{frac}(x + \alpha)$, $\operatorname{frac}(x + 2\alpha)$, $\operatorname{frac}(x + 3\alpha)$, ... *In this system, the orbit of x is finite if and only if α is rational and, when α is irrational, the orbit of x is dense in* $[0, 1)$.

2.6 Kronecker and chaos in the music of Steve Reich

Steve Reich[6], writing about his 1965 composition *"It's Gonna Rain"*, says that it

> ... was composed in San Francisco in January 1965. The voice belongs to a young black Pentecostal preacher who called himself Brother Walter. I recorded him along with the pigeons and traffic one Sunday afternoon in Union Square in downtown San Francisco. Later at home I started playing with tape loops of his voice and, by accident, discovered the process of letting two identical loops go gradually out of phase with each other.

Reich describes this process of "letting ... identical loops go gradually out of phase with each other" as follows:

2.6. Kronecker and chaos in the music of Steve Reich

> In the first part of the piece the two loops are lined up in unison, gradually move completely out of phase with each other, and then slowly move back to unison. In the second part, two much longer loops gradually begin to go out of phase with each other. This two-voice relationship is then doubled to four with two voices going out of phase with the other two. Finally, the process moves to eight voices and the effect is a kind of controlled chaos, which may be appropriate to the subject matter—the end of the world.

It seems to me that this is only an indication of how, precisely, the loops are caused to go "out of phase" with each other. Does the word "gradually" refer to a *continuous* changing of phase, or does it, as perhaps seems more likely, refer to a series of *discrete* changes? Here we will analyze a mathematical model for a continuous changing of phase, and the discrete model is left as a matter for reader investigation. We assume the two identical tape loops start off in the same position on each of two identical tape players. If the two tapes start moving at the same constant speed, the music we hear will be the same music as on any single one of the tapes, because the two tapes are exactly synchronized and play the same notes of the music at the same time—there will be no *interaction* between the music on the two tapes. However, the music on the one tape will go *out of phase* with the music on the other tape if one tape moves at a constant speed v_1 say, while the other tape moves at a constant but *different* speed v_2.

So, imagine we have two circular and identical tape loops of radius r. These circular loops of tape move about their respective centers in an counterclockwise direction with speeds v_1 and v_2 respectively. The music on the first tape commences at the point P_1 on the tape, and the music on the second tape commences at the point P_2 on the tape. As the first tape turns, there is a stable point Q_1, that is not on the tape and does not move with the tape, at which the tape player plays the notes that are recorded on the part of the tape that is at Q_1 at the time. Similarly, as the second tape runs, there is a stable point Q_2, that is not on the tape and does not move with the tape, at which the tape player plays the notes that are recorded on the tape at Q_2. Denote time by t. When $t = 0$, we assume the two tapes start to play the beginning of the music—that is, the tapes are synchronized at $t = 0$. So, when $t = 0$, the point P_1 on the tape is at the point Q_1 where the tape player reproduces the notes on the first tape, and the point P_2 on the tape is at the point Q_2 where the tape player reproduces the notes on the second tape. For convenience, we will take Q_1, Q_2 to be the extreme right-hand points on their respective circles. As the tapes are played, we hear a "music" which is the same as the music on the original tapes if $v_1 = v_2$, but which will be a possibly complicated interaction between the two identical musics on the two tapes when $v_1 \neq v_2$. We call this the *combined* music. Note that Reich says that initially the tapes " ... gradually move completely out of phase with each other, and then slowly move back to unison". Once the tapes are back in unison, the whole effect (that is the combined music) will be repeated again. Presumably that is why, in the composition, what happens next is varied by then using longer tape loops, then moving to four voices, and finally to eight. Here we shall try to understand the mathematical conditions that tell us, under our formulation of the whole procedure of the tapes moving out of phase, when the tapes will be back in unison. Alternatively, we can ask: will the combined music repeat itself after some time, starting from the beginning of the music, and if so, after how long?

Now, after the tapes have been playing for time t, the point P_1 on the first tape will have moved through a total distance $v_1 t$ around the circumference of the circle in an counterclockwise direction. If k denotes the number of times P_1 has gone all the way around its circle, we see that there is $\theta_1 \in [0, 1)$ such that

$$v_1 t = 2\pi r k + 2\pi r \theta_1. \tag{2.12}$$

Similarly, after time t, the point P_2 on the second tape will have moved through a total distance $v_2 t$ around the circumference of its circle in an counterclockwise direction. If ℓ denotes the number of times P_2 has gone all the way around its circle, we see that there is $\theta_2 \in [0, 1)$ such that

$$v_2 t = 2\pi r \ell + 2\pi r \theta_2. \tag{2.13}$$

2.6. Kronecker and chaos in the music of Steve Reich

> Since music was regarded as one of the mathematical sciences [from medieval times], *scientia mathematica*, the idealistic conception of music was that it should be based on speculative calculation and proportion together with reason and order, *unitas*, thus enabling maintenance of the total cosmic order.
>
> Music was thought of as the heir to the aestheticism of classicism, which took as its model the nature of the arts in classical Greece and Rome, and in particular in the belief that all phenomena are sublimated through order, harmony and universality; it was able to appeal subtly but forcibly to profound emotions through the avenue of the perspicacious intellect which is part of man's innate nature.
>
> Music viewed in this light was considered to have the power to transcend differences of taste, form and belief which were the products of different historical periods, and to be able to do this without sacrificing purity of expression.
>
> In other words, music as an art would not succumb to the ephemeral quirks of the age and would retain an immutable universality. Such is the essence of the music of J.S. Bach.
>
> —Masahiro Arita, in the introduction to J.S. Bach's *Musical Offering* for the Denon CD CO-78915.

When will the music repeat itself starting at the beginning again? Well, this occurs after time t if and only if, at time t, P_1 is at Q_1 and P_2 is at Q_2. It follows from (2.12) and (2.13) that this is equivalent to having

$$\theta_1 = 0 \text{ and } \theta_2 = 0. \tag{2.14}$$

Now, if (2.14) holds, it follows from (2.12) and (2.13) that $k, \ell \geq 1$ and that

$$\frac{v_1}{v_2} = \frac{k}{\ell},$$

so that v_1/v_2 is rational. On the other hand, suppose that v_1/v_2 is rational, so that $v_1/v_2 = k/\ell$ for some $k, \ell \in \mathbb{N}$. Then, observe that

$$\frac{2\pi r k}{v_1} = \frac{2\pi r k \ell}{v_1 \ell} = \frac{2\pi r k \ell}{v_2 k} = \frac{2\pi r \ell}{v_2}.$$

Put

$$t = \frac{2\pi r k}{v_1} = \frac{2\pi r \ell}{v_2}.$$

Then, it is clear from (2.12) and (2.13) that (2.14) holds for this value of t, so that the music repeats itself from the beginning after a lapse of time $2\pi rk/v_1 = 2\pi r\ell/v_2$. Incidentally, this shows that the shortest possible time after which the music can repeat itself from the beginning is obtained by writing $v_1/v_2 = k/\ell$ with k, ℓ having no common factors. In any case, we have proved the following.

Proposition 2.12. *In the formulation as described, where the tapes move respectively at the speeds v_1, v_2, there will be a time at which the music will repeat itself from the beginning if and only if v_1/v_2 is rational. Alternatively expressed, the music never repeats itself from the beginning if and only if v_1/v_2 is irrational.*

Now what happens if v_1/v_2 is irrational? This is where Kronecker's Theorem sheds some light. In this case, using (2.12) and (2.13), observe that (2.14) can be written as

$$\text{frac}\left(\frac{v_1 t}{2\pi r}\right) = 0 \quad \text{and} \quad \text{frac}\left(\frac{v_2 t}{2\pi r}\right) = 0. \tag{2.15}$$

Proposition 2.12 shows that (2.15) cannot be satisfied by any value of $t > 0$, but Kronecker's Theorem will show that the first equation in (2.15) can be satisfied by infinitely many values of t for which also the second equation is "approximately" satisfied to within any degree of approximation that we specify. To see this, observe that

$$\text{frac}\left(\frac{v_1 t}{2\pi r}\right) = 0, \quad \text{for all} \quad t = \frac{2\pi r}{v_1}, \frac{4\pi r}{v_1}, \frac{6\pi r}{v_1}, \ldots \tag{2.16}$$

We now check what happens in the second equation for these values of t. Putting t successively equal to the successive values of t in (2.16) gives that $\text{frac}(v_2 t/2\pi r)$ is successively equal to

$$\text{frac}\left(\frac{v_2}{v_1}\right), \text{frac}\left(2\left(\frac{v_2}{v_1}\right)\right), \text{frac}\left(3\left(\frac{v_2}{v_1}\right)\right), \text{frac}\left(4\left(\frac{v_2}{v_1}\right)\right), \ldots \tag{2.17}$$

But we are considering the case where v_1/v_2 is irrational. So by Kronecker's Theorem, if we are given an arbitrary $\varepsilon > 0$, there is an infinite

2.6. Kronecker and chaos in the music of Steve Reich

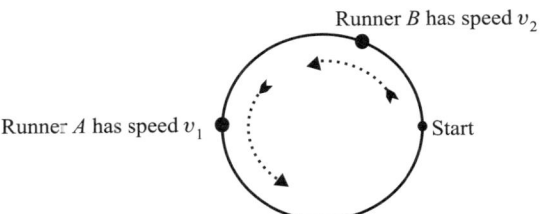

Figure 2.10. Runners A and B run counterclockwise around a circular race track of radius r, starting at the same point. Runner A has a speed v_1 which is greater than the speed v_2 of the runner B. Then A will move ahead of B and eventually overtake B. The process will then repeat itself, indefinitely. The time at which the first overtaking occurs is $2\pi r/(v_1 - v_2)$, with the subsequent overtaking times being $2\pi rk/(v_1 - v_2)$ for $k = 2, 3, \ldots$. But what about the possible points where the overtaking will occur? These are at distances of $2\pi r \,\mathrm{frac}(kv_1/(v_1 - v_2))$ for $k = 1, 2, \ldots$, going around the circle counterclockwise from the starting point. If v_1/v_2 is irrational, this set of points is *dense* in the unit circle $\mathbb{T}$. But if v_1/v_2 is rational, let's say that $v_1/v_2 = p/q$ where p, q are positive integers with no common factor and $p > q$. In this case, overtaking will occur at a finite number of points, given by the distances $2\pi r \,\mathrm{frac}\bigl(q/(p-q)\bigr), 2\pi r \,\mathrm{frac}\bigl(2q/(p-q)\bigr), \ldots, 2\pi r \,\mathrm{frac}\bigl((p-q-1)q/(p-q)\bigr)$ counterclockwise from the starting point.

number of $n \in \mathbb{N}$ such that $\mathrm{frac}\bigl((n(v_2/v_1)\bigr) < \varepsilon$. Choosing such an n, put $t = 2\pi rn/v_1$. It then follows from (2.16) and (2.17) that for this value of t,

$$\mathrm{frac}\left(\frac{v_1 t}{2\pi r}\right) = 0 \quad \text{and} \quad \mathrm{frac}\left(\frac{v_2 t}{2\pi r}\right) < \varepsilon. \qquad (2.18)$$

We can interpret (2.18) as saying: *when v_1/v_2 is irrational, the combined music can never repeat itself exactly, but it will repeat itself to within any arbitrarily assigned degree of accuracy an infinite number of times.*

There is another way of thinking about the preceding ideas. Imagine there are two runners, who run around a circular race track. When time is 0, they start running around the track in the same direction, starting from the same point. The first runner runs with a speed v_1 and the second runner runs with speed v_2. Assume that $v_1 > v_2$. As time goes on the first runner will get further and further ahead, until at last he or she will catch

52   Chapter 2. Irrational Numbers and Dynamical Systems

up to the slower runner—the slower runner will have been "lapped". Now let us ask the question: is the point at which the faster runner catches up with the slower runner ever the same as the point from which they started running? In the context of the music and the two tapes, this is equivalent to the question: does the music ever repeat itself, starting from the beginning? So, we see that the faster runner will catch up with the slower runner at the point from which they started running if and only if v_1/v_2 is rational. Thus, when v_1/v_2 is rational, the runners will run the same race over and over again, just as in the case of the music, the music repeats itself over and over again indefinitely. On the other hand, if v_1/v_2 is irrational, one runner will be at the starting point an infinite number of times while simultaneously the other runner is as close as we wish to specify to the starting point, but never *exactly at* the starting point.

2.7 The ideas in Weyl's Theorem on irrational numbers

Kronecker's Theorem shows that the terms of the sequence of fractional parts of the multiples of an irrational number are "spread everywhere" throughout the interval [0, 1). However, Kronecker's Theorem says nothing about the *nature* of this spreading. For example, do the terms of the sequence tend to appear more frequently in one part of the interval rather than another? Or are the terms of the sequence random in the sense that they favor no particular part of the interval? A clue to what happens comes from the following observation: when $frac(\ell\alpha)$ is close to 0, the terms $frac(\ell\alpha), frac(2\ell\alpha), \ldots$ increase *by equal amounts* for a period (see Figure 2.3, and Figure 2.5 for the unit circle interpretation); while if $frac(\ell\alpha)$ is close to 1, the terms $frac(\ell\alpha), frac((2\ell\alpha)), \ldots$ decrease *by equal amounts* for a period (see Figure 2.5 for the unit circle interpretation). Since each of these periods of increasing and decreasing occurs at a fixed rate, while either period lasts, it seems that the terms of the subsequence $frac(\ell\alpha), frac(2\ell\alpha), \ldots$ may be "evenly spread" or "uniformly distributed" throughout the interval. In fact, we shall see that the proportion of terms in the *original* sequence $frac(\alpha), frac(2\alpha),$

2.7. The ideas in Weyl's Theorem on irrational numbers

$frac(3\alpha), \ldots$ that lie in a subinterval J, obtained as a proportion of the first n terms that lie in J and then letting n tend to ∞, is equal to the length $\mu(J)$ of the interval J and depends only on the length. Thus, this sequence of points does not "show a preference" for one part of the interval $[0, 1)$ over any other part. Also, one way of thinking of the length $\mu(J)$ of J is that it represents the probability that a number "chosen at random" in $[0, 1)$ will belong to J. In this sense, therefore, we shall see that the terms of the sequence obtained from the fractional parts of the multiples of an irrational number are randomly distributed through the interval $[0, 1)$, but our above observations of regularities in the distribution suggest a "weak" form of random behavior. The following result was proved by Hermann Weyl[7] in 1909–10.

Theorem 2.13 (Weyl's Theorem). *Let α be an irrational number. Then the sequence $frac(\alpha), frac(2\alpha), frac(3\alpha), \ldots$ is distributed in $[0, 1)$ in such a way that for every subinterval J of $[0, 1)$, the proportion of the first n terms of the sequence that belong to J has a limit as n tends to infinity, and this limit is $\mu(J)$, the length of J.*

Let us now formulate Weyl's Theorem in more technical terms. Recall that if A is a finite set, $|A|$ denotes the number of elements in A. Let α be an irrational number. Then, for a subset J of $[0, 1)$ and for $n \in \mathbb{N}$, make the definition that $A(\alpha, J, n)$ is the set of all j such that $1 \leq j \leq n$ and $\text{frac}(j\alpha) \in J$. Thus,

$$A(\alpha, J, n) = \left\{ j : 1 \leq j \leq n \text{ and } \text{frac}(j\alpha) \in J \right\}.$$

Then, the proportion of the first n terms of the sequence $\text{frac}(\alpha)$, $\text{frac}(2\alpha)$, $\text{frac}(3\alpha), \ldots$ that belong to J is $|A(\alpha, J, n)|/n$. We see that Weyl's Theorem is equivalent to saying that for every irrational number α and for every subinterval J of $[0, 1)$,

$$\lim_{n \to \infty} \frac{|A(\alpha, J, n)|}{n} = \mu(J). \tag{2.19}$$

54   Chapter 2. Irrational Numbers and Dynamical Systems

Note that this is a refinement of Kronecker's Theorem. For, Kronecker's Theorem is equivalent to saying that if J is a subinterval of $[0, 1)$ with $\mu(J) > 0$, and if α is irrational, then

$$\lim_{n \to \infty} |A(\alpha, J, n)| = \infty,$$

and this is clearly implied by (2.19). However, note that the proof of Weyl's Theorem uses Kronecker's Theorem, so that the latter cannot be deduced from the former.

Now, there are two intervals for which (2.19) is clearly true. One is when J is the empty interval $\emptyset$. In this case $A(n, \emptyset, \alpha) = \emptyset$, so that $|A(n, \emptyset, \alpha)| = 0$ and

$$\lim_{n \to \infty} \frac{|A(\alpha, \emptyset, n)|}{n} = 0 = \mu(\emptyset).$$

Another instance is when the interval is the whole of $[0, 1)$. Since frac$(j\alpha) \in [0, 1)$ for all j, we have in this case that $A(n, [0, 1), \alpha) = \{1, 2, \ldots, n\}$. Hence, $|A(\alpha, [0, 1), n)| = n$ and so

$$\lim_{n \to \infty} \frac{|A(\alpha, [0, 1), n)|}{n} = \lim_{n \to \infty} \frac{n}{n} = 1 = \mu([0, 1)).$$

In the general case, the main idea is to show that if two intervals J, K of $[0, 1)$ have lengths $\mu(J)$, $\mu(K)$ respectively, and if $0 < \mu(J), \mu(K) < 1$, then

$$\lim_{n \to \infty} \frac{|A(\alpha, J, n)|}{|A(\alpha, K, n)|} = \frac{\mu(J)}{\mu(K)}. \tag{2.20}$$

Now, suppose we have proved (2.20) and consider the intervals $[0, 1/2)$ and $[1/2, 1)$. We have $[0, 1) = [0, 1/2) \cup [1/2, 1)$, and this union is disjoint. It follows that

$$\{1, 2, \ldots, n\} = A(\alpha, [0, 1), n) = A(\alpha, [0, 1/2), n) \cup A(\alpha, [1/2, 1), n),$$

2.7. The ideas in Weyl's Theorem on irrational numbers

where this union is disjoint. Then, considering the numbers of elements in these sets gives

$$\begin{aligned} n &= |A(\alpha, [0, 1), n)| \\ &= |A(\alpha, [0, 1/2), n)| + |A(\alpha, [1/2, 1), n)|. \end{aligned}$$

Thus, for all $n = 1, 2, \ldots$,

$$\frac{n}{|A(\alpha, [0, 1/2), n)|} = 1 + \frac{|A(\alpha, [1/2, 1), n)|}{|A(\alpha, [0, 1/2), n)|}. \tag{2.21}$$

Now the intervals $[0, 1/2)$ and $[1/2, 1)$ both have length $1/2$. So if (2.20) were true, it would follow from (2.21) that

$$\begin{aligned} \lim_{n \to \infty} \frac{n}{|A(\alpha, [0, 1/2), n)|} &= 1 + \lim_{n \to \infty} \frac{|A(\alpha, [1/2, 1), n)|}{|A(\alpha, [0, 1/2), n)|} \\ &= 1 + \frac{1/2}{1/2} \\ &= 1 + 1 \\ &= 2. \end{aligned}$$

Taking reciprocals gives

$$\lim_{n \to \infty} \frac{|A(\alpha, [0, 1/2), n)|}{n} = \frac{1}{2} = \mu([0, 1/2)). \tag{2.22}$$

This shows that (2.19) would then at least be true for the interval $[0, 1/2)$. However, we could now use (2.20) again, to deduce that (2.19) is true for every interval J having a length $\mu(J)$ with $0 < \mu(J) < 1$. For, it would

56   Chapter 2. Irrational Numbers and Dynamical Systems

follow from (2.20) and (2.22) that

$$\lim_{n\to\infty} \frac{|A(\alpha, J, n)|}{n} = \lim_{n\to\infty} \left(\frac{|A(\alpha, J, n)|}{|A(\alpha, [0, 1/2), n)|} \cdot \frac{|A(\alpha, [0, 1/2), n)|}{n} \right)$$

$$= \lim_{n\to\infty} \left(\frac{|A(\alpha, J, n)|}{|A(\alpha, [0, 1/2), n)|} \right)$$

$$\times \lim_{n\to\infty} \left(\frac{|A(\alpha, [0, 1/2), n)|}{n} \right)$$

$$= \frac{\mu(J)}{\mu([0, 1/2))} \cdot \frac{1}{2}$$

$$= \frac{\mu(J)}{1/2} \cdot \frac{1}{2}$$

$$= \mu(J).$$

This would show that (2.19) holds for every subinterval J of $[0, 1)$, thus proving Weyl's Theorem. Thus, if we can prove (2.20), Weyl's Theorem will follow.

The proof of (2.20) is done in stages, in the next section. Here we give an overview of the approach. First we consider two intervals J, K that are of equal length and have the form

$$J = [\text{frac}(p\alpha), a) \quad \text{and} \quad K = [\text{frac}(q\alpha), b), \qquad (2.23)$$

where $p, q \in \mathbb{N}$. (Actually, and more generally, we consider "arcs" rather than intervals, as described subsequently.) Because the left endpoints of J, K are terms in the sequence $(\text{frac}(n\alpha))$ an easy calculation comparing the number of elements in $A(\alpha, J, n)$ with the number in $A(\alpha, K, n)$ enables us to show that

$$\left| |A(\alpha, J, n)| - |A(\alpha, K, n)| \right| \leq |p - q|.$$

Now Kronecker's Theorem implies that $\lim_{n\to\infty} |A(\alpha, K, n)| = \infty$, and it will follow that

$$\lim_{n\to\infty} |A(\alpha, J, n)|/|A(\alpha, K, n)| = 1.$$

2.8. The proof of Weyl's Theorem

So (2.20) is then established for intervals J, K of the special type in (2.23). In the general case, we "approximate" a general interval by intervals that are like the ones J, K just considered. The fact that (2.20) holds for all the intervals used in the approximating procedure enables us to deduce that (2.20) actually holds for all intervals. We deduce that for *all* subintervals J, K of $[0, 1)$ with $\mu(K) > 0$,

$$\lim_{n\to\infty} \frac{|A(n, J, \alpha)|}{|A(n, K, \alpha)|} = \frac{\mu(J)}{\mu(K)},$$

and as we have seen above Weyl's Theorem will follow. Note that the approximating process is quite simple from the geometric point of view, but it is not so easy to describe completely in a mathematical form suitable for rigorously proving (2.20). There is a danger that the technicalities of the discussion may obscure the underlying geometric simplicity. While I have tried to avoid this, it may be a good idea to complement the main text with a study of Figure 2.16, which indicates the geometric interpretation of the approximation idea lying at the heart of the proof.

2.8 The proof of Weyl's Theorem

Weyl's Theorem 2.13 is proved by a process of "approximating" intervals by smaller intervals of a particular type, about which much more is known concerning the sequence of multiples of the given irrational number. In fact, the approximation process involves a more general notion than that of an interval, namely the notion of an arc.

Definition A subset J of $[0, 1)$ is called an *arc* if there are $a, b \in [0, 1]$ such that: either $a \leq b$ and $J = [a, b)$, or $a > b$ and $J = [0, b) \cup [a, 1)$. In each case, the arc J is denoted by $[a, b)$. Thus, for $a, b \in [0, 1]$,

$$[a, b) = \begin{cases} [a, b), & \text{if } a \leq b; \\ [0, b) \cup [a, 1), & \text{if } a > b. \end{cases}$$

Thus, in the former case, the arc J is a half-open interval $[a, b)$, closed on the left and open on the right. In this case, the *length* $\mu(J)$ of J is

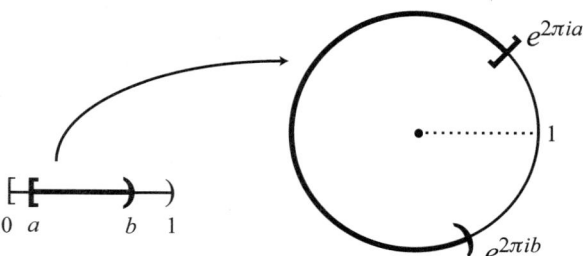

Figure 2.11. The picture on the left is of the arc $[a, b)$ when $a < b$, in which case the arc is the *interval* $[a, b)$. The function $t \mapsto e^{2\pi i t}$ maps $[a, b)$ onto an ordinary arc on the unit circle—the half open arc that goes from $e^{2\pi i a}$ to $e^{2\pi i b}$.

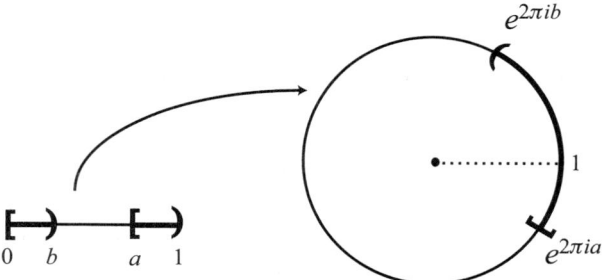

Figure 2.12. The picture on the left is of the arc $[a, b)$ when $a > b$, in which case the arc is the union of the intervals $[a, 1)$ and $[0, b)$. The function $t \mapsto e^{2\pi i t}$ maps $[a, 1)$ onto the arc in $\mathbb{T}$ going counterclockwise from $e^{2\pi i a}$ to 1, and it maps $[0, b)$ onto the arc in $\mathbb{T}$ going from 1 to $e^{2\pi i b}$. Consequently, although in this case the arc $[a, b)$ is the union of two intervals, the function $t \mapsto e^{2\pi i t}$ still maps $[a, b)$ onto an *ordinary arc* on the unit circle—namely, the half open arc that includes 1 and goes from $e^{2\pi i a}$ to $e^{2\pi i b}$. Figures 2.11 and 2.12 combine to show that an arc in $[0, 1)$ corresponds to an ordinary arc in $\mathbb{T}$.

2.8. The proof of Weyl's Theorem

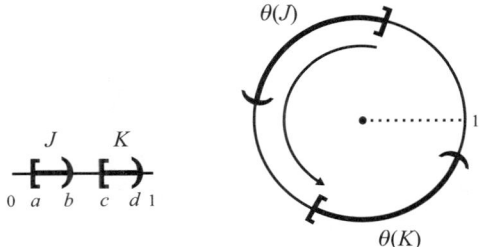

Figure 2.13. Figures 2.13 and 2.14 illustrate the proof of Lemma 2.14. The idea is that what happens on $[0, 1)$ can be interpreted equivalently on the unit circle $\mathbb{T}$, by means of the function $\theta : t \mapsto e^{2\pi i t}$ that maps $[0, 1)$ onto $\mathbb{T}$. In the figure, on the left we have two arcs $J = [a, b)$ and $K = [c, d)$ that are intervals of equal length, with $c > a$. As, J, K have equal lengths, $c - a + J = K$—which is to say that $\tau_{c-a}(J) = K$, so in this case the result follows easily without the need to go to $\mathbb{T}$. But anyway, observe that θ maps the arcs J, K respectively onto the ordinary arcs $\theta(J), \theta(K)$ in $\mathbb{T}$, as shown. On $\mathbb{T}$, τ_{c-a} corresponds to an counterclockwise rotation through an angle $2\pi(c - a)$, and this will rotate $\theta(J)$ onto $\theta(K)$.

$b - a$. In the latter case, J is the union of two disjoint intervals $[a, 1)$ and $[0, b)$ and the *length* $\mu(J)$ of J is $1 + b - a$. Note that for $0 < a \leq 1$, $[a, 0) = [a, 1)$.

In this notation, $[b, a)$ is the complement of $[a, b)$ in $[0, 1)$. The fact that $[b, a)$ is the complement of $[a, b)$ in $[0, 1)$ is consistent with the observation that

$$1 = \mu([0, 1)) = \mu([a, b)) + \mu([b, a)). \tag{2.24}$$

Using (2.24), or by showing it directly, it is easy to check that the length of an arc $[a, b)$ is equal to frac$(b - a)$.

The reason that a set of the form $[a, b)$ is called an arc, is that under the function $t \mapsto e^{2\pi i t}$ that maps $[0, 1)$ onto $\mathbb{T}$, an arc $[a, b)$ maps into an *ordinary* arc on the unit circle—namely, the arc that commences at the point $e^{2\pi i a}$ on the circle and proceeds in an counterclockwise direction to the point $e^{2\pi i b}$, but not including this latter point. This is illustrated in Figures 2.11 and 2.12.

60   Chapter 2. Irrational Numbers and Dynamical Systems

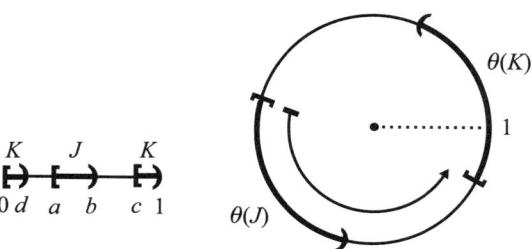

Figure 2.14. On the left we have two arcs $J = [a, b)$ and $K = [c, d)$ of equal length. J is an interval but K is the union of the two intervals $[c, 1)$ and $[0, d)$. Under the function θ, the arcs J, K respectively map to ordinary arcs $\theta(J), \theta(K)$ in $\mathbb{T}$ of equal length, as shown. As $\theta(J), \theta(K)$ have equal lengths, a counterclockwise rotation through an angle $2\pi(c - a)$ will rotate $\theta(J)$ onto $\theta(K)$. Interpreting this in terms of what happens on $[0, 1)$, we see that τ_{c-a} maps J onto K. The argument appears easier and more natural on the circle because, on the circle, there are no endpoints so that we do not have to consider separate cases. So, the approach has been to say: "on the circle the result is clear, and when we interpret this result on the interval $[0, 1)$, we get the conclusion we need".

Lemma 2.14. *Let* $J = [a, b)$ *and* $K = [c, d)$ *be two arcs of equal length. Then* $\tau_{c-a}(J) = K$.

Proof. A visual proof is illustrated in Figures 2.13 and 2.14. Alternatively, the result follows from Exercise 11. □

Lemma 2.15. *Let* $\alpha \in \mathbb{R}$ *be irrational. Let* J, K *be two arcs of* $[0, 1)$ *having equal and positive length and for which there are* $p, q \in \mathbb{N}$ *and* $a, b \in [0, 1)$ *with*

$$J = \big[\operatorname{frac}(p\alpha), a\big) \text{ and } K = \big[\operatorname{frac}(q\alpha), b\big).$$

Then,

$$\lim_{n \to \infty} \frac{|A(\alpha, J, n)|}{|A(\alpha, K, n)|} = 1.$$

Proof. Recall that

$$A(\alpha, J, n) = \big\{j : \operatorname{frac}(j\alpha) \in J \text{ and } 1 \leq j \leq n\big\}.$$

2.8. The proof of Weyl's Theorem

If $p = q$, then $J = K$ as the arcs have equal length, in which case the result is true. If $p \neq q$, either $p > q$ or $p < q$. Assume that $p < q$, and put $r = q - p \in \mathbb{N}$. Let τ denote the function $x \longmapsto \mathrm{frac}(q\alpha - p\alpha + x)$. Then, by Lemma 2.14, $x \in J \iff \tau(x) \in K$. Using the fact that $\mathrm{frac}\big(x + \mathrm{frac}(y)\big) = \mathrm{frac}(x + y)$, we now have

$$\begin{aligned}
j \in A(\alpha, J, n) &\iff \mathrm{frac}(j\alpha) \in J \text{ and } 1 \le j \le n \\
&\iff \tau(\mathrm{frac}(j\alpha)) \in K \text{ and } 1 \le j \le n \\
&\iff \mathrm{frac}(q\alpha - p\alpha + \mathrm{frac}(j\alpha)) \in K \text{ and } 1 \le j \le n \\
&\iff \mathrm{frac}((q - p + j)\alpha)) \in K \text{ and } 1 \le j \le n \\
&\iff \mathrm{frac}((j + r)\alpha) \in K \text{ and } 1 \le j \le n. \quad (2.25)
\end{aligned}$$

Thus,

$$j \in A(\alpha, J, n) \text{ and } 1 \le j \le n - r \implies j + r \in A(\alpha, K, n),$$

so that

$$\left|\{j : j \in A(\alpha, J, n) \text{ and } 1 \le j \le n - r\}\right| \le |A(\alpha, K, n)|. \quad (2.26)$$

Now, the definition of the set $A(\alpha, J, n)$ ensures that the set $\{j \in A(\alpha, J, n)$ and $1 \le j \le n - r\}$ has at least $|A(\alpha, J, n)| - r$ elements, so it follows from (2.26) that

$$|A(\alpha, J, n)| - r \le |A(\alpha, K, n)|. \quad (2.27)$$

On the other hand, we deduce using (2.25) that

$$\begin{aligned}
\ell \in A(\alpha, K, n) \text{ and } r < \ell \le n &\implies \mathrm{frac}(\ell\alpha) \in K \text{ and } r < \ell \le n \\
&\implies \mathrm{frac}\big(((\ell - r) + r)\alpha\big) \in K \\
&\qquad \text{and } 1 \le \ell - r \le n - r < n \\
&\implies \ell - r \in A(\alpha, J, n),
\end{aligned}$$

so that

$$\left|\{\ell : \ell \in A(\alpha, K, n) \text{ and } r < \ell \le n\}\right| \le |A(\alpha, J, n)|. \quad (2.28)$$

But, the set $\{\ell \in A(\alpha, K, n) \text{ and } r < \ell \leq n\}$ has at least $|A(\alpha, K, n)| - r$ elements, so from (2.28) we deduce that

$$|A(\alpha, K, n)| - r \leq |A(\alpha, J, n)|. \tag{2.29}$$

Now, (2.27) and (2.29) give

$$\Big| |A(\alpha, J, n)| - |A(\alpha, K, n)| \Big| \leq r,$$

from which we have

$$\left| 1 - \frac{|A(\alpha, J, n)|}{|A(\alpha, K, n)|} \right| \leq \frac{r}{|A(\alpha, K, n)|}. \tag{2.30}$$

But Kronecker's Theorem implies that

$$\lim_{n \to \infty} |A(\alpha, K, n)| = \infty,$$

(see Exercise 8) so it follows from (2.30) that

$$\lim_{n \to \infty} \frac{|A(\alpha, J, n)|}{|A(\alpha, K, n)|} = 1,$$

as required. □

Lemma 2.16. *Let $0 < \beta < 1$ and $p \in \mathbb{N}$. Then the arc* [frac($p\beta$), frac($p + 1)\beta$) *has length* β.

Proof. It was mentioned earlier that the length of an arc $[a, b)$ is frac($b - a$). Thus, using (3) of Proposition 2.1, the length of the arc [frac($p\beta$), frac($p + 1)\beta$) is

$$\text{frac}\big(\text{frac}(p + 1)\beta) - \text{frac}(p\beta)\big)$$
$$= \text{frac}\Big[(p + 1)\beta - p\beta - \text{int}((p + 1)\beta) + \text{int}(p\beta)\Big]$$
$$= \text{frac}(\beta) = \beta.$$

□

2.8. The proof of Weyl's Theorem

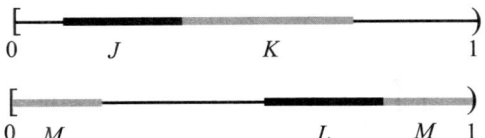

Figure 2.15. The picture above shows the arc K following the arc J. Below, the picture shows the arc M following the arc L.

Lemma 2.17. *Let $\alpha \in \mathbb{R}$. Then, for all $j, n \in \mathbb{N}$, $\mathrm{frac}\bigl(n\, \mathrm{frac}(j\alpha)\bigr) = \mathrm{frac}(nj\alpha)$.*

Proof. Using (3) of Proposition 2.1 we have

$$\mathrm{frac}(n\, \mathrm{frac}(j\alpha)) = \mathrm{frac}\bigl(nj\alpha - n\, \mathrm{int}(j\alpha)\bigr) = \mathrm{frac}(nj\alpha).$$

$\square$

Lemma 2.18. *Let J, K be subsets of $[0, 1)$ and let $n \in \mathbb{N}$. Then the following hold.*
(1) $J \subseteq K \implies A(\alpha, J, n) \subseteq A(\alpha, K, n)$.
(2) If $J \cap K = \emptyset$, then

$$A(\alpha, J \cup K, n) = A(\alpha, J, n) \cup A(\alpha, K, n).$$

(3) $A\bigl(\alpha, [0, 1), n\bigr) = \{1, 2, \ldots, n\}$.

Proof. These statements follow immediately from the definition of the set $A(\alpha, J, n)$. $\square$

The following definition has a geometric interpretation that is illustrated in Figure 2.15.

Definition Let J, K be arcs of $[0, 1)$. Then the arc K *follows* the arc J if J is of the form $[a, b)$ and K is of the form $[b, c)$. In this case, $J \cup K = [a, c)$ so that $J \cup K$ is also an arc. Also, if arcs J, K are disjoint, $\mu(J) + \mu(K) = \mu(J \cup K) \leq 1$. Note that if K follows J and L follows K, then $J \cup K$ is an arc and L follows $J \cup K$.

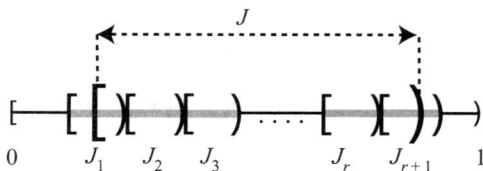

Figure 2.16. The figure illustrates an important idea in the proof of Weyl's Theorem. The intervals $J_1, J_2, \ldots, J_{r+1}$ follow on from each other and have equal length β. The intervals $J_2, \ldots, J_r$ are contained within J, so their union "approximates" J from the *inside*. The intervals J_1 and J_{r+1} "overlap" J, the former at the left-hand end-point of J, the latter at the right-hand end-point. So, the union of $J_1, J_2, \ldots, J_{r+1}$ "approximates" J from the *outside*. The idea then is to estimate the number of points of the sequence (frac$(n\alpha)$) in J by estimating the number of points of (frac$(n\alpha)$) that are in the intervals $J_1, J_2, \ldots, J_{r+1}$. This is possible because of the special form of the intervals J_j. We end up getting an estimate of the number of points of (frac$(n\alpha)$) in J from both the "inside" and the "outside" of J. A detailed analysis shows that as $\beta \to 0$, that is as the (common) length of the intervals J_j goes to 0, these estimates are comparable and, with further work, Weyl's Theorem follows.

The following result is the main one needed for Weyl's Theorem. A basic idea in the proof is that a given arc may be economically covered by a succession of small, adjacent arcs of equal length. This idea is illustrated in Figure 2.16.

Proposition 2.19. *Let α be an irrational number. Let J and K be two subintervals of $[0, 1)$ with $0 < \mu(J), \mu(K) < 1$. Then*

$$\lim_{n \to \infty} \frac{|A(\alpha, J, n)|}{|A(\alpha, K, n)|} = \frac{\mu(J)}{\mu(K)}.$$

Proof. Let $\varepsilon > 0$. If $k \in \mathbb{N}$ and we put $\beta = \text{frac}(k\alpha)$, then β is irrational, $\beta > 0$ and there are unique $r, s \in \mathbb{N}$ such that

$$r - 1 \le \frac{\mu(J)}{\beta} < r \text{ and } s - 1 \le \frac{\mu(K)}{\beta} < s, \quad (2.31)$$

2.8. The proof of Weyl's Theorem

which implies that

$$\mu(J) < r\beta, \; \frac{1}{\mu(K)} \le \frac{1}{(s-1)\beta} \; \text{ and } \; \mu(J) \ge (r-1)\beta, \; \frac{1}{\mu(K)} > \frac{1}{s\beta}. \tag{2.32}$$

Note from (2.31) that the smaller β is, the larger are r and s. Kronecker's Theorem further allows us to choose k so that β is as small as we wish, so by choosing β to be sufficiently small, both r and s will be as large as we wish. We will choose β to be sufficiently small so that, in addition to the above,

$$r > \frac{1+\mu(J)}{1-\mu(J)}, \; s > \frac{1+\mu(K)}{1-\mu(K)}, \tag{2.33}$$

$$1-\varepsilon < \left(1 - \frac{1}{s}\right)\left(\frac{1-1/r}{1+1/s}\right)^2 \; \text{ and } \; \left(\frac{1}{1-1/r}\right)\left(\frac{1+1/r}{1-1/s}\right)^2 < 1+\varepsilon. \tag{2.34}$$

This means that $r, s > 1$. The reasons for selecting r, s so that these properties hold may well seem mysterious at this juncture, but the selections are merely designed to make sure that certain inequalities, that appear in subsequent calculations, hold. Also, again for future use, note that (2.32) gives

$$\left(1 - \frac{1}{s}\right)\frac{\mu(J)}{\mu(K)} = \frac{s-1}{s} \cdot \frac{\mu(J)}{\mu(K)} < \frac{s-1}{s} \cdot \frac{r\beta}{(s-1)\beta} = \frac{r}{s}, \; \text{and}$$

$$\left(\frac{1}{1-1/r}\right)\frac{\mu(J)}{\mu(K)} = \frac{r}{r-1} \cdot \frac{\mu(J)}{\mu(K)} > \frac{r}{r-1} \cdot \frac{(r-1)\beta}{s\beta} = \frac{r}{s}.$$

Hence,

$$\left(1 - \frac{1}{s}\right)\frac{\mu(J)}{\mu(K)} < \frac{r}{s} < \left(\frac{1}{1-1/r}\right)\frac{\mu(J)}{\mu(K)}. \tag{2.35}$$

The idea is that we are going to approximate each of the intervals J, K from both the "outside" and the "inside" by a finite disjoint union of arcs each having length β. The procedure is the same for each interval, so we discuss it in detail for J only. The ideas are illustrated in Figure 2.16.

Observe that by (2.31), there are disjoint arcs $J_1, J_2, \ldots, J_{r+1}$ of $[0, 1)$ such that these arcs follow on from each other, each arc has length β and

$$\bigcup_{j=2}^{r} J_j \subseteq J \subseteq \bigcup_{j=1}^{r+1} J_j. \tag{2.36}$$

Note that whereas the arcs $J_2, J_3, \ldots, J_r$ are contained in J, it is a consequence of the choice of r as in (2.31) that the arcs J_1 and J_{r+1} merely "overlap" J, the former at the left end-point of J, the latter at the right end-point. This situation is illustrated in Figure 2.16. In fact, if we use the irrationality of β and apply Kronecker's Theorem, it follows from Lemma 2.16 and (2.36), that there is $p \in \mathbb{N}$ so that the arcs $J_1, J_2, \ldots, J_{r+1}$ may be taken to be

$$J_j = \Big[\text{frac}((p+j-1)\beta), \text{frac}((p+j)\beta)\Big), \text{ for } j = 1, 2, \ldots, r+1. \tag{2.37}$$

Now, the left-hand inequality in (2.33) gives $(r+1)\mu(J) < r - 1$. Using this and (2.31) we have

$$(r+1)\beta = (r+1)\mu(J) \cdot \frac{\beta}{\mu(J)} < (r-1)\beta\mu(J)^{-1} \leq \mu(J)\mu(J)^{-1} = 1.$$

Thus, the total length of the arcs $J_1, J_2, \ldots, J_{r+1}$ is at most the length of the whole interval $[0, 1)$. It follows that the arcs $J_1, J_2, \ldots, J_{r+1}$ are pairwise disjoint. So, by (2.36) and (2) of Lemma 2.18,

2.8. The proof of Weyl's Theorem

$$\left| A\left(\alpha, \bigcup_{j=2}^{r} J_j, n\right) \right| \leq |A(\alpha, J, n)| \leq \left| A\left(\alpha, \bigcup_{j=1}^{r+1} J_j, n\right) \right|$$

$$\implies \sum_{j=2}^{r} |A(\alpha, J_j, n)| \leq |A(\alpha, J, n)| \leq \sum_{j=1}^{r+1} |A(\alpha, J_j, n)|$$

$$\implies |A(\alpha, J_1, n)| \cdot \left(\sum_{j=2}^{r} \frac{|A(\alpha, J_j, n)|}{|A(\alpha, J_1, n)|} \right) \tag{2.38}$$

$$\leq |A(\alpha, J, n)|$$

$$\leq |A(\alpha, J_1, n)| \cdot \left(\sum_{j=1}^{r+1} \frac{|A(\alpha, J_j, n)|}{|A(\alpha, J_1, n)|} \right).$$

Now, $\beta = \text{frac}(k\alpha)$, so by Lemma 2.17 and (2.37) we have

$$J_j = \Big[\text{frac}((p+j-1)k\alpha), \text{frac}((p+j)k\alpha)\Big), \text{ for } j = 1, 2, \ldots, r+1.$$

Hence, we may apply Lemma 2.15 to the arcs of equal length $J_1, J_2, \ldots, J_{r+1}$. We deduce that

$$\lim_{n \to \infty} \frac{|A(\alpha, J_j, n)|}{|A(\alpha, J_1, n)|} = 1, \text{ for all } j = 1, 2, \ldots, r+1.$$

Hence there is $n_0 \in \mathbb{N}$ such that for all $j = 1, 2, \ldots, r+1$,

$$n > n_0 \implies 1 - \frac{1}{r} < \frac{|A(\alpha, J_j, n)|}{|A(\alpha, J_1, n)|} < 1 + \frac{1}{r}.$$

Using this, it follows from (2.38) that for all $n > n_0$,

$$(r-1)\left(1 - \frac{1}{r}\right) |A(\alpha, J_1, n)| \leq |A(\alpha, J, n)|$$

$$\leq (r+1)\left(1 + \frac{1}{r}\right) |A(\alpha, J_1, n)|.$$

That is, for all $n > n_0$,

$$r\left(1 - \frac{1}{r}\right)^2 |A(\alpha, J_1, n)| \leq |A(\alpha, J, n)|$$
$$\leq r\left(1 + \frac{1}{r}\right)^2 |A(\alpha, J_1, n)|. \quad (2.39)$$

Now, exactly the same argument can be used on the interval K that was just used on J. There is $q \in \mathbb{N}$ such that if

$$K_j = \left[\text{frac}((q + j - 1)\beta), \text{frac}(q + j)\beta)\right),$$

$$\text{for } j = 1, 2, \ldots, s + 1, \text{ then } \bigcup_{j=2}^{s} K_j \subseteq K \subseteq \bigcup_{j=1}^{s+1} K_j,$$

where $K_1, K_2, \ldots, K_{s+1}$ are chosen in the same way as the $J_1, J_2, \ldots, J_{r+1}$. In the same way that we deduced (2.39), we see there is n'_0 such that for all $n > n'_0$,

$$s\left(1 - \frac{1}{s}\right)^2 |A(\alpha, K_1, n)| \leq |A(\alpha, K, n)|$$
$$\leq s\left(1 + \frac{1}{s}\right)^2 |A(\alpha, K_1, n)|. \quad (2.40)$$

It follows from (2.39) and (2.40) that for all $n > \max(n_0, n'_0)$,

$$\frac{r}{s}\left(\frac{1 - 1/r}{1 + 1/s}\right)^2 \frac{|A(\alpha, J_1, n)|}{|A(\alpha, K_1, n)|} \leq \frac{|A(\alpha, J, n)|}{|A(\alpha, K, n)|}$$
$$\leq \frac{r}{s}\left(\frac{1 + 1/r}{1 - 1/s}\right)^2 \frac{|A(\alpha, J_1, n)|}{|A(\alpha, K_1, n)|}.$$

2.8. The proof of Weyl's Theorem

We deduce from this, using (2.35), that for all $n > \max(n_0, n_0')$,

$$\left(1 - \frac{1}{s}\right)\left(\frac{1 - 1/r}{1 + 1/s}\right)^2 \frac{\mu(J)}{\mu(K)} \leq \frac{|A(\alpha, J, n)|}{|A(\alpha, K, n)|} \cdot \frac{|A(\alpha, K_1, n)|}{|A(\alpha, J_1, n)|}$$

$$\leq \left(\frac{1}{1 - 1/r}\right)\left(\frac{1 + 1/r}{1 - 1/s}\right)^2 \frac{\mu(J)}{\mu(K)}.$$

Using (2.34) we now have that for all $n > \max(n_0, n_0')$,

$$1 - \varepsilon < \frac{|A(\alpha, J, n)|}{|A(\alpha, K, n)|} \cdot \frac{|A(\alpha, K_1, n)|}{|A(\alpha, J_1, n)|} \cdot \frac{\mu(K)}{\mu(J)} < 1 + \varepsilon.$$

So, it follows from the definition of a convergent sequence that

$$\lim_{n \to \infty} \frac{|A(\alpha, J, n)|}{|A(\alpha, K, n)|} \cdot \frac{|A(\alpha, K_1, n)|}{|A(\alpha, J_1, n)|} \cdot \frac{\mu(K)}{\mu(J)} = 1.$$

However, because of the special form of the intervals J_1, K_1, we can apply Lemma 2.15 to get

$$\lim_{n \to \infty} \frac{A(\alpha, J_1, n)}{A(\alpha, K_1, n)} = 1.$$

Hence,

$$\lim_{n \to \infty} \frac{|A(\alpha, J, n)|}{|A(\alpha, K, n)|} = \frac{\mu(J)}{\mu(K)}.$$

□

Theorem 2.20 (Weyl's Theorem II). *Let α be an irrational number, and let J be a subinterval of $[0, 1)$. Then,*

$$\lim_{n \to \infty} \frac{\left|\{j : 1 \leq j \leq n \text{ and } \operatorname{frac}(j\alpha) \in J\}\right|}{n} = \mu(J).$$

70   Chapter 2. Irrational Numbers and Dynamical Systems

Proof. If J is a single point, if J is empty, or if $J = [0, 1)$, then the result is true. So we may take J to be an interval with $0 < \mu(J) < 1$. Using Proposition 2.19 we have

$$\lim_{n\to\infty} \frac{|A(\alpha, J, n)|}{n} = \lim_{n\to\infty} \left(\frac{|A(\alpha, J, n)|}{|A(\alpha, [0, 1/2), n)|} \cdot \frac{|A(\alpha, [0, 1/2), n)|}{n} \right)$$

$$= \lim_{n\to\infty} \left(\frac{|A(\alpha, J, n)|}{|A(\alpha, [0, 1/2), n)|} \cdot \frac{|A(\alpha, [0, 1/2), n)|}{|A(\alpha, [0, 1/2), n)| + |A(\alpha, [1/2, 1), n)|} \right)$$

$$= \lim_{n\to\infty} \left(\frac{|A(\alpha, J, n)|}{|A(\alpha, [0, 1/2), n)|} \right)$$

$$\cdot \lim_{n\to\infty} \left(\frac{1}{1 + |A(\alpha, [1/2, 1), n)| \cdot |A(\alpha, [0, 1/2), n)|^{-1}} \right)$$

$$= \frac{\mu(J)}{\mu([0, 1/2))} \cdot \frac{1}{1 + \mu([1/2, 1)) \mu([0, 1/2))^{-1}} = \frac{\mu(J)}{1/2} \cdot \frac{1}{1 + 1} = \mu(J).$$

$\square$

A sequence (x_n) of points in $[0, 1)$ is *uniformly distributed* in $[0, 1)$ if, for every subinterval $[a, b)$ of $[0, 1)$,

$$\lim_{n\to\infty} \frac{\left|\{j : 1 \leq j \leq n \text{ and } x_j \in [a, b)\}\right|}{n} = b - a.$$

Thus, Weyl's Theorem says that if α is irrational, the sequence (frac$(n\alpha)$) is uniformly distributed in $[0, 1)$.

2.9 Chaos in Kronecker systems

Notions of chaos in dynamical systems have been around for a long time. There is no single generally accepted definition of a chaotic dynamical system, just as there is no single definition of a random system. However, it would seem that any mathematical notion of "chaos" in a dynamical system should have implications, or at least overtones, of "random" or "complex" behavior. The notion of chaos presented in this Section emphasizes the idea that in a chaotic system, even when the evolution of

2.9. Chaos in Kronecker systems

the system is determined by an exact rule, the future behavior of the system cannot be predicted. Technically, this is known as "sensitivity to initial conditions", and it can also be regarded as a way of describing "information loss" in a system. As we shall see, a Kronecker system $([0, 1), \tau_\alpha)$ associated with Kronecker's Theorem exhibits this behavior when α is irrational, so in such cases we can regard the Kronecker systems as chaotic[8].

Now, suppose we have a dynamical system (S, f) and suppose that this system can be "decomposed" into two subsystems. That is, suppose that there is a non-trivial partition of S into two subsets S_1, S_2 such that (S_1, f) and (S_2, f) are dynamical systems in their own right. Then, S is the disjoint union of non-empty sets S_1 and S_2, f maps S_1 into S_1 and f maps S_2 into S_2. This means that the systems (S_1, f) and (S_2, f) are entirely independent of each other. Assuming that the systems (S_1, f) and (S_2, f) are not merely trivial or negligible parts of the whole system, we would not then consider the whole system (S, f) to be "chaotic" or "random", because once the system is in a state in S_1 (say), the system will remain in a state in S_1 for evermore. Similarly, if the system is in a state in S_2, then the system will remain in S_2 for evermore. To this extent, the future of the system is predictable. The idea of the following definition is to formally identify those systems on intervals that cannot be split into subsystems.

Definition Let S be an interval and let f be a transformation on S. Then the dynamical system (S, f) is *transitive* if, whenever we are given $x, y \in S$ and $\varepsilon > 0$, there is $z \in S$ and $n \in \mathbb{N}$ such that

$$|x - z| < \varepsilon \quad \text{and} \quad |f^n(z) - y| < \varepsilon. \tag{2.41}$$

The idea is that if we take two arbitrary states x, y in a transitive system, we can "approximately" get from near the state x to near the state y if we let the system run long enough over time—we might say that over time, every state is approximately attainable from any other state, or we might say that we can evolve from any part of the system so as to arrive in the future in any other part. These ideas are expressed by saying that the state z, which is "near" x to within ε, evolves after some n iterations so as to be within a distance of ε of the other state y.

In the Kronecker system $([0, 1), \tau_\alpha)$, we will see that the irrationality of α determines the transitivity of the system.

Theorem 2.21. *Let* $\alpha \in \mathbb{R}$. *Then the system* $([0, 1), \tau_\alpha)$ *is transitive if and only if α is irrational.*

Proof. We have $\tau_\alpha(z) = \text{frac}(z + \alpha)$ and, in fact,

$$\tau_\alpha^n(z) = \text{frac}(z + n\alpha), \tag{2.42}$$

for all $z \in [0, 1)$ and $n = 0, 1, 2, \ldots$. Assume that α is irrational, let $x, y \in [0, 1)$ and let $\varepsilon > 0$. By Theorem 2.11, the sequence $(\text{frac}(x + n\alpha))$ is dense in $[0, 1)$, so by using (2.42) we see that there is $n \in \mathbb{N}$ such that

$$|\tau_\alpha^n(x) - y| = |\text{frac}(x + n\alpha) - y| < \varepsilon.$$

It follows that (2.41) holds with $z = x$, and we see that when α is irrational, the system $([0, 1), \tau_\alpha)$ is transitive.

It remains to prove that if α is rational, the system $([0, 1), \tau_\alpha)$ is not transitive. This is Exercise 18. □

Definition Let S be an interval and let f be a transformation on S. Then the dynamical system (S, f) is *sensitive to initial conditions* if there exists a number $\delta > 0$ that has the following property: given $x \in S$ and $\varepsilon > 0$, there is $z \in S$ and $n \in \mathbb{N}$ such that

$$|x - z| < \varepsilon \quad \text{and} \quad |f^n(x) - f^n(z)| > \delta. \tag{2.43}$$

The idea here is that no matter what state x the system is in, there is a state z as close to x as we wish such that if the system evolves from z in place of x, then after some number n of iterations of the system, the state of the system will be $f^n(z)$, and will differ from $f^n(x)$ by at least δ. That is, an arbitrarily small deviation in the initial conditions of the system will lead, in a certain precise sense, to a comparatively large deviation in at least some of the future states of the system—that

2.9. Chaos in Kronecker systems

is, we have "lost information" about the initial state of the system after a certain number of iterations. In fact, if the system has sensitivity to initial conditions, we have that for each x in S and $A \in (0, \infty)$, there is $z \in S$ and $n \in \mathbb{N}$ such that

$$\frac{|f^n(x) - f^n(z)|}{|x - z|} > A.$$

Whereas a statement that a system is transitive is a way of saying that the system cannot be decomposed into simpler systems, a statement that a system is sensitive to initial conditions is a markedly "chaotic" or, perhaps, a "randomness" requirement, in that it expresses an idea of the "unpredictability" of the system. Sensitive dependence upon initial conditions is generally regarded as a hallmark of chaos in dynamical systems (see [3] for further discussion).

Theorem 2.22. *Let α be an irrational number. Then the system $([0, 1), \tau_\alpha)$ is sensitive to initial conditions.*

Proof. As in Theorem 2.21, we will use the facts that for all $y \in [0, 1)$, $\tau_\alpha^n(y) = \text{frac}(y + n\alpha)$ and $(\text{frac}(n\alpha))$ is a sequence dense in $[0, 1)$. Let $x \in [0, 1)$, let $0 < \varepsilon < 1/2$ and let $z \in [0, 1)$ with $z \in (x, x + \varepsilon)$. Then $|x - z| < \varepsilon$ and, as $(\text{frac}(n\alpha))$ is a sequence dense in $[0, 1)$, there is n such that $\text{frac}(n\alpha) \in (1 - z, 1 - x)$. As $x + \text{frac}(n\alpha) < 1$, we have

$$\begin{aligned}
\tau_\alpha^n(x) &= \tau_{\text{frac}(n\alpha)}(x) \\
&= \text{frac}(x + \text{frac}(n\alpha)) \\
&= x + \text{frac}(n\alpha) > x + 1 - z > 1 - \varepsilon.
\end{aligned} \quad (2.44)$$

Also, as $1 < z + \text{frac}(n\alpha)$, we have

$$\begin{aligned}
\tau_\alpha^n(z) &= \tau_{\text{frac}(n\alpha)}(z) = \text{frac}(z + \text{frac}(n\alpha)) \\
&= z + \text{frac}(n\alpha) - 1 < z + 1 - x - 1 = z - x < \varepsilon. \quad (2.45)
\end{aligned}$$

It follows from (2.44) and (2.45) that

$$|\tau_\alpha^n(x) - \tau_\alpha^n(z)| > 1 - 2\varepsilon. \quad (2.46)$$

74   Chapter 2. Irrational Numbers and Dynamical Systems

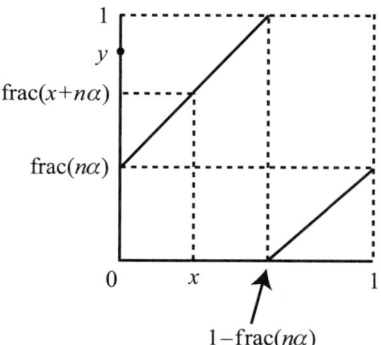

Figure 2.17. The Figure illustrates the graph of $\tau_{n\alpha} = \tau_\alpha^n$, and the proof of transitivity of the system $([0, 1), \tau_\alpha)$ for irrational α. By Kronecker's Theorem, if x, y are given, as the sequence $(\text{frac}(x + n\alpha))$ is dense in $[0, 1)$, there is $n \in \mathbb{N}$ such that $\tau_\alpha^n(x) = \text{frac}(x + n\alpha)$ is as close to y as we care to specify in advance. This gives transitivity. In the picture, where $x < y$, we see that this means selecting n so that the graph of $\tau_{n\alpha}$ has the position depicted, with $\text{frac}(x + n\alpha)$ close to y.

Thus, if δ is a number in $(0, 1)$, for all sufficiently small $\varepsilon > 0$, we see from (2.46) that for every $x \in [0, 1)$ there is $z \in (0, 1)$ with $|x - z| < \varepsilon$ such that $|\tau_\alpha^n(x) - \tau_\alpha^n(z)| > \delta$. That is, the definition of sensitive dependence upon initial conditions is satisfied, and any value of $\delta \in (0, 1)$ suffices to satisfy the definition. □

In the context of Theorems 2.21 and 2.22, Weyl's Theorem says that the inequality in (2.41) that determines transitivity will be fulfilled for values of n that have a certain asymptotic frequency, and that the inequality in (2.43) that produces sensitivity can likewise be given a certain asymptotic frequency in n. In this sense, Weyl's Theorem can be looked at as saying that "chaos" is uniformly distributed throughout the system.

Finally, note that if τ_α is an irrational translation on $[0, 1)$, the corresponding transformation on $\mathbb{T}$ is the translation $\rho_\alpha : z \longmapsto e^{2\pi i \alpha} z$. However, whereas we have seen that τ_α has sensitivity to initial condi-

Exercises

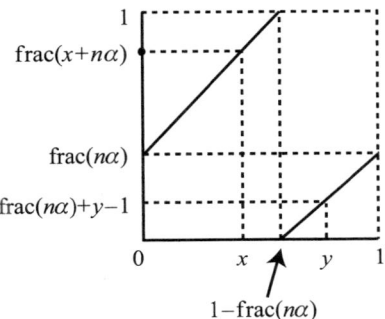

Figure 2.18. The figure illustrates the graph of $\tau_{n\alpha} = \tau_\alpha^n$, and the sensitivity of the system $([0, 1), \tau_\alpha)$ for irrational α. Given x, let y be chosen "sufficiently close" to x (in the picture, $x < y$). Again, as the sequence (frac($x + n\alpha$)) is dense, there is n such that $1 - $ frac($n\alpha$) is between x and y. Then, the difference between $\tau_\alpha^n(x)$ and $\tau_\alpha^n(y)$ is

$$|\tau_\alpha^n(x) - \tau_\alpha^n(y)| = |x + \text{frac}(n\alpha) - y - \text{frac}(n\alpha) - 1|$$
$$= |x - y - 1| \geq 1 - |x - y|, \quad (*)$$

and sensitivity follows. The discontinuity at $1 - $ frac($n\alpha$) of the transformation τ_α^n is a "point of chaos" in the system in the sense that it's the existence of such a discontinuity in the required position which makes the system sensitive. From this viewpoint, Weyl's Theorem says that the "points of chaos" are distributed uniformly throughout the system as time evolves.

tions, ρ_α does not. This is because

$$|\rho_\alpha^n(z_1) - \rho_\alpha^n(z_2)| = |z_1 - z_2|,$$

so the inequalities as in (2.43) do not hold in this complex context, and so sensitivity cannot hold.

Exercises

1. Prove that frac($2\sqrt{2}$)/ frac($\sqrt{2}$) is rational.
2. Let α be irrational and let $m, n \in \mathbb{N}$. Prove that if frac($m\alpha$)/ frac($n\alpha$) is rational then
$$\frac{\text{frac}(m\alpha)}{\text{frac}(n\alpha)} = \frac{m}{n}.$$

76   Chapter 2. Irrational Numbers and Dynamical Systems

Also, if m is the largest element of $\mathbb{N}$ such that $m \operatorname{frac}(\alpha) < 1$, prove that

$$\frac{\operatorname{frac}(j\alpha)}{\operatorname{frac}(\alpha)} = j, \text{ for } j = 1, 2, \ldots, m,$$

and that $\operatorname{frac}(j\alpha)/\operatorname{frac}(\alpha)$ is irrational for $j > m$.

3. Let α be an irrational number. Then for a given non-empty open subinterval U of $[0, 1)$, prove from the Pigeon-Hole Principle that there is $m \in \mathbb{Z}$ such that the fractional part of $m\alpha$ is in U. Note that Kronecker's Theorem is the stronger statement that the m appearing here can in fact be chosen to be in $\mathbb{N}$.

4. Let α be real. Prove that the following are equivalent.

   (i) α is rational.
   (ii) $\operatorname{frac}(\alpha)$ is rational.
   (iii) There are $m, n \in \mathbb{N}$ such that $m \neq n$ and $\operatorname{frac}(m\alpha) - \operatorname{frac}(n\alpha)$ is rational.
   (iv) For all $m, n \in \mathbb{N}$ such that $m \neq n$, $\operatorname{frac}(m\alpha) - \operatorname{frac}(n\alpha)$ is rational.

5. Let $X = \{x_1, x_2, \ldots, x_n\}$ be a given finite set, and let the sets $X_1, X_2, \ldots, X_r$ partition X. If $n > r$ but r does not divide into n, prove that there is $j \in \{1, 2, \ldots, r\}$ such that at least $\operatorname{int}(n/r) + 1$ of the points $x_1, x_2, \ldots, x_n$ lie in X_j.

6. (This exercise arises from a suggestion of David M. Bressoud.) Let $b \in \mathbb{N}$ with $b \geq 2$. Let $1 \leq n_1 < n_2 < n_3 < \cdots$ be an increasing sequence in $\mathbb{N}$ such that $\lim_{j \to \infty}(n_j - n_{j-1}) = \infty$. If $\alpha = \sum_{j=1}^{\infty} 1/b^{n_j}$, prove that α is irrational. Now, prove that the terms of $\operatorname{frac}(n\alpha)$ are "partially predictable", by showing that

$$\operatorname{frac}(b^k \alpha) \in \left[\frac{1}{b^{n_1-k}}, \frac{1}{b^{n_1-k}} + \frac{1}{(b-1)b^{n_2-k-1}} \right),$$
for $k = 1, 2, \ldots, n_1 - 1$, if $n_1 > 1$

and

$$\operatorname{frac}(b^k \alpha) \in \left[\frac{1}{b^{n_{j+1}-k}}, \frac{1}{b^{n_{j+1}-k}} + \frac{1}{(b-1)b^{n_{j+2}-k-1}} \right),$$
for $n_j \leq k \leq n_{j+1} - 1$.

Deduce that $\lim_{j \to \infty} \operatorname{frac}(b^{n_j} \alpha) = 0$.

Exercises

7. Let $0 < \alpha < 1$. Prove that for all $n \in \mathbb{N}$,
$$\text{int}((n+1)\alpha) - \text{int}(n\alpha) \in \{0, 1\}.$$
If we further assume that α is irrational, deduce from Kronecker's Theorem that
$$\text{int}((n+1)\alpha) = \text{int}(n\alpha), \quad \text{for infinitely many } n \in \mathbb{N},$$
and then deduce from Weyl's Theorem that
$$\lim_{n \to \infty} \frac{1}{n} \left| \{j : 1 \le j \le n \text{ and } \text{int}((j+1)\alpha) > \text{int}(j\alpha)\} \right| = \alpha, \text{ and}$$
$$\lim_{n \to \infty} \frac{1}{n} \left| \{j : 1 \le j \le n \text{ and } \text{int}((j+1)\alpha) = \text{int}(j\alpha)\} \right| = 1 - \alpha.$$

8. Let J be a subinterval of $[0, 1)$ having positive length and let α be irrational. For $n = 1, 2, \ldots$, let
$$A(\alpha, J, n) = \{j : 1 \le j \le n \text{ and } \text{frac}(j\alpha) \in J\}.$$
Deduce from Kronecker's Theorem that
$$\lim_{n \to \infty} |A(\alpha, J, n)| = \infty.$$

9. Consider the following and give proofs as required.

   (i) Let (a_n) be a sequence of numbers in $(0, \infty)$ such that $\lim_{n \to \infty} a_n = 0$ and $\sum_{j=1}^{\infty} a_j = \infty$. Prove that the sequence $\left(\text{frac}(\sum_{j=1}^{n} a_j) \right)$ is dense in $[0, 1]$. Hence deduce that the sequence $\left(\text{frac}\left(\sum_{j=1}^{n} 1/j \right) \right)$ is dense in $[0, 1]$.

   (ii) Let $f : [1, \infty) \longrightarrow (0, \infty)$ be a twice differentiable function such that
   $$f'(x) > 0 \quad \text{and} \quad f''(x) < 0 \text{ for all } x, \quad \lim_{n \to \infty} f'(n) = 0$$
   and
   $$\sum_{n=1}^{\infty} f'(n) = \infty.$$
   Using (i), or otherwise, prove that the sequence $\left(\text{frac}(f(n)) \right)$ is dense in $[0, 1]$. Deduce that both the sequences $\left(\text{frac } \sqrt{n} \right)$ and $\left(\text{frac}(\log n) \right)$ are dense in $[0, 1]$.

   (iii) Prove that for $0 \le a < b < 1$,
   $$\lim_{n \to \infty} \frac{1}{n} \left| \{j : 1 \le j \le n \text{ and } \text{frac}\left(\sqrt{j} \right) \in [a, b)\} \right| = b - a.$$

(iv) If $0 \le a \le b \le 1$, prove that

$$\liminf_{n\to\infty} \frac{1}{n}\left|\{j : 1 \le j \le n \text{ and } \operatorname{frac}(\log j) \in [a,b)\}\right| = \frac{e^{b-a}-1}{e-1},$$

and

$$\limsup_{n\to\infty} \frac{1}{n}\left|\{j : 1 \le j \le n \text{ and }\right.$$

$$\left.\operatorname{frac}(\log j) \in [a,b)\}\right| = e^{1-b+a}\left(\frac{e^{b-a}-1}{e-1}\right).$$

(v) Can the methods used to derive the results in (iii) and (iv) be used to prove Weyl's Theorem? If not, what are the difficulties?

10. Let $p, q \in \mathbb{N}$, let p, q have no common factors, and put $\alpha = p/q$. Let $x \in [0, 1)$ and let τ_α denote the function mapping $[0, 1)$ into $[0, 1)$ given by $x \longmapsto \operatorname{frac}(x + \alpha)$. Prove that the points

$$\operatorname{frac}\left(x + \frac{p}{q}\right), \operatorname{frac}\left(x + \frac{2p}{q}\right), \ldots, \operatorname{frac}\left(x + \frac{(q-1)p}{q}\right)$$

are distinct. Deduce that in the dynamical system $([0,1), \tau_\alpha)$, every point is a periodic point with period q.

11. Let $[a, b)$ and $[c, d)$ be two arcs in $[0, 1)$ of equal length, and let τ denote the function $x \longmapsto \operatorname{frac}(c - a + x)$ on $[0, 1)$. Prove that $\tau(J) = K$.

12. Prove that if $\alpha \in \mathbb{R}$ and $x \in [0, 1)$, then the sequence

$$\operatorname{frac}(x + \alpha), \operatorname{frac}(x + 2\alpha), \operatorname{frac}(x + 3\alpha), \ldots,$$

is dense in $[0, 1)$ if and only if α is irrational. Deduce that if the sequence $\operatorname{frac}(x + \alpha), \operatorname{frac}(x + 2\alpha), \operatorname{frac}(x + 3\alpha), \ldots$ is dense in $[0, 1)$ for a *single* value of x, it is dense in $[0, 1)$ for *all* values of x.

13. Let α be an irrational number and let τ_α be the transformation on $([0, 1)$ given by $x \longmapsto \operatorname{frac}(x + \alpha)$. Corresponding to the dynamical systems interpretation of Kronecker's Theorem, prove the dynamical systems version of Weyl's Theorem. That is, if $a, b \in [0, 1)$ with $a \le b$, and if $x \in [0, 1)$, prove that

$$\lim_{n\to\infty} \frac{1}{n}\left|\{j : 1 \le j \le n \text{ and } \tau_\alpha^n(x) \in [a,b)\}\right| = b - a.$$

Exercises

14. Let (x_n) and (y_n) be two sequences in $[0, 1)$ with $\lim_{n\to\infty} y_n = 1$.
    (i) If (x_n) is dense in $[0, 1]$, prove that $(x_n y_n)$ is dense in $[0, 1]$.
    (ii) If (x_n) is uniformly distributed in $[0, 1]$, prove that $(x_n y_n)$ is uniformly distributed in $[0, 1]$.

15. Let J be an interval and let $f : J \longrightarrow J$ be a transformation on J. If there is an orbit of f that is dense in J, prove that the system (J, f) is transitive.

16. For $x, y \in [0, 1)$, let
    $$d(x, y) = \min\{|x - y|, 1 - |x - y|\}.$$
    Thus, d is a function from $[0, 1) \times [0, 1)$ into $[0, \infty)$. Let $\alpha \in \mathbb{R}$ and let τ_α be the transformation $x \longmapsto \text{frac}(x + \alpha)$ on $[0, 1)$. Consider the following statements.
    (i) $d(x, y) \in [0, 1/2]$ for all $x, y \in [0, 1)$.
    (ii) $d(x, y) = \min\{\text{frac}(x - y), \text{frac}(y - x)\}$.
    (iii) $d(x, y) = 0 \iff x = y$.
    (iv) $d(x, y) = d(y, x)$, for all $x, y \in [0, 1)$.
    (v) $d(x, z) \le d(x, y) + d(y, z)$, for all $x, y, z \in [0, 1)$.
    (vi) $d(x, y) = d(\tau_\alpha(x), \tau_\alpha(y))$.

    We think of $d(x, y)$ as a notion of *distance* between x and y and the fact that $d \ge 0$ and the statements (iii), (iv), (v) reflect intuitive properties we would expect a notion of distance to have. The property (iv) is known as the *triangle inequality*, and is an analogue of the fact that in Euclidean geometry, the length of the side of a triangle is less than or equal to the sum of the lengths of the other two sides. Statement (vi) above shows that τ_α preserves the "distance" between points. Now, prove statements (i) to (vi) above.

    Interpret the "distance function" d by means of pictures on the unit interval. A different geometric interpretation can be given by identifying $[0, 1)$ with the unit circle $\mathbb{T}$ by means of the function $t \longmapsto (\cos 2\pi t, \sin 2\pi t)$ as in Figure 2.1. Give a description of this interpretation, including pictures.

17. Let $\mathbb{C}$ denote the set of complex numbers. The unit circle $\mathbb{T}$ is the subset of $\mathbb{C}$ given by
    $$\{z : z \in \mathbb{C} \text{ and } |z| = 1\}.$$
    Graphically, $\mathbb{T}$ is the circle of radius 1 whose center is the origin. Consider the function $\theta : [0, 1) \longrightarrow \mathbb{T}$ given by $\theta(t) = \cos 2\pi t + i \sin 2\pi t$. Alternatively, by Euler's Theorem, $\theta(t) = e^{2\pi i t}$ for $t \in [0, 1)$.

(i) Prove that θ maps $[0, 1)$ into $\mathbb{T}$.
(ii) Prove that θ is one-to-one.
(iii) Prove that θ maps $[0, 1)$ onto $\mathbb{T}$.
(iv) Let us say that an *arc* in $\mathbb{T}$ is a subset of $\mathbb{T}$ that is an arc in the usual geometric sense. If J is an arc in $[0, 1)$, prove that $\theta(J)$ is an arc in $\mathbb{T}$. If K is a subinterval of $[0, 1)$ of the form $[a, b)$, and if K is an arc in $[0, 1)$ that is not an interval, then $\theta(J)$ and $\theta(K)$ are both arcs in $\mathbb{T}$, but what is the difference, if any, between them?
(v) If J is an arc in $[0, 1)$, and using the usual notion of arc length in $\mathbb{T}$, prove that the arc $\theta(J)$ in $\mathbb{T}$ has length $2\pi\mu(J)$.
(vi) If $\alpha \in \mathbb{R}$, let ρ_α be the function on $\mathbb{T}$ given by

$$\rho_\alpha(w) = e^{2\pi i \alpha} w, \text{ for } w \in \mathbb{T}.$$

Geometrically, ρ_α is an counterclockwise rotation on $\mathbb{T}$ through an angle $2\pi\alpha$. Prove that

$$\theta \circ \tau_\alpha = \rho_\alpha \circ \theta \text{ or, equivalently, } \theta \circ \tau_\alpha \circ \theta^{-1} = \rho_\alpha.$$

This shows that the translation τ_α on $[0, 1)$ corresponds to the counterclockwise rotation ρ_α through the angle $2\pi\alpha$ on $\mathbb{T}$.
(vii) If $\alpha \in [0, 1)$, prove that α is irrational if and only if $\theta(\alpha)$ is not a root of unity.
(viii) If $\alpha \in [0, 1)$ and $z = \theta(\alpha) = e^{2\pi i \alpha}$, prove that

$$\theta(\text{frac}(n\alpha)) = z^n, \text{ for all } n \in \mathbb{Z}.$$

(ix) If $z \in \mathbb{T}$, prove that z is not a root of unity if and only if the sequence (z^n) is dense in $\mathbb{T}$. That is, prove that z is not a root of unity if and only if for each $w \in \mathbb{T}$ and each $\varepsilon > 0$, there is $n \in \mathbb{N}$ such that $|w - z^n| < \varepsilon$.

18. Let α be rational and let τ_α be the transformation $x \mapsto \text{frac}(x + \alpha)$ on $[0, 1)$. Prove that the system $([0, 1), \tau_\alpha)$ is not transitive.

Investigations

1. Let $\alpha, \beta \in \mathbb{R}$. Discuss conditions under which there is $j \in \mathbb{N}$ such that $\text{frac}(j\alpha) = \text{frac}(j\beta)$. When this occurs, for a given $j \in \mathbb{N}$, is there $k \in \mathbb{N}$ such that $\text{frac}(j\alpha) = \text{frac}(k\beta)$? More generally, discuss when there are $j, k \in \mathbb{N}$ such that $\text{frac}(j\alpha) = \text{frac}(k\beta)$. Another way of thinking

Investigations

about these questions is by means of the dynamical systems $([0, 1), \tau_\alpha)$, $([0, 1), \tau_\beta)$ corresponding to α and β, because of the fact that the sequence (frac$(n\alpha)$), for example, is the orbit of 0 in the system $([0, 1), \tau_\alpha)$. Interpret the conclusions in terms of orbits in the appropriate dynamical systems.

2. Let $J = [a, b)$ be an arc of $[0, 1)$ and let $\alpha \in [0, 1)$. Let τ_α denote the function $x \longmapsto$ frac$(x + \alpha)$. Prove that τ_α maps the arc J onto the arc $K =$ [frac$(a + \alpha)$, frac$(b + \alpha)$). Using this and Kronecker's Theorem, investigate the nature of the orbit of a general point $x \in [0, 1)$ within the dynamical system $([0, 1), \tau_\alpha)$ by considering, for example, whether successive points in the orbit of x are equally spaced, and how such an observation might vary for rational and irrational α.

3. Two runners run at constant speeds around a circular racetrack, starting at the same point at the same time, but they run in *opposite* directions. Discuss whether the runners ever run past each other at exactly the point where they started. What is the smallest time in which this could occur, if it does occur? If it never occurs, discuss whether there is an "approximate" phenomenon that occurs, and analyze it mathematically.

4. Three runners run around a circular race track in the same direction, each with her own constant speed. Investigate under what conditions the three runners simultaneously will find themselves together at some point on the race track.

5. We have three tapes of music. We start the three tapes at the same time, on three different tape players. Each tape player plays at its own constant speed, and we listen to the resulting "music". Investigate under what conditions there is a future time at which the conditions we started with are repeated, so that the music repeats itself.

6. Investigate what happens under a *discrete* model of phase shifting. In this case assume we have two circular tapes of radius r, and assume that they have a common speed v. The tapes are copies of each other. The music on the first tape commences at the point P_1 on the tape, and the music on the second tape commences at the point P_2 on the tape. As the first tape turns, there is a stable point Q_1, that is not on the tape and does not move with the tape, at which the tape player plays the notes that are recorded on the part of the tape that is at Q_1 at the time. Similarly, as the second tape runs, there is a stable point Q_2, that is not on the tape and does not move with the tape, at which the tape player plays the notes that are recorded on the tape at Q_2. Denote time by t. The discrete model of phase shifting means (say) that after the elapse of passage of time t_0, the second tape is instantaneously advanced through a distance d. So, starting at $t = 0$, both tapes will be

playing identical music, but at time t_0, the second tape is advanced by d and the tapes will then be "out of phase" and playing different music. This step is repeated at time $2t_0$, then at $3t_0$ and so on. After time t, the first tape will have moved through a distance vt and the second tape will have moved through a distance $vt + d \operatorname{int}(t/t_0)$. As in the case of the continuous phase model, discussed in Section 2.6, there will be $k, \ell \in \mathbb{Z}$ and $\theta_1, \theta_2 \in [0, 1)$ such that

$$vt = 2\pi rk + 2\pi r\theta_1 \quad \text{and} \quad vt + d \operatorname{int}\left(\frac{t}{t_0}\right) = 2\pi r\ell + 2\pi r\theta_2.$$

Now, investigate whether the music ever repeats itself, and if not, does it "approximately" repeat itself? In either case, does our knowledge of Kronecker's dynamical system and Weyl's Theorem enable us to say anything about patterns in the repetition or "approximate" repetition of the music?

7. Steve Reich has used the idea of phase shifting in several compositions. Here, in part, is how he describes the ideas in *Piano Phase*, composed in 1967:

> The musical material is a small number of repeating patterns that may be learned and memorized in a few minutes. The score then shows that two musicians begin in unison playing the same pattern over and over again and that while one of them stays put, the other gradually increases his or her tempo so as to slowly move one beat ahead of the other. This process is repeated until both players are back in unison, at which point the pattern is changed and the phasing process begins again.

Although this refers to a different composition using pianos, essentially this corresponds to a possible way of interpreting what happens in *It's Gonna Rain* (mentioned in Section 2.6) if the phase shifting is achieved by *discrete* changes. Using the conclusions from the previous investigation, discuss whether the music will always repeat itself. How would this be interpreted in terms of runners running around a circular track?

8. Recall that a sequence (x_n) in $[0, 1)$ is called *uniformly distributed* if for each subinterval J of $[0, 1)$,

$$\lim_{n \to \infty} \frac{1}{n} \left|\left\{j : 1 \leq j \leq n \text{ and } x_j \in J\right\}\right| = \mu(J).$$

Now let $\sigma : \mathbb{N} \longrightarrow \mathbb{N}$ be a one-to-one and onto function (such a function sometimes is called a *permutation* on $\mathbb{N}$).

(i) If (x_n) is uniformly distributed in $[0, 1)$, and σ is a permutation on $\mathbb{N}$, prove that both the sequences (x_n) and $(x_{\sigma(n)})$ are dense in $[0, 1)$.

(ii) If (x_n) is uniformly distributed in $[0, 1)$, give an example of a permutation σ such that the sequence $(x_{\sigma(n)})$ is *not* uniformly distributed in $[0, 1)$.

(iii) Given the result in (ii), if (x_n) is uniformly distributed in $[0, 1)$, investigate the conditions on a permutation σ that ensure that the sequence $(x_{\sigma(n)})$ *is* uniformly distributed in $[0, 1)$. For example, it may seem likely that if a permutation σ on $\mathbb{N}$ is such that there is $m \in \mathbb{N}$ with $|n - \sigma(n)| \le m$ for all $n \in \mathbb{N}$, then $(x_{\sigma(n)})$ will be uniformly distributed. But is this in fact correct and, if so, is it the best that can be said?

9. Let $f : [1, \infty) \longrightarrow (0, \infty)$ be an increasing function. Taking note of the results in Exercise 9, investigate the conditions on f that imply that the sequence $(\text{frac}(f(n)))$ is uniformly distributed in $[0, 1)$, and those that imply the same sequence is not uniformly distributed in $[0, 1)$.

Notes

1. [Page 25] Pythagoras of Samos (c.570—c.495 BCE) was an early Greek mathematician and philosopher who is noted by mathematicians and school children for the theorem named after him. We attribute to Pythagoras what is perhaps the first scientific discovery, that pure musical notes correspond to ratios of whole numbers. Concerning the Greeks and irrational numbers, Sir Thomas Heath writes [5, pp. 90–91]

   > it is certain that the incommensurability of the diagonal of a square with its side, that is the 'irrationality' of $\sqrt{2}$, was discovered by the school of Pythagoras ... and ... the Pythagoreans invented a method of obtaining an infinite series of arithmetical ratios approaching more and more closely to the value of $\sqrt{2}$.

   There is a discussion of irrational numbers in the dialogue *Theaetetus* of Plato (c. 429–347 BC). In it Theaetetus proposes that the square root of any positive integer that is not a perfect square is irrational, improving upon arguments of the geometer Theodorus who apparently had separate proofs for showing that $\sqrt{3}, \sqrt{5}, \ldots, \sqrt{17}$ are irrational (see [11, pp. 23–25 and 138–139]).

2. [Page 25] Sir Karl Popper (1902–1994) was a philosopher of science. He is noted especially for his view that the systematic use of *falsification* of

ideas and theories is the feature of science that distinguishes it *as science from other endeavors.* The falsification of a physical theory corresponds, in mathematics, to the production of a counter-example to a conjecture. The quotation is from Popper's essay "The nature of philosophical problems and their roots in science" [12, p. 75]. In the essay Popper is arguing, more generally, that genuine philosophical problems are rooted in problems lying outside philosophy.

3. [Page 26] These comments were made by Ennio De Giorgi (1928–1996) in an interview with Michele Emmer, an edited version of which appears in the *Notices of the American Mathematical Society* [2]. The same volume of the *Notices* also contains an evaluation of the work of De Giorgi, and of the man himself, by Jacques-Louis Lions and Francois Murat [7].

4. [Page 33] The pigeon-hole principle is also known as Dirichlet's principle, the drawer principle, and the letter-box principle. Another way of looking at it is the following: if we have n boxes each containing a single object, and if we place another object in any one of the boxes, then one of the boxes must contain two objects.

5. [Page 35] Leopold Kronecker (1823–1891) proved that the sequence (frac($n\alpha$)) is dense in $[0, 1)$ for irrational α in 1884 (in fact the result here is the one-dimensional version of Kronecker's more general result, which he obtained for higher dimensions as well as for the interval $[0, 1)$). Kronecker made the famous statement "God made the integers, all else is the work of Man". In [4, Chapter XXIII], Hardy and Wright present several proofs of Kronecker's Theorem and give more detailed references. Also, Apostol in [1, Chapter 7] has a discussion of Kronecker's Theorem and its applications in number theory and to the Riemann zeta function.

6. [Page 46] Steve Reich is an American composer, born in 1936. He studied philosophy at Cornell University, and is one of the earliest of the "minimalist" composers. His compositions "It's Gonna Rain" and "Piano Phase", and others, are on the Elektra/Nonesuch compact disc 979 169-2, and it is from the notes accompanying this disc that the quotations in this section are taken.

7. [Page 53] The result known as Weyl's Theorem was proved independently by P. Bohl, W. Sierpiński and H. Weyl in 1909–10 (see [6, p. 21] for precise references). The usual proof of Weyl's Theorem depends upon *Weyl's criterion*, which says that a sequence (x_n) in $[0, 1)$ is uniformly distributed if and only if

$$\lim_{n\to\infty} \frac{1}{n} \sum_{j=1}^{n} e^{2\pi k x_j} = 0, \text{ for all } k \in \mathbb{Z} \text{ with } k \neq 0.$$

The proof of Weyl's criterion rests upon arguments from real analysis that involve trigonometric approximation. The proof of Weyl's Theorem presented here is more elementary and is an adaptation of the proof given by A. Miklavc [8]. A different proof of Weyl's Theorem, using continued fractions, may be found in [4, pp. 390-392] and [10, pp. 71–75]. See also the discussion here in Chapter 5, where Weyl's Theorem is presented as an application of Birkhoff's Ergodic Theorem.

8. [Page 71] Further details concerning $([0, 1), \tau_\alpha)$ as a chaotic system may be found in [9].

Bibliography

[1] T. M. Apostol, *Modular Functions and Dirichlet Series in Number Theory*, Springer-Verlag, New York, 1976, second edition 1990.

[2] E. De Giorgi and M. Emmer, *Interview with Ennio De Giorgi*, Notices Amer. Math. Soc., **44** (1997), 1097–1101.

[3] R. Devaney, *A First Course in Chaotic Dynamical Systems*, Addison Wesley, Massachusetts, 1992.

[4] G. H. Hardy and E. M. Wright, *An Introduction to the Theory of Numbers*, Clarendon Press, Oxford, 1938, fourth edition reprinted 1968.

[5] T. L. Heath, *A History of Greek Mathematics, volume 1: from Thales to Euclid*, Clarendon Press, Oxford, 1921, reprinted 1965.

[6] L. Kuipers and H. Niederreiter, *Uniform Distribution of Sequences*, Pure and Applied Mathematics no. 29, Wiley, New York 1974.

[7] J.-L. Lions and F. Murat, Ennio De Giorgi (1928–1996), *Notices Amer. Math. Soc.*, **44** (1997), 1095-1096.

[8] A. Miklavc, Elementary proofs of two theorems on the distribution of numbers $n\theta(\mathrm{mod}\,1)$, *Proc. Amer. Math. Soc.*, **39** (1973), 279–280.

[9] R. Nillsen, Chaos and one-to-oneness, *Math. Magazine*, **72** (1999), 14–21.

[10] I. Niven, *Irrational Numbers*, Carus Mathematical Monographs no. 11, Mathematical Association of America, Washington DC, 1967.

[11] Plato, *Theaetetus*, translated by R. A. H. Waterfield, Penguin Books, Middlesex, UK, 1987.

[12] K. R. Popper, *Conjectures and Refutations*, Routledge, 1963.

Chapter 3 Flow Chart

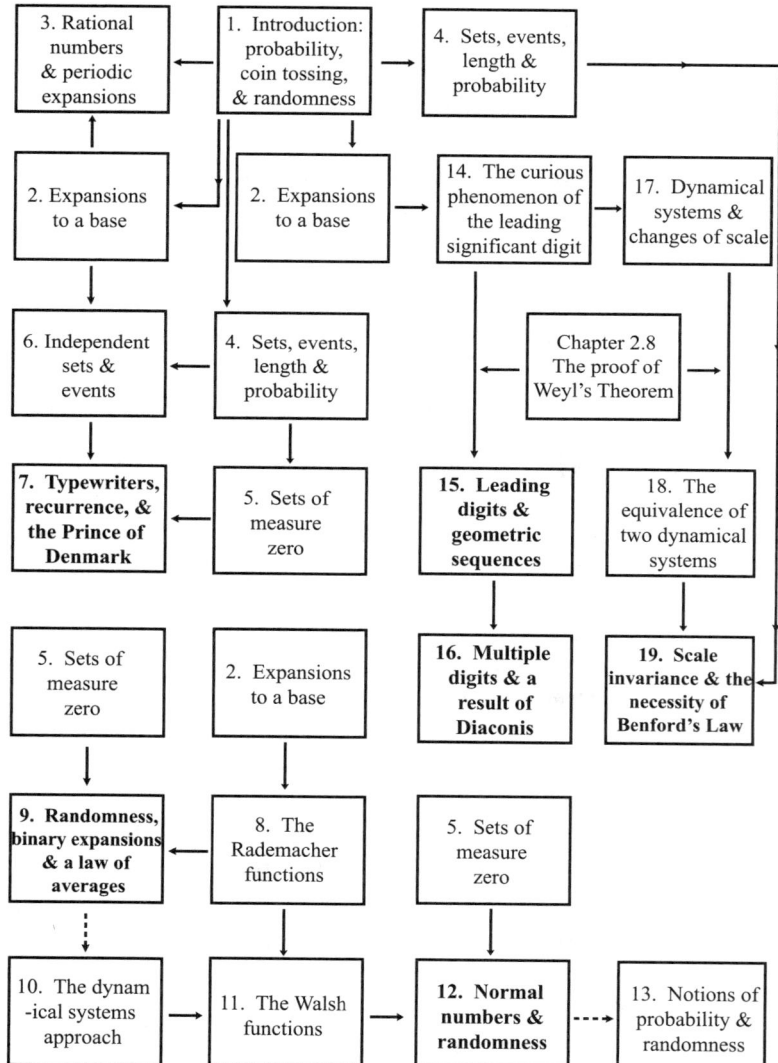

Chapter 3

Probability and Randomness

> It may be taken for granted that any attempt at defining disorder in a formal way will lead to a contradiction. This does not mean that the notion of disorder is contradictory. It is so, however, as soon as I try to formalize it. Hans Freudenthal (1905–1990)

3.1 Introduction: probability, coin tossing and randomness

A main aim in this chapter is to pursue a connection between elementary real analysis and probability theory. In probability, we imagine an experiment or a procedure whose set of all possible outcomes is denoted by S. The set S is usually called the *sample space*. Thus, a point x in S represents a possible outcome of the experiment. Now, consider a subset A of S. If the experiment is carried out and produces an outcome x, we can observe whether $x \in A$ or $x \notin A$. For this reason, A is often called an *event*. Then, if the experiment is carried out repeatedly, we can observe the proportion of times for which the outcome x was in A. As the experiment is repeated indefinitely, this proportion might be expected to approach a limit, which is called the *probability of the event A*—this approach regards probability as a measure of the frequency of an event. Alternatively, we can think of the probability of an event A as the proportion of S that is "occupied" by A, or as the ratio of the "length" or "area" of A to the "length" or "area" of S.

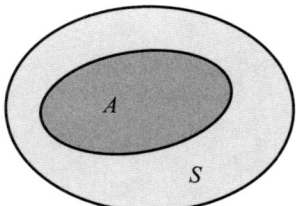

Figure 3.1. The event A is depicted as a subset of S, the set of all possible outcomes of a given experiment. The probability of A may be thought of as the proportion of S "occupied" by A. We expect it will coincide with the frequency with which the event A is observed when the experiment is carried out repeatedly. It is important to be clear as to what the experiment is and what the associated sample space is. For example, when we toss a coin twice, it can be regarded as *repetition* of the experiment of tossing a coin once, or it can be regarded as a *single* experiment in which we toss a coin *twice*. In each case the experiment and the associated sample space are different.

As an example, consider an experiment where a coin is tossed *once*. The sample space S consists of two possible outcomes, heads (H) or tails (T). Thus, $S = \{H, T\}$. If the coin is unbiased, we would expect that the probability of the event $\{H\}$ is $1/2$, and the same for $\{T\}$. Now consider an experiment where a coin is tossed *twice*. In this case, there are four possible outcomes, and the sample space is $S = \{HH, HT, TH, TT\}$ (here, HT means that a head was obtained on the first toss and a tail was obtained on the second toss, and so on for the other possible cases). If the coin is unbiased, we would expect that the probability of the event HH would be $1/4$, and the same for the other events TH, HT, TT. However, we would expect that the probability of getting one head *and* one tail, that is, the probability of the event $\{HT, TH\}$, would be $1/2$. In these two experiments, the sample space has only a finite number of points, but the ideas can be extended to discrete sample spaces—that is, to sample spaces that are finite or whose elements can be arranged as an infinite sequence of outcomes. Note that the sample space depends entirely upon the experiment that is being conducted—if the experiment is to toss a coin 100 times, there are 2^{100} possible outcomes and this is the number of elements in the sample space.

3.1. Introduction: probability, coin tossing and randomness

An important example of an infinite sample space arises from tossing a coin successively, and where the process never stops. We might say that the coin is tossed an *infinite* number of times. Because in practice we cannot toss a coin an infinite number of times, this is an idealized "mind experiment". (Such mind experiments have also played an important role in twentieth century physics, Einstein being particularly fond of them.) In this case the sample space consists of all possible sequences of heads and tails. Now, a sequence of heads and tails can be made into a sequence of 0s and 1s, simply by replacing each head by a 0 and each tail by a 1. So, in this case, the sample space can be regarded as the set of all sequences of 0s and 1s.

In Chapter 3 the main interest is in the case where the sample space is a bounded interval of real numbers, usually the unit interval $[0, 1)$, and a main technical tool for this is the relationship between $[0, 1)$ and the sample space consisting of all sequences of 0s and 1s (see the remarks in Section 3.2). Many of the notable results in Chapter 3 take the form of asserting that under certain conditions, the probability of an event, interpreted as a frequency, equals the probability of the event interpreted as a ratio of lengths or areas. To illustrate this type of result, consider a given interval S. In this case, consider the experiment of selecting a single point in S at random—this is an idealized "mind experiment", since the mechanism for selecting a point "at random" is not specified and it is something that we imagine only. By successively repeating this experiment, we obtain a sequence (x_n) of points in S. Now, let A be a subinterval of S. The number

$$\left|\left\{j : 1 \leq j \leq n \text{ and } x_j \in A\right\}\right|$$

is the number of times the event A occurs in the observation of the first n terms of (x_j). If

$$\lim_{n \to \infty} \left|\left\{j : 1 \leq j \leq n \text{ and } x_j \in A\right\}\right| = \infty, \qquad (3.1)$$

we see that the event A occurs infinitely many times, and we say that the event is *recurrent*. Also, the number

$$\frac{1}{n}\left|\left\{j : 1 \leq j \leq n \text{ and } x_j \in A\right\}\right|$$

is the proportion of times in the first n repetitions of the experiment that the event $x \in A$ occurs. Then, in the frequency interpretation of probability, the probability of the event A equals

$$\lim_{n\to\infty} \frac{1}{n} \left| \left\{ j : 1 \leq j \leq n \text{ and } x_j \in A \right\} \right|,$$

assuming that this limit exists. Thus, in this context, the assertion that the frequency notion of probability coincides with the "ratio of lengths" notion of probability would take the form of asserting that

$$\lim_{n\to\infty} \frac{1}{n} \left| \left\{ j : 1 \leq j \leq n \text{ and } x_j \in A \right\} \right| = \frac{\text{the length of } A}{\text{the length of } S}. \quad (3.2)$$

The above considerations apply to *all* sequences (x_n) of points in S, even if we do not think of them as observed results of an experiment. Note that if (3.2) holds and A has positive length, then A must be recurrent. So, the question of the coincidence of the two notions of probability, as in (3.2), is a refinement of the notion of recurrence, as expressed by (3.1).

A connection with dynamical systems comes about when the sequence of points (x_n) arises from the orbit of a point in a dynamical system. Thus, if S is a bounded interval of real numbers, if $f : S \longrightarrow S$ is a transformation on S, and if $x \in S$, for the sequence $(f^{n-1}(x))$, (3.1) takes the form

$$\lim_{n\to\infty} \left| \left\{ j : 1 \leq j \leq n \text{ and } f^{j-1}(x) \in A \right\} \right| = \infty, \quad (3.3)$$

and (3.2) takes the form

$$\lim_{n\to\infty} \frac{1}{n} \left| \left\{ j : 1 \leq j \leq n \text{ and } f^{j-1}(x) \in A \right\} \right| = \frac{\text{the length of } A}{\text{the length of } S}. \quad (3.4)$$

Now we do not expect that (3.3) or (3.4) will hold for all points in a dynamical system and all subintervals A of S. But if (3.4) holds for most points x and all subintervals A, we might consider the system to be "random" in the following sense: that whatever the initial point x is, and whatever part A of the system we choose, the points in the orbit of x spend, in the long run, the same proportion of time in A, and this

3.1. Introduction: probability, coin tossing and randomness

proportion depends only on the length of A and not on the position of A. That is, the system does not favor one part of S over another, but essentially all orbits are spread equally throughout S. Equations (3.2) and (3.4) express the idea that the observations are "unbiased", in that the frequency of the given event equals what would be expected in an unbiased or random system. Proving that equations like (3.2) and (3.4) hold in various dynamical systems is a major theme of this chapter.

Typically, treatment of the type of material in this chapter depends upon a knowledge of measure theory. One of the reasons for this is given by Feller [12, p. 197–198] as follows:

> In the present volume we are confined to the theory of discrete sample spaces, and this means a considerable loss of mathematical elegance ... we calculate the obvious limits of probabilities and clearly require no measure theory for that purpose. But only general measure theory shows that our limits are independent of the particular passage to the limit.

In the approach here, measure theory is avoided so as to minimize prerequisite knowledge, while at the same time it aims not to sacrifice any mathematical elegance or completeness. The intention is to achieve this by developing new methods of argument and by restricting the events to those that are "permissible". When S is an interval, the "permissible" events, that is the "permissible" subsets of S, are the sets that are finite unions of intervals together with the "sets of measure zero". These latter sets have a simple definition, and their consideration does not involve general measure theory.

Amongst other results, these ideas lead to the *Normal Numbers Theorem*, according to which every number x in $[0, 1)$, unless x is in some preassigned set of measure zero, has the property that every given finite sequence of 0s and 1s occurs within the binary expansion of x—that is, the expansion of x to the base 2—in the expected proportion. This shows that the binary expansion of a typical number exhibits a strong "randomness" property, and emphasizes the connection between the real numbers and randomness, which appears when we represent numbers as sequences of discrete bits of information. Other themes developed in this part include recurrence, Benford's logarithmic law on the relative frequency of the digits occurring in binary expansions and in expansions to arbitrary bases, and the uniqueness of a probability law that remains the same when the data it describes are transformed by a change of units (such a law is necessarily Benford's logarithmic law).

3.2 Expansions to a base

A familiar idea in school mathematics is the decimal representation of numbers. The idea is that there are the ten symbols $0, 1, 2, 3, \ldots, 9$, and a number in $[0, 1)$ is represented by an infinite sequence of these symbols. If x is such a number, it is usual to write $x = .c_1 c_2 c_3 \ldots$, where $c_1, c_2, c_3, \ldots$ are called the *digits* of x relative to the base 10. The equation $x = .c_1 c_2 c_3 \ldots$ sometimes is called the *expansion* of x to the base 10. In this work, there is a need to consider expansions to a general base, and especially to the base 2. There is also the problem of the *meaning* of the equation $x = .c_1 c_2 c_3 \ldots$ in the decimal expansion of x, and the meaning of the corresponding equation when x is expanded using some other base. In this section these and other matters are clarified, being considered as important in themselves, but also for the extensive use of them made in later parts of this work.

In the case of the base 2 the expansion of a number x in $[0, 1)$ is called the *binary expansion* of x and leads to writing $x = .c_1 c_2 c_3 \ldots$, where $c_1, c_2, \ldots$ is a sequence of 0s and 1s. This sequence may sometimes be called the *binary sequence* of x. Here, we can regard each 0 in the expansion as a "head" which is the result of a toss in an infinite sequence of tosses of a coin, and we can regard a 1 as representing a "tail". That is, the digits in the binary expansion of a number may be regarded as the sequence of heads and tails resulting from a coin tossing experiment corresponding to the number. A coin tossing experiment is a (presumably) random process, so this indicates the connection between the description of numbers by means of their expansions to a base and ideas of randomness. In fact, $[0, 1)$ can be essentially *identified* with the set of all sequences of 0s and 1s—that is, with the set of all sequences

3.2. Expansions to a base

of heads and tails resulting from tossing a coin repeatedly, without limit (see Exercise 1). A particular purpose of this section is to see that rational numbers have expansions which are in a sense "finite" or periodic, but irrational numbers have expansions which are "infinite", "aperiodic" or, perhaps, "random".

Definitions. Let $b \in \mathbb{N}$ with $b \geq 2$. This number is kept fixed for this section, and we shall call the number b the *base*. Now, let $x \in [0, 1)$ be given. We now describe the digits of x relative to the base b. The first digit of x relative to the base b is denoted by $d_1(x)$ and is given by

$$d_1(x) = \max\left\{j : 0 \leq j \leq b-1 \text{ and } \frac{j}{b} \leq x\right\}.$$

This definition implies that

$$\frac{d_1(x)}{b} \leq x < \frac{d_1(x)}{b} + \frac{1}{b}. \tag{3.5}$$

We proceed by induction. Suppose that the first n digits $d_1(x), d_2(x), \ldots, d_n(x)$ have been defined and that

$$\frac{d_1(x)}{b} + \frac{d_2(x)}{b^2} + \cdots + \frac{d_n(x)}{b^n}$$
$$\leq x < \frac{d_1(x)}{b} + \frac{d_2(x)}{b^2} + \cdots + \frac{d_n(x)}{b^n} + \frac{1}{b^n}. \tag{3.6}$$

Note that (3.6) is true in the case $n = 1$, because of (3.5). The $(n+1)^{th}$ digit of x relative to the base b is denoted by $d_{n+1}(x)$ and is given by

$$d_{n+1}(x) = \max\left\{j : 0 \leq j \leq b-1 \text{ and }\right.$$
$$\left. \frac{d_1(x)}{b} + \frac{d_2(x)}{b^2} + \cdots + \frac{d_n(x)}{b^n} + \frac{j}{b^{n+1}} \leq x\right\}.$$

This definition of $d_{n+1}(x)$ implies that

$$\frac{d_1(x)}{b} + \frac{d_2(x)}{b^2} + \cdots + \frac{d_n(x)}{b^n} + \frac{d_{n+1}(x)}{b^{n+1}} \leq x$$
$$< \frac{d_1(x)}{b} + \frac{d_2(x)}{b^2} + \cdots + \frac{d_n(x)}{b^n} + \frac{d_{n+1}(x)}{b^{n+1}} + \frac{1}{b^{n+1}}. \tag{3.7}$$

So, (3.7) shows that if $d_1(x), \ldots, d_n(x)$ have been defined so that (3.6) holds, then $d_1(x), \ldots, d_n(x), d_{n+1}(x)$ as defined will also satisfy (3.6), but with $n + 1$ in place of n. However, $d_1(x)$ has been defined, and (3.5) shows that (3.6) holds when $n = 1$. So, by induction, $d_n(x)$ has been defined for all $n \in \mathbb{N}$, and (3.6) holds for all $n \in \mathbb{N}$. The digit $d_n(x) \in \{0, 1, 2, \ldots, b-1\}$ is called the nth *digit of x relative to the base b*.

As well as defining the digits of x relative to a base b, the above remarks, as expressed in (3.6), have proved the first part of the following result.

Theorem 3.1. *Let $b \in \mathbb{N}$ with $b \geq 2$. Let $x \in [0, 1)$ and let $d_1(x), d_2(x), d_3(x), \ldots$ denote the sequence of digits of x relative to the base b. Then, for all $n \in \mathbb{N}$, $d_n(x) \in \{0, 1, 2, \ldots, b-1\}$ and*

$$\sum_{j=1}^{n} \frac{d_j(x)}{b^j} \leq x < \sum_{j=1}^{n} \frac{d_j(x)}{b^j} + \frac{1}{b^n}. \tag{3.8}$$

The sequence $(d_n(x)/b^n)$ is summable, and

$$x = \sum_{n=1}^{\infty} \frac{d_n(x)}{b^n}. \tag{3.9}$$

Also, there is no $r \in \mathbb{N}$ such that $d_n(x) = b-1$ for all $n \geq r+1$.

Proof. Owing to (3.6), (3.8) holds, so that $|\sum_{j=1}^{n} d_j(x)/b^j - x| < 1/b^n$. As $\lim_{n \to \infty} 1/b^n = 0$, it follows that

$$x = \lim_{n \to \infty} \sum_{j=1}^{n} \frac{d_j(x)}{b^j} = \sum_{n=1}^{\infty} \frac{d_j(x)}{b^j}.$$

Finally, assume that $r \in \mathbb{N}$ is such that $d_n(x) = b-1$ for all $n \geq r+1$.

3.2. Expansions to a base

Then, we have from (3.9) that

$$x = \sum_{n=1}^{\infty} \frac{d_n(x)}{b^n}$$
$$= \sum_{n=1}^{r} \frac{d_n(x)}{b^n} + \sum_{n=r+1}^{\infty} \frac{b-1}{b^n} = \sum_{n=1}^{r} \frac{d_n(x)}{b^n} + \frac{b-1}{b^{r+1}}\left(\frac{1}{1-1/b}\right)$$
$$= \sum_{n=1}^{r} \frac{d_n(x)}{b^n} + \frac{1}{b^r}. \qquad (3.10)$$

However, the way the digits $d_1(x), d_2(x), \ldots$ have been defined ensures, as seen in (3.6), that $x < \sum_{n=1}^{r} \frac{d_n(x)}{b^n} + \frac{1}{b^r}$. This contradicts (3.10), so there cannot be $r \in \mathbb{N}$ such that $d_n(x) = b - 1$ for all $n \geq r + 1$. □

Definitions. The equation $x = \sum_{n=1}^{\infty} d_n(x)/b^n$ in (3.9) is called the *expansion of x to the base b* and $d_n(x)$ is called the nth *digit of x relative to the base b*. When $b = 2$, the expansion of x to the base 2 is called the *binary expansion* of x, the sequence $(d_n(x))$ may be called the *binary sequence of x*, $d_n(x)$ may be called the nth *binary digit*, and 0 and 1 may be referred to as *binary symbols*.

These ideas are most familiar to us when the base is 10, in which case we get the decimal expansion. Thus, when we write $13/99 = 0.1313131313\cdots$, we mean that $13/99$ is equal to the (infinite) sum

$$\frac{1}{10} + \frac{3}{10^2} + \frac{1}{10^3} + \frac{3}{10^4} + \frac{1}{10^5} + \frac{3}{10^6} + \cdots .$$

The idea in calculating the digits in the expansion of a number to a base is to *approximate* the number by a sum of multiples of powers of the reciprocal of the base. That is why $d_{n+1}(x)$ was defined so that (3.7) would hold, with the result that (3.8) holds for all n. The sequence $d_1(x), d_2(x), \ldots$ may be thought of as a "transform" of x, which represents x as an infinite sequence of discrete bits of of data, and the formula (3.9) shows how x may be recovered, in principle, from these bits of data.

96   Chapter 3. Probability and Randomness

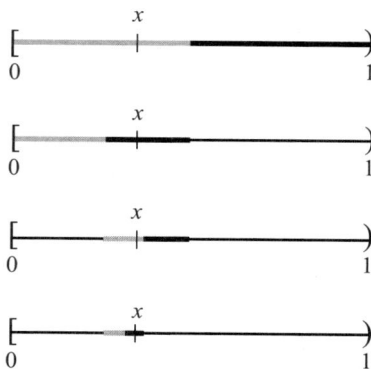

Figure 3.2. The figure illustrates the successive calculation of the first four binary digits of a given number $x \in [0, 1)$. At each stage, if x is in a grey interval that digit is 0, but if x is in a black interval that digit is 1. So, in the figure, the first digit of x is 0. Having observed the first digit, we observe the next digit by carrying out the same process on the grey interval in which x was first observed. That is, this interval is subdivided equally into grey and black intervals as shown, and according as to which of them x is in, the second digit is assigned. Here it is 1. The process is repeated indefinitely. The first four digits of x in the case illustrated by the figure are 0, 1, 0, 1. The digits of x relative to a base are calculated by an iterative process, where each stage of the process is strictly analogous to the one before, but on a different scale.

In general with a given base b, to get $d_1(x)$, we carry out an approximating step on the interval $[0, 1)$ by subdividing the interval into b equal subintervals, getting an approximation $d_1(x)/b$ to x, accurate to within $1/b$. Then, to get $d_2(x)$, we carry out the *same* conceptual approximating step on the interval $[d_1(x)/b, d_1(x)/b + 1/b)$, by subdividing *this* interval into b equal subintervals, getting an approximation $d_1(x)/b + d_2(x)/b^2$ to x, accurate to within $1/b^2$. We continue to calculate the digits of x relative to b by carrying out successive similar approximations on a sequence of decreasing intervals and at stage n, we have an approximation $\sum_{j=1}^{n} d_j(x)/b^j$ to x, accurate to within $1/b^n$. So, the successive approximations are accurate to within $1/b, 1/b^2, 1/b^3, \ldots$, and so on, which are terms in a geometric sequence.

Now, we examine how the calculation of the digits of numbers relative to a given base is related to *inequality* between numbers.

3.2. Expansions to a base

Lemma 3.2. *Let $b, r, n \in \mathbb{N}$ with $b \geq 2$ and $n \geq r$, and let $a_r, a_{r+1}, \ldots, a_n \in \{0, 1, 2, \ldots, b-1\}$ with $a_r \neq 0$. Then,*

$$\sum_{j=r}^{n} \frac{a_j}{b^j} \leq \frac{a_r}{b^r} + \frac{1}{b^r} - \frac{1}{b^n},$$

with equality holding precisely when $a_r = a_{r+1} = \cdots = a_n = b-1$.

Proof. We have

$$\sum_{j=r}^{n} \frac{a_j}{b^j} = \frac{a_r}{b^r} + \sum_{j=r+1}^{n} \frac{a_j}{b^j} \leq \frac{a_r}{b^r} + (b-1) \sum_{j=r+1}^{n} \frac{1}{b^j}$$
$$= \frac{a_r}{b^r} + (b-1) \left(\frac{1/b^{r+1} - 1/b^{n+1}}{1 - 1/b} \right) = \frac{a_r}{b^r} + \frac{1}{b^r} - \frac{1}{b^n}.$$

As well, it follows from this argument that the inequality is an equality precisely when $a_j = b-1$ for all $j = r+1, r+2, \ldots, n$. □

Lemma 3.3. *Let $b \in \mathbb{N}$ with $b \geq 2$, let $n \in \mathbb{N}$ and let $a_1, a_2, \ldots, a_n$ and $a'_1, a'_2, \ldots, a'_n$ be two sequences of elements in $\{0, 1, 2, \ldots, b-1\}$ such that $a_j \neq a'_j$ for at least one value of j. Let*

$$r = \min\{j : 1 \leq j \leq n \text{ and } a_j \neq a'_j\}.$$

Then,

$$\sum_{j=1}^{n} \frac{a_j}{b^j} < \sum_{j=1}^{n} \frac{a'_j}{b^j} \iff a_r < a'_r.$$

Also, when this occurs,

$$\sum_{j=1}^{n} \frac{a_j}{b^j} + \frac{1}{b^n} \leq \sum_{j=1}^{n} \frac{a'_j}{b^j}.$$

98   Chapter 3. Probability and Randomness

Proof. If $a_r < a'_r$, $a_r + 1 \leq a'_r$, and then,

$$\sum_{j=1}^{n} \frac{a_j}{b^j} = \sum_{j=1}^{r-1} \frac{a'_j}{b^j} + \sum_{j=r}^{n} \frac{a_j}{b^j}$$

$$\leq \sum_{j=1}^{r-1} \frac{a'_j}{b^j} + \frac{a_r}{b^r} + \frac{1}{b^r} - \frac{1}{b^n}, \text{ by Lemma 3.2,}$$

$$= \sum_{j=1}^{r-1} \frac{a'_j}{b^j} + \frac{a_r + 1}{b^r} - \frac{1}{b^n}$$

$$\leq \sum_{j=1}^{r-1} \frac{a'_j}{b^j} + \frac{a'_r}{b^r} - \frac{1}{b^n}$$

$$= \sum_{j=1}^{r} \frac{a'_j}{b^j} - \frac{1}{b^n}$$

$$\leq \sum_{j=1}^{n} \frac{a'_j}{b^j} - \frac{1}{b^n}$$

$$< \sum_{j=1}^{n} \frac{a'_j}{b^j}. \tag{3.11}$$

The converse statement follows by assuming that $a'_r < a_r$, and applying the same calculations. The latter also proves, together with (3.11), the final statement in the lemma. □

Note that Lemma 3.3 implies the familiar fact that if, for example, we have $x = 0.137693211$ and $y = 0.137793211$, where x, y are here in decimal form, then $x < y$. The following result expresses the uniqueness of a finite expansion to the base b. Its proof is immediate from Lemma 3.3.

3.2. Expansions to a base

Corollary 1. *Let* $b \in \mathbb{N}$ *with* $b \geq 2$ *and let* $a_1, a_2, \ldots, a_n, a'_1, a'_2, \ldots, a'_n \in \{0, 1, 2, \ldots, b-1\}$ *be such that*

$$\sum_{j=1}^{n} \frac{a_j}{b^j} = \sum_{j=1}^{n} \frac{a'_j}{b^j}.$$

Then $a_1 = a'_1, a_2 = a'_2, \ldots, a_n = a'_n$.

The following result is important for future applications, as it establishes a precise link between the first n digits in the expansion of a number to some base, and the location of the number within a subinterval of $[0, 1)$.

Theorem 3.4. *Let* $b \in \mathbb{N}$ *with* $b \geq 2$. *Let* $x \in [0, 1)$ *and let* $d_1(x), d_2(x), d_3(x), \ldots$ *denote the sequence of digits of* x *relative to the base* b. *Let* $c_1, c_2, \ldots, c_n \in \{0, 1, \ldots, b-1\}$ *be given. Then,*

$$d_1(x) = c_1, d_2(x) = c_2, \ldots, d_n(x) = c_n$$
$$\iff x \in \left[\sum_{j=1}^{n} \frac{c_j}{b^j}, \sum_{j=1}^{n} \frac{c_j}{b^j} + \frac{1}{b^n} \right).$$

Proof. If $d_1(x) = c_1, d_2(x) = c_2, \ldots, d_n(x) = c_n$, we have $x \in [\sum_{j=1}^{n} c_j/b^j, \sum_{j=1}^{n} c_j/b^j + 1/b^n)$ by (3.8), which proves one implication.

On the other hand, for each choice of $a_1, a_2, \ldots, a_n \in \{0, 1, 2, \ldots, b-1\}$, let

$$J(a_1, a_2, \ldots, a_n) = \left[\sum_{j=1}^{n} \frac{a_j}{b^j}, \sum_{j=1}^{n} \frac{a_j}{b^j} + \frac{1}{b^n} \right).$$

There are b^n intervals of the form $J(a_1, a_2, \ldots, a_n)$ and each has length $1/b^n$. Moreover, Lemma 3.3 shows that these intervals are disjoint. Now we know that $x \in J(d_1(x), d_2(x), \ldots, d_n(x))$ by (3.8), so that if also $x \in J(c_1, c_2, \ldots, c_n)$ it must be that $J(d_1(x), d_2(x), \ldots, d_n(x)) = J(c_1, c_2, \ldots, c_n)$. But Corollary 1 then implies that $d_1(x) = c_1, d_2(x) = c_2, \ldots$ and $d_n(x) = c_n$. □

Finally, there is the question of the *uniqueness* of the expansion of a number to a given base. The final part of Theorem 3.4 showed that the sequence $d_1(x), d_2(x), \ldots$ of digits to the base b never ended in an infinite sequence of terms $b - 1$. However, in the case of base 10, for example, we might say that $0.5999999\ldots$ represents the number $3/5$, since

$$\frac{3}{5} = \frac{5}{10} + \frac{9}{10^2} + \frac{9}{10^3} + \cdots, \qquad (3.12)$$

whereas with the definition of the digits $d_n(3/5)$ we have the usual expansion

$$\frac{3}{5} = \frac{6}{10} + \frac{0}{10^2} + \frac{0}{10^3} + \cdots.$$

Thus, the way we have defined the sequence of digits of a number excludes the "repeating decimals" in (3.12). The general situation is clarified by the following result.

Proposition 3.5. *Let $b \in \mathbb{N}$ with $b \geq 2$. Let $a_1, a_2, \ldots$ and $a'_1, a'_2, \ldots$ be two distinct sequences such that $a_n, a'_n \in \{0, 1, \ldots, b-1\}$ for all $n \in \mathbb{N}$. Put $r = \min\{n : a_n \neq a'_n\}$, and assume that $a_r > a'_r$. Then*

$$\sum_{n=1}^{\infty} \frac{a_n}{b^n} = \sum_{n=1}^{\infty} \frac{a'_n}{b^n} \iff a_r = a'_r + 1 \text{ and } a'_n = b - 1 \text{ for all } n \geq r + 1.$$

Proof. Let $\sum_{n=1}^{\infty} a_n/b^n = \sum_{n=1}^{\infty} a'_n/b^n$. Then, using the definition of r we have

$$\sum_{n=r+1}^{\infty} \frac{a'_n}{b^n} = -\frac{a'_r}{b^r} + \sum_{n=r}^{\infty} \frac{a'_n}{b^n} = -\frac{a'_r}{b^r} + \sum_{n=r}^{\infty} \frac{a_n}{b^n} \geq -\frac{a'_r}{b^r} + \frac{a_r}{b^r} = \frac{a_r - a'_r}{b^r}.$$

Thus,

$$\frac{1}{b^r} = \sum_{n=r+1}^{\infty} \frac{b-1}{b^n} \geq \sum_{n=r+1}^{\infty} \frac{a'_n}{b^n} \geq \frac{a_r - a'_r}{b^r}, \qquad (3.13)$$

and it follows that $a_r - a'_r \leq 1$. As $a_r > a'_r$ we deduce that $a_r = a'_r + 1$. But then, (3.13) takes the form

$$\frac{1}{b^r} = \sum_{n=r+1}^{\infty} \frac{b-1}{b^n} \geq \sum_{n=r+1}^{\infty} \frac{a'_n}{b^n} \geq \frac{1}{b^r},$$

which makes it clear that $a'_n = b - 1$ for all $n \geq r + 1$. This proves the statement of equivalence in one direction, and the other direction is readily checked. □

Definition Let $b \in \mathbb{N}$ with $b \geq 2$ and let $x \in [0, 1)$. If $a_1, a_2, a_3, \ldots$ is a sequence with $a_n \in \{0, 1, 2, \ldots, b - 1\}$ for all n, and if $x = \sum_{n=1}^{\infty} a_n/b^n$, we write this as $x = 0.a_1 a_2 a_3 \ldots$ and we say that $a_1, a_2, a_3, \ldots$ *represents x to the base b.*

Proposition 3.5 shows that the only way that two distinct sequences of digits can represent the same number to the base b is when one of the sequences has the property that from some point on, all digits must equal $b - 1$. However, Theorem 3.4 shows that this latter situation is specifically excluded for the digits $d_1(x), d_2(x), \ldots$, relative to the base b. So, in this sense, the expansion of a number in $[0, 1)$ to a base b is unique. Note the dynamical system on $[0, 1)$ whose transformation is given by $x \mapsto \text{frac}(bx)$ is naturally associated with the expansion of numbers to the base b. In the case $b = 2$, the idea is developed in Section 3.10. For the general case, see Exercise 10.

3.3 Rational numbers and periodic expansions

Let us now turn to the question of the expansions of rational and irrational numbers. Expanding to the base 10 we have

$$\frac{1}{3} = 0.3333333\ldots, \quad \frac{1}{12} = 0.083333333\ldots, \text{ and } \frac{3}{7} = 0.428571428571\ldots.$$

These rational numbers have led to "periodic" expansions, where after a few steps a pattern of some finite number of digits repeats itself.

Definition Let $d_1, d_2, \ldots$ be a sequence of real numbers. This sequence is called *periodic* if there is $k \in \mathbb{N}$ such that

$$d_{n+k} = d_n, \text{ for all } n \in \mathbb{N}.$$

The sequence is called *eventually periodic* if there are $i, k \in \mathbb{N}$ such that

$$d_{n+k} = d_n, \text{ for all } n \geq i.$$

If $x \in [0, 1)$ and b is a base, we may say that the expansion of x relative to b is *periodic*, or *eventually periodic*, if the sequence of digits of x relative to b has the corresponding property.

Let $p, q \in \mathbb{N}$ with $p < q$ and consider the rational number $p/q \in [0, 1)$. This number is kept fixed for the time being. There are functions $d : \mathbb{Z}_+ \longrightarrow \mathbb{Z}_+$ and $r : \mathbb{Z}_+ \longrightarrow \{0, 1, \ldots, q-1\}$ such that

$$n = d(n)q + r(n), \quad \text{for all } n \in \mathbb{Z}_+. \tag{3.14}$$

Here, $r(n)$ is the remainder obtained upon division of n by q.

Let $b \in \mathbb{N}$ with $b \geq 2$. We proceed to construct the digits of p/q relative to the base b, using a standard and well-known argument. Put

$$d_1 = d(bp) \text{ and } r_1 = r(bp). \tag{3.15}$$

Using (3.14), we have

$$bp = d_1 q + r_1,$$

so that

$$\frac{p}{q} = \frac{d_1}{b} + \frac{r_1}{bq}. \tag{3.16}$$

As $p/q < 1$, $d_1/b < 1$ and we see that $d_1 \in \{0, 1, 2, \ldots, b-1\}$. As $r_1 \in \{0, 1, \ldots, q-1\}$, we have from (3.16) that

$$\frac{p}{q} \in \left[\frac{d_1}{b}, \frac{d_1}{b} + \frac{1}{b}\right),$$

and it follows from (3.5) that d_1 is the first digit of p/q relative to b.

Now, having defined d_1 and r_1 by (3.15), we define the full sequences $d_1, d_2, \ldots$ and $r_1, r_2, \ldots$ inductively by putting

$$d_{n+1} = d(br_n) \text{ and } r_{n+1} = r(br_n), \text{ for all } n \geq 1.$$

Note that from (3.14) we have

$$br_n = d_{n+1}q + r_{n+1}, \text{ for all } n \geq 1. \tag{3.17}$$

As $br_n < bq$, we have $d_{n+1} < b$. As $d_1 < b$, we deduce that

$$d_j \in \{0, 1, \ldots, b-1\}, \text{ for all } j \in \mathbb{N}.$$

3.3. Rational numbers and periodic expansions

Now, from (3.16) we have

$$\begin{aligned}
\frac{p}{q} &= \frac{d_1}{b} + \frac{r_1}{bq} \\
&= \frac{d_1}{b} + \frac{br_1}{b^2 q} \\
&= \frac{d_1}{b} + \frac{d_2 q + r_2}{b^2 q}, \text{ by (3.17)}, \\
&= \frac{d_1}{b} + \frac{d_2}{b^2} + \frac{r_2}{b^2 q} \\
&= \frac{d_1}{b} + \frac{d_2}{b^2} + \frac{br_2}{b^3 q} \\
&= \frac{d_1}{b} + \frac{d_2}{b^2} + \frac{d_3}{b^3} + \frac{r_3}{b^3 q}, \text{ by (3.17)}, \\
&\cdots \cdots \\
&= \sum_{j=1}^{n} \frac{d_j}{b^j} + \frac{r_n}{b^n q}.
\end{aligned} \qquad (3.18)$$

(Strictly speaking, this conclusion requires an argument by induction.) As $r_n \in \{0, 1, \ldots, q-1\}$, it follows from (3.18) that

$$\frac{p}{q} \in \left[\sum_{j=1}^{n} \frac{d_j}{b^j}, \sum_{j=1}^{n} \frac{d_j}{b^j} + \frac{1}{b^n} \right).$$

Now, Theorem 3.4 shows that $d_1, d_2, \ldots, d_n$ are the first n digits of p/q relative to b.

Now we turn to the periodic behavior of the digits of p/q relative to b. All the terms of the sequence $r_1, r_2, \ldots$ are in $\{0, 1, 2, \ldots, q-1\}$, a finite set having q elements. So, if we consider the $q+1$ numbers $r_1, r_2, \ldots, r_q, r_{q+1}$, two of these must be equal (by the Pigeonhole Principle, discussed in 2.3 of Chapter 2). That is, there are $i, j \in \{1, 2, \ldots, q+1\}$ such that $i < j$ and $r_i = r_j$. Put $k = j - i \in \{1, 2, \ldots, q\}$. Then, we claim that

$$r_{n+k} = r_n, \quad \text{for all } n \geq i. \tag{3.19}$$

To prove this we proceed by induction. We have

$$r_{i+k} = r_{i+j-i} = r_j = r_i,$$

so that (3.19) is true for $n = i$. Now assume that (3.19) holds for some $n \geq i$. Then,

$$r_{n+1+k} = r(br_{n+k}) = r(br_n) = r_{n+1},$$

so (3.19) is true for $n + 1$. Thus, if the equation in (3.17) is true for some $n \geq i$, it is also true for $n + 1$. By induction, (3.19) is true for all $n \geq i$. Now, this has implications for the sequence $d_1, d_2, \ldots$ of digits of p/q. For, if $n \geq i + 1$, we have $n - 1 \geq i$ and (3.19) gives

$$d_{n+k} = d(br_{n-1+k}) = d(br_{n-1}) = d_n.$$

This shows that the sequence $d_1, d_2, \ldots$ is eventually periodic. So, we have seen that for every rational number, its sequence of digits relative to every base is eventually periodic. The following result summarizes this and shows that the converse statement is also true.

Theorem 3.6. *Let $b \in \mathbb{N}$ with $b \geq 2$ and let $x \in [0, 1)$. Then x is rational if and only if the sequence of digits of x relative to b is eventually periodic. That is, if $d_1, d_2, \ldots$ is the sequence of digits of x relative to b, x is rational if and only if there are $i, k \in \mathbb{N}$ such that $d_{n+k} = d_n$ for all $n \geq i$.*

Proof. It has been shown above that if x is rational, then the sequence of digits is indeed eventually periodic. Conversely, if the sequence $d_1, d_2, \ldots$ of digits of x relative to b is eventually periodic, let $i, k \in \mathbb{N}$ be such that $d_{n+k} = d_n$ for all $n \geq i$. Using this, we have

3.3. Rational numbers and periodic expansions

$$x = \sum_{n=1}^{\infty} \frac{d_n}{b^n} = \sum_{n=1}^{i-1} \frac{d_n}{b^n} + \sum_{n=i}^{\infty} \frac{d_n}{b^n}$$

$$= \sum_{n=1}^{i-1} \frac{d_n}{b^n} + \sum_{\ell=0}^{\infty} \left(\sum_{n=i+\ell k}^{i+(\ell+1)k-1} \frac{d_n}{b^n} \right)$$

$$= \sum_{n=1}^{i-1} \frac{d_n}{b^n} + \sum_{\ell=0}^{\infty} \left(\sum_{j=0}^{k-1} \frac{d_{j+i+\ell k}}{b^{j+i+\ell k}} \right) \quad (3.20)$$

$$= \sum_{n=1}^{i-1} \frac{d_n}{b^n} + \left(\sum_{j=0}^{k-1} \frac{d_{j+i}}{b^{j+i}} \right) \left(\sum_{\ell=0}^{\infty} \frac{1}{b^{\ell k}} \right)$$

$$= \sum_{n=1}^{i-1} \frac{d_n}{b^n} + \left(\sum_{n=i}^{i+k-1} \frac{d_n}{b^n} \right) \left(\frac{b^k}{b^k - 1} \right).$$

Here, the right-hand side is rational, so that x must be rational. □

Note that in the circumstances of (3.20), it follows that x is of the form $r/b^{i+k-1}(b^k - 1)$ for some $r \in \mathbb{Z}_+$.

A special case of (3.20) is when $i = 1$, which is to say that the digits of x relative to b are periodic. In this case, (3.20) takes the form

$$x = \frac{b^k}{b^k - 1} \left(\sum_{n=1}^{k} \frac{d_n}{b^n} \right) = \frac{1}{b^k - 1} \left(\sum_{n=1}^{k} d_n b^{k-n} \right),$$

and we see that x equals $r/(b^k - 1)$ for some $r \in \mathbb{Z}_+$.

Theorem 3.6 could be expressed by saying that a number in $[0, 1)$ is irrational if and only if the sequence of digits in the expansion of the number to the base b obtained is *not* eventually periodic. Since a number is rational or irrational regardless of the base, we see that if a number has an eventually periodic expansion in one base, it will have an eventually periodic expansion in every base. So, we can also say that for a given number in $[0, 1)$, either the sequence of digits in its expansion *is* eventually periodic in every base, or the sequence of digits in its expansion *is not* eventually periodic in every base.

Theorem 3.6 complements the result of Proposition 2.2, where it was seen that α is rational if and only if the sequence (frac$(n\alpha)$) in [0, 1) is periodic. Again, we have seen that rational numbers are associated with finite, repetitive and predictable behavior, while irrationals are associated with infinite and non-repetitive behavior. Recent ideas and results on rational numbers and their expansions are in [2, 16, 41, 43].

3.4 Sets, events, length and probability

We need to deal with various operations on subsets of $\mathbb{R}$. This arises, in part, from the connection between sets and functions. If A is a subset of $\mathbb{R}$, the *characteristic function* of A is given by

$$\chi_A(x) = \begin{cases} 1, & \text{if } x \in A; \\ 0, & \text{if } x \in A. \end{cases}$$

Then, if A, B are two subsets of $\mathbb{R}$, taking the intersection of A, B corresponds to multiplying these functions, owing to the identity

$$\chi_A \chi_B = \chi_{A \cap B}.$$

On the other hand, if we add these functions, we get

$$\chi_A + \chi_B = \chi_{A \cap B^c} + \chi_{A^c \cap B} + 2\chi_{A \cap B} = \chi_{A \cup B} + \chi_{A \cap B}.$$

Thus, adding the characteristic functions corresponding to the sets A, B leads us to consider the sets $A \cup B$ and $A \cap B$. Furthermore,

$$1 - \chi_A = \chi_{A^c},$$

so that subtracting a characteristic function from 1 corresponds to taking the complement of a set.

A different viewpoint comes from regarding each set as an "event", an interpretation of particular importance in probability and randomness.

3.4. Sets, events, length and probability

The statement "$x \in A$" can be regarded as an event or observation. The idea is that we obtain an observation x, and then either $x \in A$ or $x \notin A$. We regard "$x \in A$" as an occurrence of the "event" A. It is customary, in fact, to regard the set A itself as the "event" in this case. We need to understand how different properties of events are affected by taking combinations of events, or by constructing new events from given ones. Constructing new events comes from taking unions, intersections and complements of given events. So, it seems we need the family of events to be stable under these operations. On the real line, it is the intervals which seem to be the simplest events, but although intervals are stable under intersection, they are not stable under unions or complements. Specifically, if A, B are intervals, then $A \cap B$ is an interval, but $A \cup B$ is not generally an interval, and neither is the complement of an interval generally an interval. We introduce a family of sets that contains the intervals and has the properties that every finite union or intersection of sets in the family is also in the family, and that the complement of every set in the family is also in the family.

Definition A subset A of $\mathbb{R}$ is called *basic* if it is the union of a finite number of intervals (these intervals may be open, closed, half open or half closed). The family of all basic subsets of some given interval generally is denoted by $\mathcal{B}$.

The following result establishes the fundamental properties of the basic sets.

Proposition 3.7. *Let S be an interval and let $\mathcal{B}$ denote the family of all basic subsets of S. Then $\mathcal{B}$ has the following properties.*

(1) *If $A, B \in \mathcal{B}$, then $A \cup B \in \mathcal{B}$.*

(2) *If $A, B \in \mathcal{B}$, then $A \cap B \in \mathcal{B}$.*

(3) *Every finite union of sets in $\mathcal{B}$ is also in $\mathcal{B}$, and every finite intersection of sets in $\mathcal{B}$ is also in $\mathcal{B}$.*

(4) *If $A \in \mathcal{B}$, then $A^c \in \mathcal{B}$.*

(5) *A subset of S is in $\mathcal{B}$ if and only if it is a finite, disjoint union of intervals.*

Proof. (1) Let $A, B \in \mathcal{B}$. Observe that because each of the sets A, B is a finite union of intervals, the union $A \cup B$ is also a finite union of intervals and hence $A \cup B$ is in $\mathcal{B}$.

(2) As $A, B \in \mathcal{B}$, there are intervals $J_1, J_2, \ldots, J_r$ and $K_1, K_2, \ldots, K_s$ such that

$$A = \bigcup_{j=1}^{r} J_j \text{ and } B = \bigcup_{k=1}^{s} K_k.$$

Then,

$$A \cap B = \left(\bigcup_{j=1}^{r} J_j\right) \cap \left(\bigcup_{k=1}^{s} K_k\right) = \bigcup_{j=1}^{r} \bigcup_{k=1}^{s} J_j \cap K_k. \quad (3.21)$$

As each set $J_j \cap K_k$ is an intersection of intervals, each $J_j \cap K_k$ is an interval. Thus, we see in (3.21) that $A \cap B$ is a finite union of intervals and so $A \cap B$ is in $\mathcal{B}$.

(3) This is more or less obvious, but strictly it follows by an induction argument, applied first to the conclusion in (1) and then to the conclusion in (2). Here is the gist of it in the case of taking finite unions. Let $A_1, A_2, \ldots, A_n \in \mathcal{B}$ and consider the statement that $A_1 \cup A_2 \cup \cdots \cup A_r \in \mathcal{B}$. This statement is true for $r = 1$ because A_1 is basic, and the statement is true for $r = 2$ by (1), already proved. However, if the statement is true for r where $r < n$, it follows from (1) that it is also true for $r + 1$. By induction, it is true for all $r = 1, 2, \ldots, n$.

(4) Let $A \in \mathcal{B}$. Then $A = \bigcup_{j=1}^{r} J_j$ where each J_j is an interval. Observe that

$$A^c = J_1^c \cap J_2^c \cap \cdots \cap J_r^c.$$

Now as each set J_j is an interval, each J_j^c is either an interval or is the union of two intervals. Hence, each set J_j^c is basic. Thus, A^c is a finite intersection of basic sets and so must be basic, by part (3) just proved.

(5) First, we make an observation about intervals. If J, K are two intervals and J is not a subset of K, then $K \cap J^c$ is an interval. Now,

3.4. Sets, events, length and probability

we proceed by induction. The inductive assumption is that a set that is a union of n intervals is a finite and disjoint union of intervals. This assumption is clearly true when $n = 1$, so we assume it is true for some $n \in \mathbb{N}$. Now, let A be a basic set that is the union of $n + 1$ intervals $A_1, A_2, \ldots, A_{n+1}$. By the inductive assumption, there are disjoint intervals $J_1, J_2, \ldots, J_r$ such that $\bigcup_{j=1}^{n} A_j = \bigcup_{j=1}^{r} J_j$. So we have

$$A = \left(\bigcup_{j=1}^{n} A_j\right) \cup A_{n+1} = (J_1 \cup J_2 \cup \cdots \cup J_r) \cup A_{n+1}. \quad (3.22)$$

Now in (3.22), if $J_j \subseteq A_{n+1}$ for all $j \in \{1, 2, \ldots, r\}$, then J_j can be omitted from the sets in the union on the right of (3.22), and the validity of (3.22) is not affected. So, in (3.22) we can assume that for no $j \in \{1, 2, \ldots, n\}$ do we have $J_j \subseteq A_{n+1}$. By our opening observation on intervals, this means that for all $j \in \{1, 2, \ldots, r\}$,

$$A_{n+1} \cap J_j^c \text{ is an interval.} \quad (3.23)$$

Now from (3.22) we have

$$A = \left(\bigcup_{j=1}^{r} J_j\right) \cup A_{n+1} = \left(\bigcup_{j=1}^{r} J_j\right) \cup \left(A_{n+1} \cap \left(\bigcup_{j=1}^{r} J_j\right)^c\right)$$

$$= \left(\bigcup_{j=1}^{r} J_j\right) \cup \left(\bigcap_{j=1}^{r} A_{n+1} \cap J_j^c\right). \quad (3.24)$$

We deduce from (3.23) that $\bigcap_{j=1}^{r}(A_{n+1} \cap J_j^c)$ is an interval, so the right-hand side of (3.24) expresses A as a finite, disjoint union of intervals. Thus, if the inductive assumption holds for n it also holds for $n + 1$. □

From the viewpoint of analysis, the basic sets are the ones needed for forming the step functions. On the other hand, in the present context, and from the viewpoint of probability, the basic sets are the events that may legitimately occur. In elementary calculus, one operation which is carried out on step functions is to integrate them, and this needs the notion of the *length* of an interval. In probability, a vital notion is that of the *probability of an event*. Now, an important connection between real analysis and probability comes from the fact that for a subinterval J of $[0, 1)$, the length of J is the *same* as the probability of the event J. This comes about because the length of $[0, 1)$ is 1, the length of J represents the proportion of $[0, 1)$ which is occupied by J. On the other hand, if we assume that the outcome or event "$x \in J$" is a random process which does not favor one part of the interval $[0, 1)$ over any other, the probability of the outcome "$x \in J$", that is to say the probability of the event J, will also be the proportion of $[0, 1)$ which is occupied by J. Thus, the length of J and the probability of the event J are equal! However, note that this conclusion depends on the assumption that the process "$x \in J$" does not favor any one part of $[0, 1)$ over any other part. The length of J being equal to the probability of the event J expresses the idea that the process "$x \in J$" does not favor any one part of $[0, 1)$ over any other.

Definition Let S be an interval. A function $v : S \longrightarrow \mathbb{R}$ is called *non-decreasing* if, for all $x, y \in S$,

$$x \geq y \implies v(x) \geq v(y).$$

If S is an interval, its *closure* $\overline{S}$ is the interval S together with its endpoints. This means that if S includes its endpoints, $\overline{S} = S$.

Now, assume that an interval S and a non-decreasing function $v : \overline{S} \longrightarrow \mathbb{R}$ are given. Then if A is a bounded subinterval of J there are $a, b \in S$ with $a \leq b$ such that A equals $[a, b], (a, b), [a, b)$ or $(a, b]$. We then make the definition that

$$\mu_v(A) = v(b) - v(a). \tag{3.25}$$

Note that the domain of v is taken to be $\overline{S}$ to allow for the possibility, for example, that (a, b) may be a subinterval of S but a or b might be an endpoint of S. The interval (a, a) is empty, so the empty set $\emptyset$ is an interval and $\mu_v(\emptyset) = 0$. Also, a single point $\{a\}$ equals the interval $[a, a]$, so $\mu_v(\{a\}) = 0$. If the subinterval A is unbounded, either $A = \mathbb{R}$

3.4. Sets, events, length and probability

or there is $a \in \mathbb{R}$ such that A is $[a, \infty)$, (a, ∞), $(-\infty, a)$ or $(-\infty, a]$. If $A = \mathbb{R}$, make the definition that

$$\mu_v(A) = \lim_{x \to \infty} v(x) - \lim_{x \to -\infty} v(x).$$

If $A = [a, \infty)$ or $A = (a, \infty)$, make the definition that

$$\mu_v(A) = \lim_{x \to \infty} v(x) - v(a).$$

If $A = (-\infty, a)$ or $A = (-\infty, a]$, make the definition that

$$\mu_v(A) = v(a) - \lim_{x \to -\infty} v(x).$$

This extends the definition of $\mu_v(A)$ to unbounded subintervals A of S, if any. Allowing for the assumption that v is non-decreasing, we see that $\mu_v(A) \geq 0$ for all subintervals A of S, and it may happen that $\mu_v(A) = \infty$ for some subintervals A of S.

Note that $\mu_v(A)$ may be called the *v-length* of A. The case of most interest to us is when v is the function given by $v(x) = x$ for all $x \in S$. In this case, μ_v is denoted simply by μ. We see from (3.25) that if $a, b \in \mathbb{R}$, and A is $[a, b]$, (a, b), $[a, b)$ or $(a, b]$, then $\mu(A) = b - a$. But if the interval A is unbounded, then $\mu(A) = \infty$. Thus, $\mu(A)$ is the usual length of the interval A, and may simply be called the *length of A*.

When S is an interval and J is a subinterval of S, $\mu(J)/\mu(S)$ is, in terms of length, the proportion of S which is occupied by J, and the probability of the event J is also $\mu(J)/\mu(S)$. This emphasizes that whereas length is an "absolute" concept and is independent of the size of the enclosing interval, probability is a relative concept, and compares the size of the event compared with the size of the largest possible event, the event that consists of all possible outcomes.

Now, the idea is to extend the concept of v-length from intervals to all of the basic sets. Since Proposition 3.7 has shown that every basic set is a disjoint union of intervals, it seems reasonable to say that the v-length of a basic set is the sum of the v-lengths of the intervals that express the set as a disjoint union. This is what we do, but there is a problem in that we must show that such a sum is the same, regardless of the manner in which the basic set is expressed as a disjoint union of intervals.

Lemma 3.8. *Let S be an interval and let $J_1, J_2, \ldots, J_n$ be disjoint intervals such that $J = \bigcup_{j=1}^{n} J_j$. Also, let $v : \overline{J} \longrightarrow \mathbb{R}$ be a non-decreasing function. Then,*

$$\mu_v(J) = \sum_{j=1}^{n} \mu_v(J_j).$$

Proof. The proof proceeds by an inductive argument on the number n of disjoint intervals. If $n = 1$, $J = J_1$ and the result is true because $\mu_v(J) = \mu_v(J_1)$.

Now, consider when $n = 2$. Let $J = J_1 \cup J_2$ where J_1, J_2 are non-empty disjoint intervals. As J is an interval, we may assume that the right-hand endpoint of J_1 equals the left-hand endpoint of J_2. If J is bounded, we will have a, b, c such that $a \leq b \leq c$, the left-hand endpoint of J and J_1 is a, the right-hand endpoint of J_1 and the left-hand endpoint of J_2 equal b, and the right-hand endpoint of J and J_2 is c. Then,

$$\mu_v(J) = v(c) - v(a) = v(c) - v(b) + v(b) - v(a) = \mu_v(J_1) + \mu_v(J_2). \tag{3.26}$$

If J is unbounded, at least one of J_1, J_2 is unbounded. Let's assume that J_1 is bounded and that J_2 is unbounded and of the form (b, ∞). Then, $J = (a, \infty)$ or $[a, \infty)$ and (3.26) takes the form

$$\mu_v(J) = \lim_{x \to \infty} v(x) - v(a) = \lim_{x \to \infty} v(x) - v(b) + v(b) - v(a)$$
$$= \mu_v(J_1) + \mu_v(J_2).$$

A similar argument works for all the possibilities of J_1 and J_2 being bounded or unbounded. So the result is true whenever $n = 2$.

Now, let $n \geq 2$ and assume that the result is true for every interval that is the union of n disjoint intervals. Consider intervals $J, J_1, J_2, \ldots, J_n, J_{n+1}$ such that

3.4. Sets, events, length and probability

$$J = \bigcup_{j=1}^{n+1} J_j \text{ and } J_1, \ldots, J_n, J_{n+1} \text{ are disjoint.}$$

We may assume, by rearranging the intervals J_j if necessary, that the right hand endpoint of J_j is the left hand endpoint of the interval J_{j+1} for $j = 1, 2, \ldots, n$. Put $K = \bigcup_{j=1}^{n} J_j$. Then, K is an interval, $J = K \cup J_{n+1}$, and this expresses J as a union of two disjoint intervals. Because the result is true when $n = 2$, and because the result is assumed to be true for n, we now have

$$\begin{aligned}
\mu_v(J) &= \mu_v(K \cup J_{n+1}) \\
&= \mu_v(K) + \mu_v(J_{n+1}) \\
&= \mu_v\left(\bigcup_{j=1}^{n} J_j\right) + \mu_v(J_{n+1}) \\
&= \sum_{j=1}^{n} \mu_v(J_j) + \mu_v(J_{n+1}) \\
&= \sum_{j=1}^{n+1} \mu_v(J_j).
\end{aligned}$$

Hence, if the result is true for n, it is also true for $n + 1$. By induction, the result is true for all n, as required. □

Lemma 3.9. *Let S be a given interval, and let $v : \overline{S} \longrightarrow \mathbb{R}$ be a non-decreasing function. Let $J_1, J_2, \ldots, J_m$ be disjoint intervals, and let $K_1, K_2, \ldots, K_n$ be disjoint intervals such that*

$$\bigcup_{j=1}^{m} J_j = \bigcup_{k=1}^{n} K_k \subseteq S.$$

Then,

$$\sum_{j=1}^{m} \mu_v(J_j) = \sum_{k=1}^{n} \mu_v(K_k).$$

114   Chapter 3. Probability and Randomness

Proof. Using the given information and Lemma 3.8 gives

$$\mu_v(J_j) = \mu_v\left(J_j \cap \left(\bigcup_{k=1}^n K_k\right)\right) = \mu_v\left(\bigcup_{k=1}^n J_j \cap K_k\right)$$

$$= \sum_{k=1}^n \mu_v(J_j \cap K_k).$$

Hence,

$$\sum_{j=1}^m \mu_v(J_j) = \sum_{j=1}^m \sum_{k=1}^n \mu_v(J_j \cap K_k).$$

Now, the situation is symmetric in j, k. So, if we interchange the roles of J_j, K_k in the above argument, we will get

$$\sum_{k=1}^n \mu_v(K_k) = \sum_{k=1}^n \sum_{j=1}^m \mu_v(K_k \cap J_j)$$

$$= \sum_{j=1}^m \sum_{k=1}^n \mu_v(J_j \cap K_k)$$

$$= \sum_{j=1}^m \mu_v(J_j),$$

and the result follows. □

Definition Let S be an interval and let $v : \overline{S} \longrightarrow \mathbb{R}$ be a non-decreasing function. Let A be a basic set with $A \subseteq S$ and let $J_1, J_2, \ldots, J_n$ be a finite number of disjoint intervals such that $A = \bigcup_{j=1}^n J_j$. Then the v-*length* $\mu_v(A)$ of A is defined to be the sum $\sum_{j=1}^n \mu_v(J_j)$. In the case when $v(x) = x$ for all $x \in S$, μ_v is denoted by μ, and $\mu(A)$ is called simply the *length* of the basic set A.

Note that a basic set may always be expressed as a finite union of disjoint intervals by Proposition 3.7. Also, by Lemma 3.9, the definition of the v-length of a basic set is independent of the manner in which A is expressed as a disjoint union of intervals—we say that the concept of the

3.4. Sets, events, length and probability

v-length of a basic set is *well defined*. Note also that for a basic set that is an interval, the v-length of the set as an interval is equal to its v-length as a basic set – we say that the concept of v-length has been *extended* from intervals to basic sets. The following result describes the essential properties of v-length, considered as a function on the basic sets.

Proposition 3.10. *Let S be an interval and let $\mathcal{B}$ be the family of basic subsets of S. Let $v : \overline{S} \longrightarrow \mathbb{R}$ be a non-decreasing function. Then the following hold.*

(1) $\mu_v(\emptyset) = 0$ and $0 \leq \mu_v(A) \leq \infty$ for all $A \in \mathcal{B}$.

(2) Let $A \in \mathcal{B}$ and let $A_1, A_2, \ldots, A_n$ be mutually disjoint sets in $\mathcal{B}$ such that $A = \bigcup_{j=1}^{n} A_j$. Then

$$\mu_v(A) = \sum_{j=1}^{n} \mu_v(A_j).$$

(3) If $A, B \in \mathcal{B}$ and $A \subseteq B$, then $\mu_v(A) \leq \mu_v(B)$.
(4) Let $A, A_1, A_2, \ldots, A_n \in \mathcal{B}$. Then,

$$A \subseteq \bigcup_{j=1}^{n} A_j \implies \mu_v(A) \leq \sum_{j=1}^{n} \mu_v(A_j).$$

Proof. (1) If $a \in S$ we have $\emptyset = (a, a)$, so that $\mu_v(\emptyset) = \mu_v\bigl((a, a)\bigr) = a - a = 0$.

(2) For all $j = 1, 2, \ldots, n$, A_j is a finite union of disjoint intervals, and A is the disjoint union of all the intervals so obtained. Thus, the sum of the v-lengths of the totality of all these intervals is equal to both $\sum_{j=1}^{n} \mu_v(A_j)$ and $\mu_v(A)$.

(3) Let $A, B \in \mathcal{B}$ and $A \subseteq B$. Then, $A \cup (B \cap A^c) = B$ so that by using (1) and (2),

$$\mu_v(A) \leq \mu_v(A) + \mu_v(B \cap A^c) = \mu_v(B).$$

(4) Observe that $\bigcup_{j=1}^{n} A_j \in \mathcal{B}$. So, if $A \in \mathcal{B}$ with $A \subseteq \bigcup_{j=1}^{n} A_j$ we have by (3) that

$$\mu_v(A) \leq \mu_v\left(\bigcup_{j=1}^{n} A_j\right) = \mu_v\Big(A_1 \cup (A_2 \cap A_1^c) \cup (A_3 \cap (A_1 \cup A_2)^c) \cup$$
$$\cdots \cup \big(A_n \cap (A_1 \cup A_2 \cup \cdots \cup A_{n-1})^c\big)\Big),$$

where the union on the right is a union of mutually disjoint sets in $\mathcal{B}$. Using (2) and (3), already proved, we have

$$\mu_v(A) \leq \mu_v\left(\bigcup_{j=1}^{n} A_j\right)$$
$$= \mu_v(A_1) + \sum_{j=2}^{n} \mu_v\Big(A_j \cap (A_1 \cup A_2 \cup \cdots \cup A_{j-1})^c\Big)$$
$$\leq \sum_{j=1}^{n} \mu_v(A_j).$$

□

Definition Let S be an interval and let $\mathcal{B}$ be the family of basic subsets of S. Then a function $v : \mathcal{B} \longrightarrow [0, \infty]$ is called an *additive set function* on $\mathcal{B}$ (or on S) if for all $A, B \in \mathcal{B}$ such that $A \cap B = \emptyset$, $v(A \cup B) = v(A) + v(B)$. (We adopt the usual convention that for $x \in [0, \infty]$, $x + \infty = \infty$.) If $v(A) < \infty$ for some $A \in \mathcal{B}$, we must have $v(\emptyset) = 0$. The additive set function v is *bounded* if $v(S) < \infty$.

A case of particular interest to us in the later Sections of this Chapter is when $S = [1, b)$ for some $b > 1$ and $v(x) = \log_b(x)$ for $x \in [1, b)$. In this case the function $\mu_{\log_b}$ is an additive set function, and $\mu_{\log_b}([c, d)) = \log_b d - \log_b c$. The number $\mu_{\log_b}(A)$ is called the *logarithmic length of A to the base b*. Note that $\mu_{\log_b}([1, b)) = 1$. Note also that there are finitely additive set functions on $\mathcal{B}$ that are not of the form μ_v for any non-decreasing function v on S—in fact, if v is a bounded additive set function, there is a non-decreasing function v and a

3.5. Sets of measure zero

"discrete" additive set function ν_d such that $\nu = \mu_v + \nu_d$ (see Exercises 4 and 5).

Proposition 3.11. *Let S be an interval, let $\mathcal{B}$ be the family of basic subsets of S, and let $v : \overline{S} \longrightarrow \mathbb{R}$ be a non-decreasing function. Then μ_v is an additive set function on $\mathcal{B}$. In particular the function μ, which ascribes to an interval its usual length, is an additive set function on $\mathcal{B}$. Also, if $b > 1$, the logarithmic length $\mu_{\log_b}$ to the base b, which ascribes to an subinterval $[c, d)$ of $[1, b)$ the value $\log_b d - \log_b c$, is an additive set function on the family of basic subsets of $[1, b)$.*

Proof. This is immediate from the definitions and Proposition 3.10. □

3.5 Sets of measure zero

A special role is played in analysis and probability by the so-called sets of measure zero, or null sets[1]. There are various results in which a certain statement is true for all values of x, except for those x in some set of measure zero. The idea is that a set of measure zero is an exceptional set that is "small" in the sense of the precise definition below. If we think of a set as a possible event, then a set of measure zero corresponds to an event that has probability zero. For example, consider an experiment in which an unbiased coin is tossed over and over again, indefinitely, and the observed outcome is some particular infinite sequence of heads and tails. The set of all possible outcomes in such an experiment is the set of all possible sequences of heads and tails. Consider the set of all the outcomes of this hypothetical experiment that end in an infinite sequence of "heads". Then this set or event has probability zero, as we would expect that, for an unbiased coin, beyond any point in the tossing, "tails" will occur at some future time and there will not be an infinite run of "heads".

Definition A subset B of the real numbers is called a *set of measure zero* if, for each $\varepsilon > 0$, there is a corresponding sequence (J_n) of intervals such that

$$B \subseteq \bigcup_{n=1}^{\infty} J_n \quad \text{and} \quad \sum_{n=1}^{\infty} \mu(J_n) < \varepsilon. \tag{3.27}$$

This says that a set of measure zero is one which can be covered by an infinite sequence of intervals, the sum of whose lengths is as small as any positive number we have specified in advance. It follows from the definition that if B has measure zero, and if $C \subseteq B$, then C also has measure zero.

Which sets are sets of measure zero? A single point will have measure zero. For, let $x \in [0, 1]$ and let $\varepsilon > 0$. Then since $\{x\} = [x, x]$, $\{x\} \subseteq [x, x]$ and $\mu([x, x]) = 0 < \varepsilon$, so that $\{x\}$ is of measure zero. This still does not say much about the sets of measure zero, but more such sets can be built up from individual points, as we now see.

Suppose that a sequence $B_1, B_2, B_3 \ldots$ of sets of measure zero is given, and let $\varepsilon > 0$. Then, as each B_n is of measure zero, it follows from the definition that if in (3.27) we take B_n in place of A and $\varepsilon/2^n$ in place of ε, there is a sequence $(J_{nj})_{j=1}^{\infty}$ of intervals such that

$$B_n \subseteq \bigcup_{j=1}^{\infty} J_{nj} \quad \text{and} \quad \sum_{j=1}^{\infty} \mu(J_{nj}) < \varepsilon/2^n. \tag{3.28}$$

Now, the family $\{J_{nj} : j, n = 1, 2, \ldots\}$ of intervals may be arranged in a sequence. For example, we could write down those intervals J_{nj} with $n + j = 2$, then those with $n + j = 3$, then those with $n + j = 4$, and so on. Then, we would obtain a sequence of intervals that could be written as

$$J_{11}, J_{12}, J_{21}, J_{31}, J_{22}, J_{13}, J_{41}, \ldots.$$

Denote this sequence of intervals by (T_n). As the intervals in (T_n) are simply the intervals in $\{J_{nj} : j, n \in \mathbb{N}\}$ written in a definite order we have, using (3.28),

$$\bigcup_{n=1}^{\infty} B_n \subseteq \bigcup_{n=1}^{\infty} \bigcup_{j=1}^{\infty} J_{nj} = \bigcup_{n=1}^{\infty} T_n. \tag{3.29}$$

3.5. Sets of measure zero

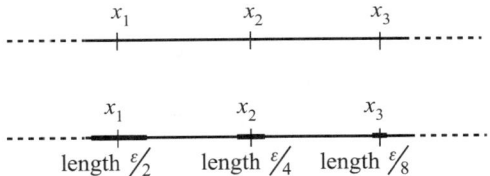

Figure 3.3. The top picture illustrates a countable set $\{x_1, x_2, \ldots, x_n, \ldots\}$. Such a set has measure zero. For, a set $\{x\}$ consisting of single point is an interval, and $\mu(\{x\}) = 0$. Now, observe that $\{x_1, x_2, \ldots, x_n, \ldots\} \subseteq \bigcup_{n=1}^{\infty} \{x_n\}$ and that $\sum_{n=1}^{\infty} \mu(\{x_n\}) = 0$, so a countable set has measure zero, from the definition. More generally, to prove a set is of measure zero, we must cover it by a sequence of intervals, the sum of whose lengths is as small as we specify in advance. The lower picture illustrates again that a countable set has measure zero, but this time by covering it using intervals of positive length. Given a number $\varepsilon > 0$ an interval of length $\varepsilon/2$, indicated in black, is placed about x_1. Then further intervals of lengths $\varepsilon/2^2, \varepsilon/2^3, \ldots$ and so on, also indicated in black, are placed successively about $x_2, x_3, \ldots$. Although it is not indicated in the figure, note that some of these intervals may overlap. Clearly, the union of all these intervals contains the set $\{x_1, x_2, \ldots, x_n, \ldots\}$. Also, the totality of the lengths of all these intervals, if they are disjoint as shown in the picture, or even if they overlap, equals $\sum_{n=1}^{\infty} \varepsilon/2^n = \varepsilon$. So, the set $\{x_1, x_2, \ldots, x_n, \ldots\}$ has measure zero, and again we see that a countable set has measure zero.

Now, if $r \in \mathbb{N}$ there will be $p, q \in \mathbb{N}$ such that

$$\sum_{n=1}^{r} \mu(T_n) \leq \sum_{n=1}^{p} \sum_{j=1}^{q} \mu(J_{nj}) \leq \sum_{n=1}^{\infty} \left(\sum_{j=1}^{\infty} \mu(J_{nj}) \right) < \sum_{n=1}^{\infty} \varepsilon/2^j = \varepsilon,$$

where we have used (3.28). As this is true for all $r \in \mathbb{N}$, we deduce that

$$\sum_{n=1}^{\infty} \mu(T_n) < \varepsilon. \tag{3.30}$$

Now (3.29) and (3.30) together show that for each $\varepsilon > 0$, $\bigcup_{n=1}^{\infty} B_n$ is contained in a sequence (T_n) of intervals whose lengths have a sum that is less than ε. It follows that $\bigcup_{n=1}^{\infty} B_n$ is a set of measure zero. This argument demonstrates that the union of a sequence of sets of measure zero is also a set of measure zero.

Finally, we see that if B is a set of measure zero, and if $x \in \mathbb{R}$, then $x + B$ is also a set of measure zero. Note that a set of the form $x + B$ is called a *translate* of B, because it results from "sliding" or "translating" B along the real axis. Thus, we are saying that a translate of a set of measure zero is a set of measure zero. To see this, first observe that if W is an interval and if $x \in \mathbb{R}$, $x + W$ is an interval and $\mu(x + W) = \mu(W)$. Now, let B have measure zero and let $\varepsilon > 0$. Then there is a sequence (W_n) of intervals such that

$$B \subseteq \bigcup_{n=1}^{\infty} W_n \quad \text{and} \quad \sum_{n=1}^{\infty} \mu(W_n) < \varepsilon.$$

Then, if $x \in \mathbb{R}$,

$$x + B \subseteq \bigcup_{n=1}^{\infty}(x + W_n) \quad \text{and} \quad \sum_{n=1}^{\infty} \mu(x + W_n) < \varepsilon.$$

Thus, $x + B$ is a set of measure zero, by definition. We summarize the above observations as follows.

Proposition 3.12. *The following hold.*

(1) A set consisting of a single point has measure zero.

(2) The union of a sequence of sets, each of which is a set of measure zero, is also a set of measure zero.

(3) A subset of a set of measure zero is a set of measure zero.

(4) A translate of a set of measure zero is also a set of measure zero.

In particular, it follows from Proposition 3.12 that a set whose elements may be arranged as a sequence of points is a set of measure zero. Such sets deserve special attention.

Now, a set is called *countable* if its elements may be arranged as a sequence, and it has been mentioned in Section 1.3 of Chapter 1 and in Figure 3.3 that the set $\mathbb{Q}$ of all rational numbers is countable. Thus, Proposition 3.12 gives the following result.

Proposition 3.13. *The following hold.*

(1) Every countable subset of $\mathbb{R}$ has measure zero.

(2) The set $\mathbb{Q}$ of all rational numbers is a set of measure zero.

3.5. Sets of measure zero

Figure 3.4. The illustrations depict the construction of Cantor's ternary set. If we are given an interval J, we can divide this interval into three subintervals of equal length. If we omit the middle of these intervals, we obtain a left-hand interval $J(\ell)$ and a right-hand interval $J(r)$, where each of these is taken to be closed. We can repeat the process on $J(\ell)$ to obtain a closed interval $J(\ell)(\ell)$, denoted by $J(\ell\ell)$, and an interval $J(\ell)(r)$, denoted by $J(\ell r)$, and so on. Now in the top picture, we start with the closed unit interval $J = [0, 1]$ and then, at the next step, the grey interval has been omitted, and we have the black intervals $J(\ell)$ on the left and $J(r)$ on the right, having a total length of $2/3$. We repeat this step on each of the intervals $J(\ell)$ and $J(r)$, obtaining at the next stage the four intervals indicated by black, namely $J(\ell\ell)$, $J(\ell r)$, $J(r\ell)$ and $J(rr)$, having a total length of $4/9$. After n steps we will obtain 2^n intervals consisting of all those of the form $J(\theta_1\theta_2\ldots\theta_n)$, where $\theta_j \in \{\ell, r\}$ for all $j = 1, 2, \ldots, n$. These 2^n intervals will have a total length of $2^n/3^n$. The *Cantor set* is defined to be the set C where

$$C = \bigcap_{n=1}^{\infty} \Big(\bigcup_{\theta_1,\ldots,\theta_n \in \{\ell,r\}} J(\theta_1\theta_2\ldots\theta_n) \Big).$$

The set C has measure zero because C is contained in $\bigcup_{\theta_1,\ldots,\theta_n \in \{\ell,r\}} J(\theta_1\theta_2\ldots\theta_n)$ for all n, the total length of the intervals in this union is $2^n/3^n$, and $\lim_{n \to \infty} 2^n/3^n = 0$. We see that if we neglect the endpoints of the intervals $J(\theta_1\theta_2\ldots\theta_n)$, C consists of those numbers x in $[0, 1]$ that have an expansion to the base 3 that is of the form $x = 0.x_1x_2x_3\ldots$, where $x_j \in \{0, 2\}$ for all j. But the set of all sequences of 0s and 2s is in one-to-one correspondence with the set of all sequences of 0s and 1s, and the latter is in an essentially one-to-one correspondence with the points of the interval $[0, 1]$, as consideration of the binary expansion shows (see Exercise 1). It can be shown from this that the Cantor set is uncountable (see Exercise 7).

It is important to realize that not all sets are sets of measure zero. No interval of positive length is a set of measure zero (see Exercise 6). In fact, if J is an interval and (J_n) is a sequence of intervals such that $J \subseteq \bigcup_{n=1}^{\infty} J_n$, then $\mu(J) \le \sum_{n=1}^{\infty} \mu(J_n)$. Also, there are sets that have measure zero, but which are not countable, such as Cantor's ternary set discussed in Figure 3.4. The sets of measure zero should be thought of as "negligible", "small" or "sparse". In terms of probability, a set of measure zero, even if it is not a basic set, may be thought of as an event with probability zero. Finally, if a statement holds for all values of x outside some particular set of measure zero, we may say that the statement holds for *almost all* x, or we may say that the statement holds *almost everywhere*.

3.6 Independent sets and events

Consider an experiment that has as sample space the interval $[0, 1)$. Then, the interval $[0, 1)$ denotes the set of all possible outcomes of the experiment. Consider also two basic subsets A, B of $[0, 1)$ that have positive lengths $\mu(A), \mu(B)$. We can also think of A, B as events which respectively have probabilities $\mu(A), \mu(B)$. In this context, independence means something like "being unaffected by previous events". So, if we say that the event B is independent of the event A, it should mean that the probability of the outcome $x \in B$ is not affected by whether or not the outcome $x \in A$ has occurred. Now, the probability that $x \in B$ occurs is $\mu(B)$. On the other hand, the probability that $x \in B$ occurs, *given* that $x \in A$ has occurred, is $\mu(A \cap B)/\mu(A)$, because $\mu(A \cap B)/\mu(A)$ represents the proportion of points in A that are also in B. Independence of the event B from the event A should mean that these probabilities are equal. Thus, independence of A, B should mean that

$$\frac{\mu(A \cap B)}{\mu(A)} = \mu(B).$$

That is, independence of B from A should mean

$$\mu(A \cap B) = \mu(A)\,\mu(B). \tag{3.31}$$

3.6. Independent sets and events

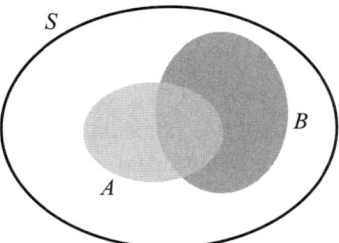

Figure 3.5. The Figure illustrates the idea of independent events in a general sample space. The events A, B are depicted as subsets of S, the "set of all possible outcomes". As S is the set of "all possible outcomes", its probability is 1. If D is a suitable subset of S, let $v(D)$ be the "area" of D. The probability of S is then $v(S) = 1$, and the probability of an event D is $v(D)$. Now, if we are given that the event A has occurred, then A becomes the "set of all possible outcomes" instead of S. So, given that A has occurred, the probability that an event B has also occurred will be the proportion of B which lies within A. That is, the probability of B, given A, is $v(A \cap B)/v(A)$. But if B is *independent* of A, the probability of B occurring will be the same, regardless of whether or not A has occurred. That is, independence of B from A means $v(A \cap B) = v(A)v(B)$.

Note that (3.31) is symmetric in A, B. This means that if (3.31) is taken as the definition of the independence of B from A, then B is independent from A if and only if A is independent from B. That is, independence is a symmetric condition on the two events.

Motivated by (3.31), we make the following formal definition of probabilistic independence for a finite number of basic sets or events.

Definition The basic subsets $A_1, A_2, \ldots, A_n$ of $[0, 1)$ are *probabilistically independent* if for all choices of $1 \le r \le n$ and $1 \le n_1 < n_2 < \cdots < n_r \le n$,

$$\mu\left(\bigcap_{j=1}^{r} A_{n_j}\right) = \prod_{j=1}^{r} \mu(A_{n_r}).$$

Usually, we shall simply refer to *independent* sets or events, rather than *probabilistically independent* sets or events—however, note that the word "independence" appears in various mathematical contexts, and its meaning may depend on the context.

When $r = 2$, the definition simply gives (3.31) for the independence of two basic sets A, B. Also, it is immediate from this definition that if the sets $A_1, A_2, \ldots, A_n$ are independent, a rearrangement of these sets also gives an independent family of sets. It is also evident that any subfamily of an independent family of basic sets will also be independent. In line with our intuitive idea of independence, there is the following result that will be used in the next section.

Proposition 3.14. *Let the basic subsets $A_1, A_2, \ldots, A_n$ of $[0, 1)$ be independent. Let $B_1, B_2, \ldots, B_n$ be sets such that for each $j = 1, 2, \ldots, n$, either $B_j = A_j$ or $B_j = A_j^c$. Then the sets $B_1, B_2, \ldots, B_n$ are also independent.*

Proof. Consider the case when $B_1 = A_1^c$, $B_2 = A_2$, $B_3 = A_3, \ldots, B_n = A_n$, and let $1 \le n_1 < n_2 < \ldots < n_r \le n$. Then if $n_1 > 1$,

$$\mu\left(\bigcap_{j=1}^r B_{n_j}\right) = \mu\left(\bigcap_{j=1}^r A_{n_j}\right) = \prod_{j=1}^r \mu(A_{n_j}) = \prod_{j=1}^r \mu(B_{n_j}). \tag{3.32}$$

When $n_1 = 1$, $B_{n_1}^c = A_1$, so that

$$\prod_{j=2}^r \mu(A_{n_j}) = \mu\left(\bigcap_{j=2}^r A_{n_j}\right) = \mu\left(\bigcap_{j=2}^r B_{n_j}\right)$$

$$= \mu\left(B_{n_1} \cap \left(\bigcap_{j=2}^r B_{n_j}\right)\right) + \mu\left(B_{n_1}^c \cap \left(\bigcap_{j=2}^r B_{n_j}\right)\right)$$

$$= \mu\left(\bigcap_{j=1}^r B_{n_j}\right) + \mu\left(A_1 \cap \left(\bigcap_{j=2}^r A_{n_j}\right)\right)$$

$$= \mu\left(\bigcap_{j=1}^r B_{n_j}\right) + \mu\left(\bigcap_{j=1}^r A_{n_j}\right) = \mu\left(\bigcap_{j=1}^r B_{n_j}\right) + \prod_{j=1}^r \mu(A_{n_j}).$$

Hence,

$$\mu\left(\bigcap_{j=1}^{r} B_{n_j}\right) = \prod_{j=2}^{r} \mu(A_{n_j}) - \prod_{j=1}^{r} \mu\left(A_{n_j}\right)$$
$$= \left(1 - \mu(A_{n_1})\right) \prod_{j=2}^{r} \mu(A_{n_j})$$
$$= \mu(A_{n_1}^c) \prod_{j=2}^{r} \mu(A_{n_j})$$
$$= \prod_{j=1}^{r} \mu(B_j). \qquad (3.33)$$

We see from (3.32) and (3.33) that $B_1, B_2, \ldots, B_n$ are independent. It follows that if a single one of the sets B_j is A_j^c, while $B_k = A_k$ for all $k \neq j$, then $B_1, B_2, \ldots, B_n$ are independent. That is, when *one* of the original sets is changed from A_j to A_j^c, independence is not affected. But this argument may now be repeated on any family of sets obtained in this way, and it follows that if *two* of the sets A_j are changed from A_j to A_j^c, then independence is not affected. This argument may be repeated up to n times, so that if $k \in \{1, 2, \ldots, n\}$, and if $B_j = A_j$ for k values of j but $B_j = A_j^c$ for the remaining values of j, then $B_1, B_2, \ldots, B_n$ are independent. □

3.7 Typewriters, recurrence, and the Prince of Denmark

Suppose we have an unbiased coin that is tossed over and over again. Then, at least eventually, we would expect to obtain a "head". Not only that, but with repeated tossing we would expect to obtain two consecutive heads. But, there is no reason to stop at two—we would expect, with repeated tossing, eventually to obtain three consecutive heads. If we reflect upon this, it seems that with repeated tossing, we should expect eventually to obtain 100 consecutive heads, or 1,000 consecutive heads, or even

10,000,000,0000 consecutive heads. In fact, for every $n \in \mathbb{N}$, it seems we should expect eventually to get n consecutive heads. Of course, the catch is that the more consecutive heads we are looking for, the longer we would expect to have to toss the coin. However, we would not expect to get an *infinite* sequence of consecutive heads as we continue to toss, since this would mean we would get heads at every toss, thus showing that the coin is biased.

Now if we tossed a coin 100 times and obtained "heads" every time, we would think that the coin was biased, so this seems to be inconsistent with the preceding thoughts. This apparent paradox is resolved by realizing that if the coin is unbiased, the tossing of the coin produces random results which nevertheless will, from time to time, exhibit order. After all, if the outcome is "random", it cannot be expected that the continued outcomes will arrange themselves in such a way that any form of order will be consistently avoided. Here, obtaining 100 consecutive heads gives to the observer a very strong impression of order rather than randomness, but if we think of randomness as a form of mindlessness, or as a type of forgetfulness of what has previously occurred, we would not expect a random process to show an awareness that it must avoid any appearance of order so as to reveal itself as random. Thus, we should expect a random process to produce appearances of order from time to time. But note that when we use the phrase "from time to time", we must realize that we may have to wait for enormously long periods, before we can reasonably expect to observe an event with a very high degree of order—such as obtaining 100 consecutive heads. So, if we pick up a coin and toss it 100 times and observe 100 heads, it is overwhelmingly likely that the coin is biased. But although an unbiased coin must produce events with a high degree of order from time to time, the *proportion* of time that we expect to observe such an event will be the lower the higher the degree of order the event requires. The fact that a temporary order is observed for a while when a coin is tossed many times does not necessarily mean that there is an order underlying the whole tossing process, nor that the outcomes of the process can be predicted.

In this section we look at these questions from the point of view of the binary expansion of numbers. The main question considered is the

3.7. Typewriters, recurrence, and the Prince of Denmark

> Is it not a wonder that anyone can bring himself to believe that a number of solid and separate particles by their chance collisions and moved only by the force of their own weight could bring into being so marvelous and beautiful a world ? If anybody thinks that this is possible, I do not see why he should not think that if an infinite number of examples of the twenty-one letters of the alphabet, made of gold or what you will, were shaken together and poured onto the ground it would be possible for them to fall so as to spell out, say, the whole text of the *Annals* of Ennius. In fact I doubt whether chance would permit them to spell out a single verse!
> —Marcus Tullius Cicero (106–43 BCE)
> *The Nature of the Gods*

extent to which the binary expansion of numbers, when thought of as producing a succession of tosses of a coin, produces "random" outcomes. The aim is to do this in a precise and complete way, using only ideas from elementary analysis.

Definition Let $\theta_1, \theta_2, \ldots, \theta_s$ be a finite sequence of 0s and 1s, and let $\alpha_1, \alpha_2, \ldots, \alpha_n, \ldots$ be an infinite sequence of 0s and 1s. That is, for each $j \in \{1, 2, \ldots, s\}$, $\theta_j \in \{0, 1\}$; and for each $j \in \mathbb{N}$, $\alpha_j \in \{0, 1\}$. We shall say that the sequence $\theta_1, \theta_2, \ldots, \theta_s$ *occurs* in the sequence $\alpha_1, \alpha_2, \ldots, \alpha_n, \ldots$ if there is $r \in \mathbb{N}$ such that

$$\theta_1 = \alpha_r, \theta_2 = \alpha_{r+1}, \theta_3 = \alpha_{r+2}, \ldots, \theta_s = \alpha_{r+s-1}.$$

In this case we shall say that the sequence $\theta_1, \theta_2, \ldots, \theta_s$ occurs in the sequence $\alpha_1, \alpha_2, \ldots, \alpha_n, \ldots$ *starting at position r*.

We shall use the notation that if $x \in [0, 1)$ and $j \in \mathbb{N}$, then $d_j(x)$ is the jth binary digit of x and $d(x)$ denotes the binary sequence of x, $d_1(x), d_2(x), \ldots$.

Lemma 3.15. *Let $\theta_1, \theta_2, \ldots, \theta_r \in \{0, 1\}$. Then the set*

$$\left\{x : x \in [0, 1) \text{ and } d_j(x) = \theta_j \text{ for all } j = 1, 2, \ldots, r\right\}$$

is a subinterval of $[0, 1)$ of length 2^{-r}.

Proof. It is proved in Theorem 3.4 that the set is the interval

$$\left[\sum_{j=1}^{r}\frac{\theta_j}{2^j},\ \sum_{j=1}^{r}\frac{\theta_j}{2^j}+\frac{1}{2^r}\right),$$

and this interval has length $1/2^r$. □

Let $r \in \mathbb{N}$ be given, let $\theta_1, \theta_2, \ldots, \theta_r \in \{0, 1\}$ be given, and let $n_1, n_2, \ldots, n_r \in \mathbb{N}$ be such that $1 \leq n_1 < n_2 < \cdots < n_r$. Note that $r \leq n_r$. Put

$$A = \Big\{x : x \in [0, 1) \text{ and } d_{n_j}(x) = \theta_j \text{ for all } j = 1, 2, \ldots, r\Big\}.$$

Let $\alpha_1, \alpha_2, \ldots, \alpha_{n_r} \in \{0, 1\}$ be such that $\alpha_{n_j} = \theta_j$ for all $j = 1, 2, \ldots, r$. That is, r of the α_j are specified in advance but $n_r - r$ of the α_j may be chosen at will from $\{0, 1\}$. Thus there are $2^{n_r - r}$ ways of choosing $\alpha_1, \alpha_2, \ldots, \alpha_{n_r}$ subject to the given constraints. For such a given choice of $\alpha_1, \alpha_2, \ldots, \alpha_{n_r}$ consider the set

$$B = \Big\{x : x \in [0, 1) \text{ and } d_j(x) = \alpha_j, \text{ for all } j = 1, 2, \ldots, n_r\Big\}.$$

By Lemma 3.15, B is an interval of length 2^{-n_r}. Moreover, A is the union of all the $2^{n_r - r}$ intervals B that may be obtained in this way, so that A is basic. Also, two such intervals B obtained from two different choices of the α_j are disjoint. Thus, it follows that the length of A equals $2^{n_r - r} \times 2^{-n_r} = 2^{-r}$. This has proved the following result.

Proposition 3.16. *Let $n_1, n_2, \ldots, n_r \in \mathbb{N}$ with $n_1 < n_2 < \cdots < n_r$, and let $\theta_1, \theta_2, \ldots, \theta_r \in \{0, 1\}$. Then the set*

$$\Big\{x : x \in [0, 1) \text{ and } d_{n_j}(x) = \theta_j, \text{ for all } j = 1, 2, \ldots, r\Big\}$$

is a finite union of intervals of equal length, and the length of this set is 2^{-r}.

Now, let $s \in \mathbb{N}$ be given and let $\theta_1, \theta_2, \ldots, \theta_s$ be a given finite sequence of 0s and 1s. So, each $\theta_j \in \{0, 1\}$. For each $k = 0, 1, 2, \ldots$ let

$$J_k = \{ks + 1, ks + 2, \ldots, (k + 1)s\}.$$

3.7. Typewriters, recurrence, and the Prince of Denmark

Note that the intervals $J_0, J_1, J_2, \ldots, J_k, \ldots$, express $\mathbb{N}$ as a union of disjoint intervals, each one having s elements. We may call J_k the kth interval of length s. Now for $k = 0, 1, 2, 3, \ldots$ let D_k denote those numbers $x \in [0, 1)$ such that $\theta_1, \theta_2, \ldots, \theta_s$ occurs in the binary expansion of x starting at $sk + 1$. That is,

$$D_k = \left\{ x : x \in [0, 1) \text{ and } d_{sk+j}(x) = \theta_j \text{ for all } j = 1, 2, 3, \ldots, s \right\}.$$

Alternatively,

$$D_k = \left\{ x : x \in [0, 1) \text{ and } d_\ell(x) = \theta_{\ell-ks} \text{ for all } \ell \in J_k \right\}.$$

Now the usefulness of Proposition 3.16 lies in the fact that it can be used to deduce that for every $n \in \mathbb{N}$ the sets $D_0, D_1, D_2, \ldots, D_n$ are independent. For, if $0 \le j_1 < j_2 < j_3 < \cdots < j_r \le n$, from the definition of the sets D_k and Proposition 3.16,

$$\mu\left(\bigcap_{k=1}^{r} D_{j_k}\right) = \mu\left(\left\{x : x \in [0, 1) \text{ and for all } \right.\right.$$
$$\left.\left. k = 1, 2, \ldots, r, d_\ell(x) = \theta_{\ell-j_k s} \text{ for all } \ell \in J_{j_k} \right\}\right)$$
$$= 2^{-rs}$$
$$= \prod_{k=1}^{r} 2^{-s}$$
$$= \prod_{k=1}^{r} \mu\left(\left\{x : x \in [0, 1) \text{ and }\right.\right.$$
$$\left.\left. d_\ell(x) = \theta_{\ell-j_k s} \text{ for all } \ell \in J_{j_k} \right\}\right)$$
$$= \prod_{k=1}^{r} \mu(D_{j_k}),$$

which shows that $D_0, D_1, D_2, \ldots, D_n$ are independent.

Now since the sets $D_0, D_1, D_2, \ldots, D_n$ are independent, it follows from Proposition 3.14 that for each $n \in \mathbb{N}$, the sets $D_0^c, D_1^c, D_2^c, \ldots, D_n^c$

are also independent. Thus,

$$\mu\left(\bigcap_{j=0}^{n} D_j^c\right) = \prod_{j=0}^{n} \mu(D_j^c) = \prod_{j=0}^{n}(1 - \mu(D_j))$$

$$= \prod_{j=0}^{n}(1 - 2^{-s}) = (1 - 2^{-s})^{n+1}.$$

As $0 < 1 - 2^{-s} < 1$, it follows that

$$\lim_{n\to\infty} \mu\left(\bigcap_{j=0}^{n} D_j^c\right) = \lim_{n\to\infty}(1 - 2^{-s})^{n+1} = 0. \qquad (3.34)$$

Now let D be the set consisting precisely of those points $x \in [0, 1)$ such that $\theta_1, \theta_2, \ldots, \theta_s$ does *not* occur in the binary sequence of x. Since, from the definition of D_k, the set D_k^c consists precisely of those $x \in [0, 1)$ such that $\theta_1, \theta_2, \ldots, \theta_s$ does not occur in the binary sequence of x starting at the point sk, it is clear that for all $n \in \mathbb{N}$,

$$D \subseteq \bigcap_{j=0}^{\infty} D_j^c \subseteq \bigcap_{j=0}^{n} D_j^c,$$

and it follows from (3.34) that D is a set of measure zero. This means we have proved the following.

Proposition 3.17. *Let $\theta_1, \theta_2, \ldots, \theta_s$ be a given finite sequence of 0s and 1s, and let D be the set consisting of those points $x \in [0, 1)$ such that $\theta_1, \theta_2, \ldots, \theta_s$ does not occur in the binary sequence of x. Then D has measure zero.*

We are now in a position to prove that, with the exception of those points that are in some fixed set of measure zero, every point x in $[0, 1)$ has the property that *every* finite sequence of symbols occurs in the binary sequence of x.

Theorem 3.18 (The Recurrence Theorem[2]). *There is a subset $\mathcal{Z}$ of $[0, 1)$ that has measure zero and has the following property: if $x \notin \mathcal{Z}$, every*

3.7. Typewriters, recurrence, and the Prince of Denmark

finite sequence of binary symbols occurs in the sequence of binary digits of x. That is, if $\theta_1, \theta_2, \ldots, \theta_s$ is a given sequence of 0s and 1s, for every $x \notin \mathcal{Z}$ there is $r \in \mathbb{N}$ such that

$$d_{r+j}(x) = \theta_j, \text{ for all } j = 1, 2, \ldots, s.$$

Furthermore, if $x \notin \mathcal{Z}$, every finite sequence of binary symbols occurs in the sequence of binary digits of x infinitely many times.

Proof. Let $n \in \mathbb{N}$. There are 2^n possible sequences of the form $\theta_1, \theta_2, \ldots, \theta_n$ where $\theta_j \in \{0, 1\}$ for each j. If these 2^n sequences, each of which has n terms, are placed adjacent to each other, a sequence of 0s and 1s having $n2^n$ terms is obtained and we denote this sequence by Θ_n. Note that the manner in which the 2^n sequences of n terms are placed adjacently to each other is not significant—all that is required is to choose one manner of doing it for each n. Now for $n = 1, 2, \ldots$ let

$$\mathcal{Z}_n = \Big\{ x : x \in [0, 1) \text{ and } \Theta_n \text{ does not occur in the sequence of binary digits of } x \Big\}.$$

Then it follows from Proposition 3.17 that $\mathcal{Z}_n$ has measure zero. Put $\mathcal{Z} = \cup_{n=1}^{\infty} \mathcal{Z}_n$ and note that by (2) of Proposition 3.12, $\mathcal{Z}$ has measure zero. Now, let $x \notin \mathcal{Z}$ and let $\phi_1, \phi_2, \ldots, \phi_s$ be a given finite sequence of symbols. Then because Θ_s consists of the juxtaposition of *all* sequences of s symbols, $\phi_1, \phi_2, \ldots, \phi_s$ occurs within Θ_s. But as $x \notin \mathcal{Z}$, $x \notin \mathcal{Z}_s$, so that Θ_s occurs in the sequence of binary digits of x. Hence the sequence $\phi_1, \phi_2, \ldots, \phi_s$ occurs in the sequence of binary digits of x. Thus, $\mathcal{Z}$ has measure zero, and every point x not in $\mathcal{Z}$ has the property that every given finite sequence of symbols occurs in $d(x)$.

Concerning the infinite occurrence of $\theta_1, \theta_2, \ldots, \theta_s$, let $p \in \mathbb{N}$ and consider the finite sequence ϕ_p of symbols $\theta_1, \ldots, \theta_s, \theta_1, \ldots, \theta_s, \ldots, \theta_1, \ldots, \theta_s$, where $\theta_1, \ldots, \theta_s$ has been repeated p times. Then, if $x \notin \mathcal{Z}$, we have seen already that ϕ_p must occur in the sequence of binary digits of x. But, each occurrence of ϕ_p is an occurrence p times of $\theta_1, \ldots, \theta_s$. So, for every $p \in \mathbb{N}$, $\theta_1, \ldots, \theta_s$ occurs at least p times in the sequence of binary digits of x. Thus, $\theta_1, \ldots, \theta_s$ occurs infinitely many times in the sequence of binary digits of x. □

Hamlet, Prince of Denmark, is perhaps the best known of all the plays of Shakespeare. At a very mundane level, we can regard this play as a sequence of symbols, which includes lower and upper case letters, symbols to denote punctuation, blank spaces, and so on. It remains a mystery as to how Shakespeare found the particular combination of symbols that produces the effect of *Hamlet*, but I think we can say that he did not merely generate random sequences of symbols to get his results. However, we can imagine these symbols being randomly produced, for example by thinking of hypothetical monkeys randomly hitting the keys of a typewriter. Although there will be more symbols on the keyboard than the two we have considered in this section, the principle of Theorem 3.18 still applies, and says in effect that a preassigned finite sequence of symbols will be produced by the monkeys provided they type for a sufficiently long time—that is, if they type for long enough, and if they type randomly, they will produce an exact copy of *Hamlet*!

Questions such as the average time it would take the monkeys to produce *Hamlet* will be considered in Chapter 4. For the time being, let us consider a simpler question suggested by Theorem 3.18. This simpler question is: *if we repeatedly toss an unbiased coin, on the average, how many tosses does it take to get the first "head"?* This question can be phrased equivalently as: if a number x is chosen at random in $[0, 1)$, on the average, what is the position in which the first 1 appears in the binary expansion of x? We know there is a set $\mathcal{Z}$ of measure zero such that if $x \in [0, 1)$ but $x \notin \mathcal{Z}$, then every finite sequence of 0s and 1s will appear in the binary digits of x. In particular, if $x \notin \mathcal{Z}$, there will be a smallest value of n in the sequence $(d_n(x))$ such that $d_n(x) = 1$. In fact, because this single digit 1 is such a simple finite sequence, for *every* non-zero x in $[0, 1)$ there is a smallest value of n in the sequence $(d_n(x))$ such that $d_n(x) = 1$. Denote this smallest value of n by $\phi(x)$. Then, the average value of ϕ is the number of tosses of an unbiased coin that is required, on the average, to obtain the first "head". We now indicate an

3.7. Typewriters, recurrence, and the Prince of Denmark

> An experiment to test the theory that a group of monkeys armed with typewriters will eventually produce the works of Shakespeare has been abandoned after the primates failed to write even one recognizable word.
>
> Lecturers and students from Plymouth University, who received 2,000 Pounds of Arts Council sponsorship for the project, installed a computer in a zoo enclosure to monitor the literary output of six monkeys.
>
> But after a month the Sulawesi crested macaques succeeded only in partially destroying the machine, using it as a lavatory, and filling five pages of text, primarily with the letter S.
>
> The students, from the MediaLab Arts course, concluded that their subjects at Paignton Zoo, in Devon, would never achieve literary greatness. Geoff Cox, the lecturer who devised the experiment, said: "The aim of the project was to show that animals cannot be reduced to the level of random processes, or, indeed, to the level of a computer. The joke, if there is one, is not on the monkeys, but on the theory itself."
>
> The conceit that monkeys might type Shakespeare, often cited in arguments about evolution, is thought to have been coined by Thomas Huxley, the foremost scientific supporter of Charles Darwin's theories.
>
> The participants at Paignton Zoo have done little to help Huxley's cause, however.
>
> Having first tried to destroy the computer by chewing the cover, the macaques eventually produced a little text. Their output improved slightly towards the end, with the letters A, J, L and M also being employed, but the monkeys failed to come up with anything remotely resembling a word.
>
> "We weren't particularly surprised that the monkeys didn't write a great deal," Dr Vicky Melfi, a research associate, said. "They are extremely intelligent, but have evolved to a completely different niche where they don't need Shakespeare. To be honest, they weren't very interested in the computer at all. They spent most of the time sitting on it, or jumping up and down ... "
>
> The results of the experiment, part of a larger project developed by i-DAT, the Institute of Digital Arts and Technology at the university, are available in a limited edition book entitled Notes Towards The Complete Works of Shakespeare.
>
> The five-page edition duly credits its authors: Elmo, Gum, Heather, Holly, Mistletoe and Rowan.
>
> The London Times, May 4th, 2003

argument which can be refined to calculate the average value of ϕ, and thus answers the above questions.

Note that by Theorem 3.4,

$$\{x : x \in [0, 1) \text{ and } d_1(x) = 1\} = \left[\frac{1}{2}, 1\right), \text{ and}$$

$$\{x : x \in [0, 1), \ d_1(x) = \cdots = d_{n-1}(x) = 0 \text{ and } d_n(x) = 1\} = \left[\frac{1}{2^n}, \frac{1}{2^{n-1}}\right). \tag{3.35}$$

Now for $n = 1, 2, \ldots$, let

$$A_n = \{x : x \in [0, 1) \text{ and } \phi(x) = n\} = \left[\frac{1}{2^n}, \frac{1}{2^{n-1}}\right), \tag{3.36}$$

using (3.35). Thus, $A_1, A_2, A_3, \ldots$, is a sequence of disjoint basic sets and $\bigcup_{n=1}^{\infty} A_n = (0, 1)$. Also, ϕ takes the constant value n on each of the sets A_n. We take the average value of ϕ to be $\sum_{n=1}^{\infty} n\mu(A_n)$, as the length of $[0, 1)$ is 1 and n is the value of ϕ on the set whose "size" is $\mu(A_n)$. Note from (3.36) that $\mu(A_n) = 2^{-n}$. The sequence $(n\mu(A_n))$ is summable and its sum is

$$\sum_{n=1}^{\infty} n\mu(A_n) = \sum_{n=1}^{\infty} n 2^{-n} = 2. \tag{3.37}$$

How do we interpret this sum? In terms of coin-tossing, it means that if we toss repeatedly an unbiased coin, the average number of tosses required before we obtain the first "head" is 2. We could also interpret this result as telling us the average number of tosses between outcomes of obtaining "heads", a value which is called an average "recurrence time". Note that (3.37) shows that the average value of the time to get "heads" is finite—this does not seem to be evident in advance.

Apart from the question of the *average* number of tosses required to get "heads", we might ask a question such as: what is the most *common* number required to get "heads"? That is, what is the most probable value of ϕ? In the above formulation, this corresponds to calculating n_0, if possible, where

$$\mu(A_{n_0}) = \max\{\mu(A_n) : n \in \mathbb{N}\}.$$

3.7. Typewriters, recurrence, and the Prince of Denmark 135

As $\mu(A_n) = 2^{-n}$ by (3.36), we see that the most probable value of ϕ is $n_0 = 1$. We might also ask: is there a *median* value for ϕ? That is, is there $m \in \mathbb{R}$ such that the probability that ϕ is less than m equals the probability that ϕ is greater than m? Such a value m, if there is one, satisfies

$$\mu(\{x : \phi(x) < m\}) = \mu(\{x : \phi(x) > m\}). \qquad (3.38)$$

In this case, we have $m > 1$, for otherwise the left-hand side of (3.38) is 0 and the right-hand side is at least $1/2$. So, assume that $m > 1$ and that (3.38) holds. Let m_0 be the maximum integer in $\mathbb{N}$ with $m_0 < m$. Note that $1 \leq m_0 < m \leq m_0 + 1$. Using (3.36), we see that

$$\begin{aligned}
\frac{2^{m_0} - 1}{2^{m_0}} &= \frac{1}{2} + \frac{1}{2^2} + \cdots + \frac{1}{2^{m_0}} \\
&= \mu\left(\cup_{n=1}^{m_0} A_n\right) \\
&= \mu(\{x : \phi(x) \leq m_0\}) \\
&= \mu(\{x : \phi(x) < m\}) \\
&= \mu(\{x : \phi(x) > m\}) \\
&= \begin{cases} \mu\left(\cup_{n=m_0+1}^{\infty} A_n\right), \text{if } m_0 < m < m_0 + 1; \\ \mu\left(\cup_{n=m_0+2}^{\infty} A_n\right), \text{if } m_0 + 1 = m. \end{cases} \\
&= \begin{cases} \mu\left(\left(0, \frac{1}{2^{m_0}}\right)\right), \text{if } m_0 < m < m_0 + 1; \\ \mu\left(\left(0, \frac{1}{2^{m_0+1}}\right)\right), \text{if } m_0 + 1 = m. \end{cases} \\
&= \begin{cases} \frac{1}{2^{m_0}}, \text{if } m < m_0 + 1; \\ \frac{1}{2^{m_0+1}}, \text{if } m_0 + 1 = m. \end{cases}
\end{aligned}$$

Thus,

$$2^{m_0} - 1 = \begin{cases} 1, \text{if } m < m_0 + 1; \\ \frac{1}{2}, \text{if } m_0 + 1 = m, \end{cases}$$

and we see that $m_0 = 1$ and that (3.38) holds if and only if $1 < m < 2$. In these circumstances, *the median* of ϕ is usually taken to be the average of the extreme values 1 and 2. That is, the median of ϕ is $3/2$, which is different from the average value of ϕ. Generally the median and the average will be different.

The above arguments become more difficult once we consider the recurrence times for finite sequences of heads and tails, rather than just the recurrence time for a single head or tail. Questions such as these will be discussed in Chapter 4.

3.8 The Rademacher functions

The digits of a number x in $[0, 1)$ relative to a base b were denoted in Section 3.2 by $d_1(x), d_2(x), \ldots$. In this Section we only consider the base 2. Thus, $d_n(x) \in \{0, 1\}$ for all $n \in \mathbb{N}$. If n is given and $x \in [0, 1)$, then $x \in [(k-1)/2^n, k/2^n)$ for a unique $k \in \{1, 2, \ldots, 2^n\}$. Then, it follows from the definitions in Section 3.2 (see also Exercise 2) that

$$d_n(x) = \begin{cases} 0, \text{ if } x \in \left[\dfrac{k-1}{2^n}, \dfrac{k}{2^n}\right) \text{ and } k \text{ is odd;} \\ 1, \text{ if } x \in \left[\dfrac{k-1}{2^n}, \dfrac{k}{2^n}\right) \text{ and } k \text{ is even.} \end{cases} \quad (3.39)$$

The *Rademacher function*[3] r_0 is given by $r_0 = 1$. If $n \in \mathbb{N}$, the Rademacher function r_n is defined to be $r_n = 2d_n - 1$. If $x \in [0, 1)$, $x \in [(k-1)/2^n, k/2^n)$ for a unique $k \in \{1, 2, \ldots, 2^n\}$. Then,

$$r_n(x) = \begin{cases} -1, \text{ if } x \in \left[\dfrac{k-1}{2^n}, \dfrac{k}{2^n}\right) \text{ and } k \text{ is odd;} \\ 1, \text{ if } x \in \left[\dfrac{k-1}{2^n}, \dfrac{k}{2^n}\right) \text{ and } k \text{ is even.} \end{cases}$$

Thus, the Rademacher function r_n alternately takes on values of -1 and 1 along a sequence of adjacent intervals of equal length 2^{-n}, so that each Rademacher function is a step function. Also, as we go from n to $n + 1$, each interval upon which r_n is constant is divided into two equal subintervals, and r_{n+1} is 1 on one of these intervals and -1 on the other. In general, if $j \leq n$, r_j is constant upon each interval of the form

3.8. The Rademacher functions

$[(k-1)/2^n, k/2^n]$ for $k \in \{1, 2, \ldots, 2^n\}$. This observation is refined in the following result.

Lemma 3.19. *Let* $n \in \mathbb{N}$, *let* $k \in \{1, 2, \ldots, 2^n - 1\}$, *and put*

$$J = \left[\frac{k-1}{2^n}, \frac{k+1}{2^n}\right).$$

Then the following hold.
 (1) $\int_J r_n \, d\mu = 0$.
 (2) *If furthermore k is odd, then for all* $j \in \{1, 2, \ldots, n-1\}$, r_j *is constant on* J.

Proof. (1) J is the disjoint union of the adjacent intervals $[(k-1)/2^n, k/2^n)$ and $[k/2^n, (k+1)/2^n)$. Then r_n is 1 on one of these intervals and -1 on the other, and the intervals have equal lengths. Thus, $\int_J r_n d\mu = 0$. (This proof can be readily formalized.)

(2) Let $j \in \{1, 2, \ldots, n-1\}$. As $[0, 1)$ is the disjoint union of the intervals $[(\ell-1)/2^j, \ell/2^j)$ for $\ell \in \{1, 2, \ldots, 2^j\}$, for some such ℓ,

$$\frac{\ell-1}{2^j} \leq \frac{k-1}{2^n} < \frac{\ell}{2^j}.$$

Then,
$$(\ell-1)2^{n-j} \leq k-1 < \ell 2^{n-j}. \quad (3.40)$$

But $n - j \geq 1$, so that 2^{n-j} is even and $\ell 2^{n-j}$ is even. Also, as k is odd, $k - 1$ is even. So (3.40) may be strengthened to obtain

$$(\ell-1)2^{n-j} \leq k-1 < k+1 \leq \ell 2^{n-j}.$$

Thus,
$$\frac{\ell-1}{2^j} \leq \frac{k-1}{2^n} < \frac{k+1}{2^n} \leq \frac{\ell}{2^j},$$

So, as r_j is constant on $[(\ell-1)/2^j, \ell/2^j)$, r_j must also be constant on $J = [(k-1)/2^n, (k+1)/2^n)$. □

The preceding properties lead to the following property of the Rademacher functions, which is an important one for proving that except for the points in some set of measure zero, every point of $[0, 1)$ has in a limiting sense the same number of 0s and 1s in its binary sequence.

Proposition 3.20. *Let* $n_1, n_2, \ldots, n_k \in \mathbb{N}$ *with* $n_1 < n_2 < \cdots < n_k$. *Then,*

$$\int_{[0,1)} r_{n_1} r_{n_2} \cdots r_{n_k} \, d\mu = 0.$$

Proof. Let, for $j = 1, 2, \ldots, 2^{n_k-1}$,

$$J_j = \left[\frac{2j-2}{2^{n_k}}, \frac{2j}{2^{n_k}} \right).$$

Then, as $2j - 1$ is odd, by (2) of Lemma 3.19, $r_{n_1} r_{n_2} \cdots r_{n_{k-1}}$ is a constant on each interval J_j. Thus,

$$\int_{J_j} r_{n_1} r_{n_2} \cdots r_{n_{k-1}} r_{n_k} \, d\mu = \text{(constant)} \cdot \int_{J_j} r_{n_k} \, d\mu = 0, \quad (3.41)$$

where we have used (1) of Lemma 3.19. However, as $[0, 1)$ is the disjoint union of the intervals $J_1, J_2, \ldots, J_{2^{n_k-1}}$, (3.41) gives

$$\int_{[0,1)} r_{n_1} r_{n_2} \cdots r_{n_k} \, d\mu = \sum_{j=1}^{2^{n_k-1}} \int_{J_j} r_{n_1} r_{n_2} \cdots r_{n_{k-1}} r_{n_k} \, d\mu = 0.$$

$\square$

The property that the integral of every product of distinct Rademacher functions is zero is very strong. For later use, the following result is stated so that it applies not just for the Rademacher functions, but also for step functions whose distinct products have a zero integral.

Proposition 3.21. *Let* (g_n) *be a sequence of step functions on* $[0, 1)$ *such that*

(1) $|g_n| = 1$ *for all* $n \in \mathbb{N}$.
(2) For all $n_1, n_2, \ldots, n_k \in \mathbb{N}$ *with* $n_1 < n_2 < \cdots < n_k$,

$$\int_{[0,1)} g_{n_1} g_{n_2} \cdots g_{n_k} \, d\mu = 0.$$

3.8. The Rademacher functions

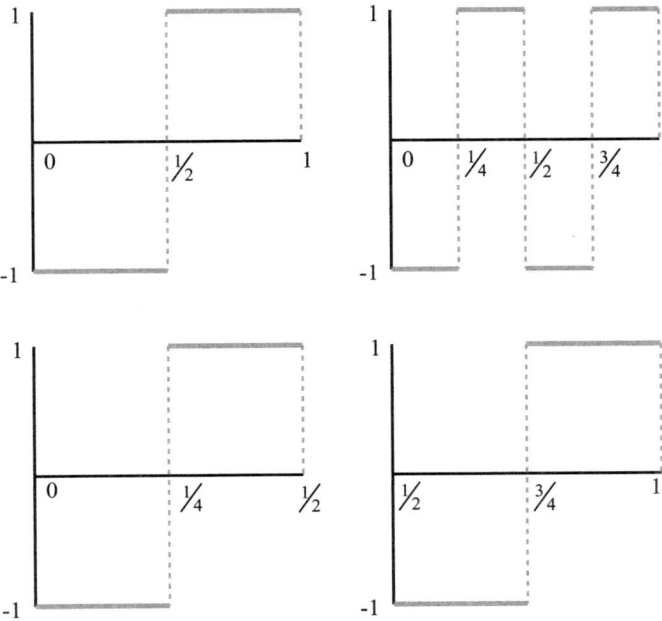

Figure 3.6. The graph on the upper left is of the Rademacher function r_1, while on the upper right is the graph of r_2. The graph of r_2 can be "split" into a "left-hand part" and a "right-hand part", as shown in the lower pictures. Note that the left-hand part of the graph of r_2, from 0 to $1/2$, looks just like the graph of r_1 on the whole of $[0, 1)$, except that the scale on the horizontal axis has been changed, with $[0, 1)$ being replaced by $[0, 1/2)$. A similar comment applies to the right-hand part of the graph of r_2, where on the horizontal axis $[0, 1)$ has been replaced by $[1/2, 1)$. We could say that the graph of r_2 is obtained from the graph of r_1 by *placing two copies of r_1 adjacent to each other*, but changing the horizontal scaling so that the new graph has domain $[0, 1)$. In other words, r_1 "replicates itself twice at a different scale" to form r_2. This is precisely expressed by the identities $r_2(x) = r_1(2x)$ for all $0 \le x < 1/2$, and $r_2(x) = r_1(2x - 1)$ for all $1/2 \le x < 1$. Similarly, the Rademacher function r_n is obtained by r_{n-1} replicating itself, and also by r_1 replicating itself $n - 1$ times. In a different context, the idea of self replication leads to the mathematical theory of fractals. The upper pictures also illustrate (2) of Lemma 3.19 in that r_1 is constant on the interval $[0, 1/2)$ of length $1/2$ (corresponding to $n = 2$ and k is odd), but r_1 is not constant on $[1/4, 3/4)$ that also has length $1/2$ (corresponding to $n = 2$ and k is even). See also the comments in Section 1.2 of Chapter 1 on self replication and composition of functions.

Then, for all $n \in \mathbb{N}$,

$$\int_{[0,1)} \left(\frac{g_1 + g_2 + \cdots + g_n}{n}\right)^2 d\mu = n,$$

and

$$\int_{[0,1)} \left(\frac{g_1 + g_2 + \cdots + g_n}{n}\right)^4 d\mu = \frac{3n-2}{n^3}.$$

Proof. Note that as $|g_n| = 1$, we have $g_n^2 = 1$, a fact which is used repeatedly in the following calculations. Since $\int_{[0,1)} g_j g_k d\mu = 0$ when $j \neq k$, and since

$$\left(\frac{g_1 + g_2 + \cdots + g_n}{n}\right)^2 = \frac{1}{n^2} \sum_{j=1}^{n} \sum_{k=1}^{n} g_j g_k$$

$$= \frac{1}{n^2} \sum_{j=1}^{n} g_j^2 + \frac{1}{n^2} \sum_{\substack{j,k=1 \\ j \neq k}}^{n} g_j g_k$$

$$= \frac{1}{n} + \frac{1}{n^2} \sum_{\substack{j,k=1 \\ j \neq k}}^{n} g_j g_k,$$

it follows that

$$\int_0^1 \left(\frac{g_1 + g_2 + \cdots + g_n}{n}\right)^2 d\mu = \frac{1}{n}.$$

Now when it comes to the second statement in the proposition, observe that

$$\left(\frac{g_1 + g_2 + \cdots + g_n}{n}\right)^4 = \frac{1}{n^4} \left(\sum_{j,k,\ell,m=1}^{n} g_j g_k g_\ell g_m\right). \quad (3.42)$$

In the summation on the right of (3.42), there are n^4 terms. These terms may be arranged into five groups, according as to how many of the indices i, j, k, ℓ are equal and how the equalities between these indices are distributed.

3.8. The Rademacher functions

(Group I.) This consists of those terms $g_j g_k g_\ell g_m$ such that all four of the indices i, j, k, ℓ are distinct. There are

$$n(n-1)(n-2)(n-3) = n^4 - 6n^3 + 11n^2 - 6n$$

such terms, and for each such term $\int_{[0,1)} g_j g_k g_\ell g_m = 0$.

(Group II.) This consists of those terms $g_j g_k g_\ell g_m$ such that two of the indices i, j, k, ℓ are equal and the other two are distinct from each other and from the two equal ones. There are

$$6n(n-1)(n-2) = 6n^3 - 18n^2 + 12n$$

such terms. For each such term there are indices s, t, u such that

$$g_j g_k g_\ell g_m = g_s^2 g_t g_u = g_t g_u,$$

where $t \neq u$. For each such term, $\int_{[0,1)} g_j g_k g_\ell g_m = 0$, which also occurred for Group I terms.

(Group III.) This consists of those terms $g_j g_k g_\ell g_m$ such that three of the indices i, j, k, ℓ are equal and the other index is distinct from from the three equal ones. There are $4n(n-1) = 4n^2 - 4n$ such terms. For every such term there are indices s, t such that $g_j g_k g_\ell g_m = g_s^3 g_t = g_s g_t$, where $s \neq t$. For every such term $\int_{[0,1)} g_j g_k g_\ell g_m = 0$, which also occurred for Group I and Group II terms.

(Group IV.) This consists of those terms $g_j g_k g_\ell g_m$ such that two of the indices i, j, k, ℓ are equal and the other indices are also equal but are distinct from the other two equal ones. There are $3n(n-1) = 3n^2 - 3n$ such terms. For each such term there are indices s, t such that $g_j g_k g_\ell g_m = g_s^2 g_t^2 = 1$. Thus, for every such term, a different type of behavior from the preceding groups occurs and $\int_{[0,1)} g_j g_k g_\ell g_m d\mu = \int_{[0,1)} 1 d\mu = 1$.

(Group V.) This consists of those terms $g_j g_k g_\ell g_m$ such that all four of the indices i, j, k, ℓ are equal. There are n such terms. For each such term there is an index s such that $g_j g_k g_\ell g_m = g_s^4 = 1$. Thus, for every such term, the same behavior occurs as for the Group IV case, and $\int_{[0,1)} g_j g_k g_\ell g_m d\mu = \int_{[0,1)} 1 = 1$.

Now it is immediate from (3.42) that

$$\int_{[0,1)} \left(\frac{g_1 + g_2 + \cdots + g_n}{n}\right)^4 d\mu = \frac{1}{n^4} \sum_{j,k,\ell,m=1}^{n} \left(\int_{[0,1)} g_j g_k g_\ell g_m d\mu\right).$$

In the sum on the right-hand side, the above discussion shows that the only non-zero contribution comes from the integrations of Group IV and Group V terms, and each such integration equals 1. Since there are $3n^2 - 3n$ Group IV terms and n Group V terms, the sum of all such integrations is $3n^2 - 3n + n = 3n^2 - 2n$. Thus,

$$\int_{[0,1)} \left(\frac{g_1 + g_2 + \cdots + g_n}{n}\right)^4 d\mu = \frac{3n^2 - 2n}{n^4} = \frac{3n - 2}{n^3}.$$

□

Incidentally, it may as a check be worth noting that the number of terms in the totality of Groups I, II, III, IV and V is

$$(n^4 - 6n^3 + 11n^2 - 6n) + (6n^3 - 18n^2 + 12n) + (4n^2 - 4n) \\ + (3n^2 - 3n) + n = n^4,$$

which is, as expected, the total number n^4 of terms in the expansion as given in (3.42).

3.9 Randomness, binary expansions and a law of averages

Again, let us turn to coin tossing, considered as producing random sequences of outcomes. If the coin is unbiased, as discussed in the preceding section, it was argued that, given any number, as the coin tossing proceeded we would expect to get at least that number of consecutive heads. Furthermore, if there is a particular finite sequence of "heads" and "tails" that is given in advance, the repeated coin tossing, when carried out for a sufficiently long time, should eventually produce the given finite sequence.

3.9. Randomness, binary expansions and a law of averages

However, this production of order from randomness is not the only type of behavior a genuinely random process would be expected to produce. If the coin is unbiased, we should expect the number of "heads" and "tails" which accumulate as the tossing proceeds to become approximately equal. That is, if the coin is unbiased, the tossing procedure should not favor "heads" over "tails" *in the long run*. While it may be possible for the process to apparently favor "heads" in the short term, in the sense that at some time a sequence of 10 consecutive heads may be observed; nevertheless, as tossing continues, we should expect that the proportion of heads which has been observed since tossing began will be very close to the proportion of tails which has been observed since tossing began. In this section, we put these ideas in a precise mathematical framework, in a manner corresponding to the earlier results on recurrence in Section 3.7, by formulating them in terms of binary expansions of numbers.

Let the function $d_n : [0, 1) \longrightarrow \{0, 1\}$ be the one that assigns the nth binary digit to each $x \in [0, 1)$. Also, for $n \in \mathbb{N}$, let $S_n : [0, 1) \longrightarrow \mathbb{N}$ be given by

$$S_n(x) = d_1(x) + d_2(x) + \cdots + d_n(x).$$

Thus, $S_n(x)$ is the *total* number of 1s in the first n binary digits of x, and $S_n(x)/n$ is the *proportion* of 1s in the first n binary digits of x.

Now, each $x \in [0, 1)$ may be thought of as corresponding to the outcome to a specific experiment of coin-tossing, as discussed in Section 3.2, and $d_n(x)$ may be thought of as giving "0" if there is a "head" on the nth toss; while $d_n(x)$ gives "1" if there is a "tail" on the nth toss. So, $S_n(x)$ can also be regarded as giving the number of "tails" that have been observed in the first n tosses. Then the proportion of "tails" that has been observed in the first n tosses is given by the number $S_n(x)/n$, and the proportion of "heads" that has been observed in the first n tosses is given by the number $1 - S_n(x)/n$. Note that these proportions will differ for different values of x, since different values of x are thought of as corresponding to different outcomes of a coin-tossing experiment. Since $\lim_{n \to \infty} S_n(x)/n = 1/2$ is equivalent to having $\lim_{n \to \infty} (1 - S_n(x)/n) = 1/2$, we see that the equation $\lim_{n \to \infty} S_n(x)/n = 1/2$ can be regarded as saying that for the coin-tossing experiment corresponding to the number x, with repeated tossing, the proportion of "heads" obtained gets closer and closer to the ratio $1/2$ and the proportion of "tails" also gets closer to the same ratio, $1/2$. This is, of course, precisely what one would expect when repeatedly tossing at random an unbiased coin. Thus, any statement that asserts that for some subset of $[0, 1)$, every number x in the set has the property that $\lim_{n \to \infty} S_n(x)/2 = 1/2$ can be regarded as expressing a "law of averages". The following result was stated in 1909 in [7] by the French Mathematician Émile Borel, and says that such a "law of averages" holds in a rather strong sense; namely that $\lim_{n \to \infty} S_n(x)/2 = 1/2$ for all numbers x in $[0, 1)$ except for those numbers in some set of measure zero.

Theorem 3.22. *(Borel's Theorem) There is a subset $\mathcal{Z}$ of $[0, 1)$ which is a set of measure zero and which has the following property:*

$$\lim_{n \to \infty} \frac{S_n(x)}{n} = \frac{1}{2}, \text{ for all } x \notin \mathcal{Z}.$$

The numbers x that have the property that $\lim_{n \to \infty} S_n(x)/n = 1/2$ are called simply normal numbers to the base 2.

Ideas in the Proof. The proof of Borel's Theorem[4] given here depends upon the results derived for the Rademacher functions in the previous section. Borel's Theorem is formulated as an equivalent statement in terms of the Rademacher functions. This approach is attributed by Riesz [38, p. 221], to Khinchine, and in it everything ultimately hinges upon the fact, which we deduce from Proposition 3.21, that

$$\sum_{n=1}^{\infty} n^{1/2} \left(\int_{[0,1)} \left(\frac{r_1 + r_2 + \cdots + r_n}{n} \right)^4 d\mu \right) < \infty.$$

(Actually, here the term $n^{1/2}$ can be replaced by n^{α} for any $0 < \alpha < 1$, but $n^{1/2}$ suffices.)

3.9. Randomness, binary expansions and a law of averages

Proof. (Borel's Theorem) As before, for each $n \in \mathbb{N}$, let r_n denote the corresponding Rademacher function. That is,

$$r_n(x) = 2d_n(x) - 1 = \begin{cases} 1, \text{ if } d_n(x) = 1; \\ -1, \text{ if } d_n(x) = 0. \end{cases}$$

Then,

$$\frac{S_n(x)}{n} - \frac{1}{2} = \frac{2S_n(x) - n}{2n} = \frac{1}{2n}\sum_{j=1}^{n}(2d_j(x) - 1) = \frac{1}{2n}\left(\sum_{j=1}^{n} r_j(x)\right),$$

and it follows that

$$\left\{x : x \in [0, 1) \text{ and } \lim_{n \to \infty} \frac{S_n(x)}{n} = \frac{1}{2}\right\}$$

$$= \left\{x : x \in [0, 1) \text{ and } \lim_{n \to \infty} \frac{1}{n}\left(\sum_{j=1}^{n} r_j(x)\right) = 0\right\}. \quad (3.43)$$

Now, let $\mathcal{Z}$ denote the set of all points x in $[0, 1)$ such that the sequence $\left(\sum_{j=1}^{n} r_j(x)/n\right)$ does *not* converge to 0. Let x be a given element of $\mathcal{Z}$. Then the sequence $\left(\sum_{j=1}^{n} r_j(x)/n\right)$ will not satisfy the definition of a convergent sequence with limit 0. Thus, there is $\varepsilon > 0$ such that $|\sum_{j=1}^{n} r_j(x)|/n > \varepsilon$ for an infinite number of $n \in \mathbb{N}$. Since $1/n^{1/8} < \varepsilon$ for all sufficiently large n, it follows that $|\sum_{j=1}^{n} r_j(x)|/n > 1/n^{1/8}$ for an infinite number of n. Now define, for $n = 1, 2, 3, \ldots$,

$$\mathcal{A}_n = \left\{ x : x \in [0,1) \text{ and } \frac{1}{n}\left|\sum_{j=1}^n r_j(x)\right| > \frac{1}{n^{1/8}} \right\}. \quad (3.44)$$

The remarks above show that if $x \in \mathcal{Z}$ and $n \in \mathbb{N}$, there is $k \in \mathbb{N}$ with $k > n$ such that $x \in \mathcal{A}_k$. So, when $x \in \mathcal{Z}$, for every n, $x \in \bigcup_{k=n}^\infty \mathcal{A}_k$. Thus,

$$\mathcal{Z} \subseteq \bigcup_{k=n}^\infty \mathcal{A}_k, \text{ for all } n \in \mathbb{N}. \quad (3.45)$$

Note that because each function r_n is a step function, the function $|\sum_{j=1}^n r_j(x)/n|$ is also a step function. Consequently, the set $\mathcal{A}_n$ is a finite union of intervals, so it is a basic set with length $\mu(\mathcal{A}_n)$. The aim now is to estimate $\mu(\mathcal{A}_n)$. To this end observe that (3.44) gives

$$\frac{1}{n^{1/2}} \leq \left(\frac{1}{n}\sum_{j=1}^n r_j(x)\right)^4, \text{ for all } x \in \mathcal{A}_n. \quad (3.46)$$

Now,

$$\mu(\mathcal{A}_n) = \int_{\mathcal{A}_n} 1 \, d\mu$$

$$= n^{1/2} \int_{\mathcal{A}_n} \frac{1}{n^{1/2}} \, d\mu$$

$$\leq n^{1/2} \int_{\mathcal{A}_n} \left(\frac{\sum_{j=1}^n r_j}{n}\right)^4 d\mu, \text{ by (3.46)},$$

$$\leq n^{1/2} \int_{[0,1)} \left(\frac{\sum_{j=1}^n r_j}{n}\right)^4 d\mu$$

$$= n^{1/2} \left(\frac{3n-2}{n^3}\right), \text{ by Propositions 3.20 and 3.21},$$

$$\leq \frac{3}{n^{3/2}}.$$

3.9. Randomness, binary expansions and a law of averages

Thus,

$$\sum_{n=1}^{\infty} \mu(\mathcal{A}_n) \leq 3 \sum_{n=1}^{\infty} \frac{1}{n^{3/2}} < \infty,$$

where we have used the fact that sequence $\{n^{-3/2}\}$ is summable (see the discussion in Ross [40, p. 60], for example). Hence,

$$\lim_{n \to \infty} \sum_{k=n}^{\infty} \mu(\mathcal{A}_k) = 0. \tag{3.47}$$

Now, let $\varepsilon > 0$. By (3.47) there is $n \in \mathbb{N}$ such that $\sum_{k=n}^{\infty} \mu(\mathcal{A}_k) < \varepsilon$. Using this n and (3.45) we now have

$$\mathcal{Z} \subseteq \bigcup_{k=n}^{\infty} \mathcal{A}_k \text{ and } \sum_{k=n}^{\infty} \mu(\mathcal{A}_k) < \varepsilon,$$

where each $\mathcal{A}_k$ is a finite, disjoint union of subintervals of $[0, 1)$. This conclusion can be obtained for all sufficiently small $\varepsilon > 0$, and we deduce that $\mathcal{Z}$ is a set of measure zero, by definition. But $\mathcal{Z}$ is by definition the set of all points x in $[0, 1)$ such that the sequence $((\sum_{j=1}^{n} r_j(x))/n)$ does not converge to 0. Using (3.43) we conclude that $\mathcal{Z}$ is a set of measure zero and

$$\lim_{n \to \infty} \frac{S_n(x)}{n} = \frac{1}{2}, \text{ for all } x \notin \mathcal{Z},$$

and Borel's Theorem has been proved. □

It is of interest to note the apparent contrast between Borel's Theorem and the result of the Recurrence Theorem 3.18 that there is a set of measure zero such that, for every given finite sequence of binary symbols, this finite sequence will appear in the binary sequence of every number that is outside of this set of measure zero—in fact, the finite sequence will appear infinitely often outside of such a set. However, this means at the least that such a binary sequence will include 100 consecutive zeros, for example. By contrast, Borel's Theorem says that in the long run, the *total number* of zeros and ones will "balance out". These two phenomena, which may seem initially to be rather contradictory, are reconciled by realizing that although most binary sequences will include finite sequences such as 100 consecutive zeros (infinitely many times, in fact), this must be balanced against what happens in between the appearance of such uncharacteristic finite sequences—for example, we shall see in Chapter 4 that as a result of Kac's formula for recurrence times, on the average we would have to observe 2^{100} binary digits of x before we would observe 100 consecutive zeros starting to occur! These comments illustrate distinct aspects of what we would expect in behavior that is really random. On the one hand, if the coin-tossing process is random, it cannot be expected to "remember" to avoid striking patterns from time to time; but on the other hand, if the coin is unbiased, it cannot be expected that in the long run certain outcomes will be favored over others.

3.10 The dynamical systems approach

Dynamical systems provide an interpretation of some earlier results; namely, the Recurrence Theorem 3.18, and Borel's Theorem 3.22. This interpretation arises from the fact that the binary digits of a number are obtained by carrying out the same calculation over and over again, at successively smaller scales. This indicates a situation analogous to taking successive compositions of a transformation in a dynamical system. Specifically, if $d_1(x), d_2(x), \ldots$ denotes the sequence of binary digits of a number x in $[0, 1)$, and if f denotes the transformation $x \longmapsto \text{frac}(2x)$ on $[0, 1)$, then

$$d_n(x) = (d_1 \circ f^{n-1})(x), \text{ for all } x \in [0, 1).$$

In view of the fact that $d_1 = \chi_{[1/2,1)}$, this means that Borel's Theorem may be formulated as the statement that

$$\lim_{n \to \infty} \frac{1}{n} \sum_{j=1}^{n} \chi_{[1/2,1)}\left(f^{j-1}(x)\right) = \lim_{n \to \infty} \frac{1}{n} \sum_{j=1}^{n} \left(d_1 \circ f^{j-1}(x)\right)$$

$$= \lim_{n \to \infty} \frac{1}{n} \sum_{j=1}^{n} d_j(x) = \frac{1}{2}, \quad (3.48)$$

3.10. The dynamical systems approach

for all $x \in [0, 1)$ except for those x in a certain set of measure zero. The idea of interpreting Borel's Theorem in this way seems to derive from the work of D. D. Wall in 1949 [45]. The expression on the left-hand side of (3.48) may be interpreted as the asymptotic proportion of time which the points in the orbit $x, f(x), f^2(x), \ldots$ of x spend in the set $[1/2, 1)$. That is, it represents an *average* of time spent in a certain part of the system. The transformation f is also described by

$$f(x) = \begin{cases} 2x, & \text{if } 0 \leq x < 1/2; \\ 2x - 1, & \text{if } 1/2 \leq x < 1. \end{cases} \quad (3.49)$$

Note that this transformation is linear and one-to-one on each of the intervals $[0, 1/2)$ and $[1/2, 1)$. Its range on each of $[0, 1/2)$ and $[1/2, 1)$ is the whole interval $[0, 1)$. Also, recall that

$$d_n(x) = \begin{cases} 0, & \text{if } x \in \left[\dfrac{k-1}{2^n}, \dfrac{k}{2^n}\right) \text{ and } k \text{ is odd}; \\ 1, & \text{if } x \in \left[\dfrac{k-1}{2^n}, \dfrac{k}{2^n}\right) \text{ and } k \text{ is even.} \end{cases}$$

We now calculate $d_n \circ f$. When restricted to $[0, 1/2)$, f is linear and its range is $[0, 1)$. So we would expect that in considering $d_n \circ f$, the behavior of d_n over the whole interval $[0, 1)$ will be "copied" by the behavior of $d_n \circ f$ on $[0, 1/2)$. That is, on $[0, 1/2)$, $d_n \circ f$ should look like d_n does on $[0, 1)$, but at half the scale. In fact, the restriction of $d_n \circ f$ to $[0, 1/2)$ is a formal copy of d_n on $[0, 1)$, in the sense of Section 1.2 in Chapter 1. Similarly, on $[1/2, 1)$, $d_n \circ f$ should look like d_n on $[0, 1)$, but at half the scale and, in fact, $d_n \circ f$ restricted to $[1/2, 1)$ is a formal copy of d_n. So, on the whole of $[0, 1)$, $d_n \circ f$ should consist of two "copies" of d_n, one copy on each of $[0, 1/2)$ and $[1/2, 1)$. However, the function d_{n+1} has precisely the property that its graph on $[0, 1)$ consists of two "copies" of d_n, one on each of the intervals $[0, 1/2)$ and $[1/2, 1)$. So, we expect that $d_{n+1} = d_n \circ f$. The following result and its proof give a precise expression to these ideas.

Proposition 3.23. *Let $f : [0, 1) \longrightarrow [0, 1)$ be given as in (3.49), and let d_n be the function that assigns the nth binary digit to numbers in $[0, 1)$.*

150   Chapter 3. Probability and Randomness

Then for all $k \in \mathbb{Z}_+$ and $n \in \mathbb{N}$,

$$d_n \circ f^k = d_{n+k}.$$

In particular, $d_n = d_1 \circ f^{n-1}$.

Proof. Let $n \in \mathbb{N}$ be given. Observe that each x in $[0, 1)$ belongs to exactly one of the intervals $[(k-1)/2^{n+1}, k/2^{n+1})$, $k = 1, 2, \ldots, 2^{n+1}$. Now we have from (3.39) that

$$x \in \left[\frac{k-1}{2^{n+1}}, \frac{k}{2^{n+1}}\right) \implies d_{n+1}(x) = \begin{cases} 0, & \text{if } k \text{ is odd}; \\ 1, & \text{if } k \text{ is even}. \end{cases} \quad (3.50)$$

(CASE I.) $x \in [0, 1/2)$. Then, $x \in [(k-1)/2^{n+1}, k/2^{n+1})$ for some $k \in \{1, 2, \ldots, 2^n\}$. As, $x \in [0, 1/2)$, we have $f(x) = 2x$. These observations give

$$x \in \left[\frac{k-1}{2^{n+1}}, \frac{k}{2^{n+1}}\right) \implies 2x \in \left[\frac{k-1}{2^n}, \frac{k}{2^n}\right)$$
$$\implies f(x) \in \left[\frac{k-1}{2^n}, \frac{k}{2^n}\right)$$
$$\implies d_n(f(x)) = \begin{cases} 0, \text{if } k \text{ is odd}; \\ 1, \text{if } k \text{ is even}. \end{cases} \quad (3.51)$$

It is clear from (3.50) and (3.51) that

$$d_{n+1}(x) = (d_n \circ f)(x) \text{ for all } x \in [0, 1/2). \quad (3.52)$$

(CASE II.) $x \in [1/2, 1)$. Then, $x \in [(k-1)/2^{n+1}, k/2^{n+1})$ for some $k \in \{2^n + 1, 2^n + 2, \ldots, 2^{n+1}\}$. As $x \in [1/2, 1)$, we have $f(x) = 2x - 1$.

3.10. The dynamical systems approach

These observations give

$$x \in \left[\frac{k-1}{2^{n+1}}, \frac{k}{2^{n+1}}\right) \implies 2x - 1 \in \left[\frac{2k-2}{2^{n+1}} - 1, \frac{2k}{2^{n+1}} - 1\right)$$

$$\implies f(x) \in \left[\frac{2k - 2 - 2^{n+1}}{2^{n+1}}, \frac{2k - 2^{n+1}}{2^{n+1}}\right)$$

$$\implies f(x) \in \left[\frac{k - 2^n - 1}{2^n}, \frac{k - 2^n}{2^n}\right)$$

$$\implies d_n(f(x)) = \begin{cases} 0, \text{ if } k \text{ is odd;} \\ 1, \text{ if } k \text{ is even.} \end{cases} \tag{3.53}$$

It is clear from (3.50), (3.52) and (3.53) that

$$d_{n+1} = d_n \circ f. \tag{3.54}$$

Now, if $k, n \in \mathbb{N}$, we can repeatedly use (3.54) to obtain

$$d_{n+k} = d_{n+k-1} \circ f = d_{n+k-2} \circ f^2 = d_{n+k-3} \circ f^3 = \cdots \cdots = d_n \circ f^k.$$

□

The identity $d_n = d_1 \circ f^{n-1}$ leads to a corresponding identity for the Rademacher functions. The nth Rademacher function is $r_n = 2d_n - 1$, so that

$$r_n = r_1 \circ f^{n-1} \text{ and, more generally, } r_{n+k} = r_n \circ f^k. \tag{3.55}$$

This observation turns out to be pertinent when we consider a more general question than the one answered by Borel's Theorem. Figure 3.7 illustrates the iterates of the transformation f given by (3.48).

Now, let $d(x) = (d_n(x))$ be the sequence of binary digits of a number $x \in [0, 1)$, and let Σ denote the set of all sequences (x_n) of zeros and ones such that for no n_0 is it the case that $x_n = 1$ for all $n \geq n_0$. Note that Theorem 3.4 and Proposition 3.5 show that $d : [0, 1) \longrightarrow \Sigma$ and that d is one-to-one and maps $[0, 1)$ onto Σ. If $x \in \Sigma$ define $\sigma(x) \in \Sigma$ by

$$\sigma(x)_n = x_{n+1}, \text{ for all } n \in \mathbb{N}. \tag{3.56}$$

That is, if $x = (x_1, x_2, x_3, \ldots)$, $\sigma(x) = (x_2, x_3, x_4, \ldots)$. Thus, σ is called the *left shift* on Σ. Note that (3.54) can be written equivalently as

$$\sigma(d(x)) = d(f(x)), \text{ for all } x \in [0, 1). \tag{3.57}$$

We are going to use these observations to give a dynamical systems interpretation of the Recurrence Theorem 3.18 and Borel's Theorem 3.22.

The Recurrence Theorem says that there is a set $\mathcal{Z}$ of measure zero in $[0, 1)$ such that if $x \notin \mathcal{Z}$, every given finite sequence $c_1, c_2, \ldots, c_s$ of terms in $\{0, 1\}$ occurs infinitely many times in the sequence of binary digits in the expansion of x. But if $c_1, c_2, \ldots, c_s$ occurs in $d(x)$ starting at position $r \in \mathbb{N}$,

$$d(x) = (d_1(x), \ldots, d_{r-1}(x), c_1, c_2, \ldots, c_s, d_{r+s}(x), d_{r+s+1}(x), \ldots),$$

which gives from (3.56) and (3.57) that

$$d(f^{r-1}(x)) = \sigma^{r-1}(d(x)) = (c_1, c_2, \ldots, c_s, d_{r+s}(x), d_{r+s+1}(x), \ldots). \tag{3.58}$$

Thus, if the finite sequence $c_1, c_2, \ldots, c_s$ of terms in $\{0, 1\}$ occurs infinitely many times in $d(x)$, then (3.58) holds for infinitely many $r \in \mathbb{N}$. But by Theorem 3.4, this is equivalent to having

$$f^{r-1}(x) \in \left[\sum_{j=1}^{s} \frac{c_j}{2^j}, \sum_{j=1}^{s} \frac{c_j}{2^j} + \frac{1}{2^s} \right), \text{ for infinitely many } r \in \mathbb{N}. \tag{3.59}$$

Now, it is easy to see that if U is an open subinterval of $[0, 1)$, there are $c_1, c_2, \ldots, c_s \in \{0, 1\}$ such that $\left[\sum_{j=1}^{r} c_j/2^j, \sum_{j=1}^{r} c_j/2^j + 1/2^r \right) \subseteq U$. Consequently we deduce the following result, which is a dynamical systems interpretation of the Recurrence Theorem 3.18.

Theorem 3.24. *Let $f : [0, 1) \longrightarrow [0, 1)$ be the function given by $f(x) = 2x$ for $0 \le x < 1/2$ and $f(x) = 2x - 1$ for $1/2 \le x < 1$. Then, there is a subset $\mathcal{Z}$ of $[0, 1)$ which is of measure zero and has the following property: if U is a non-empty open subinterval of $[0, 1)$ and $x \notin \mathcal{Z}$, $f^r(x) \in U$ for infinitely many $r \in \mathbb{N}$.*

3.10. The dynamical systems approach   153

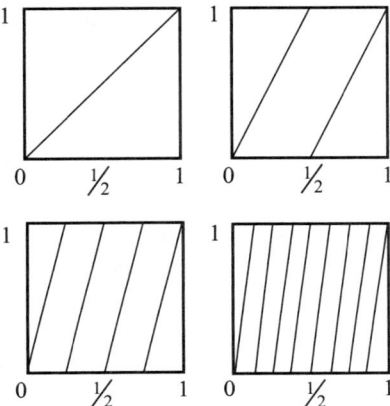

Figure 3.7. The top left graph is of the identity function $x \longmapsto x$, and the top right is of the function f where $f(x) = 2x$, for $0 \le x < 1/2$ and $f(x) = 2x - 1$ for $0 \le x < 1$. We now iterate f to get the functions f^2 and f^3 depicted in the bottom left and right pictures respectively. If $x \in [(k-1)/4, k/4)$ for some $k \in \{1, 2, 3, 4\}$, then $f^2(x) = 4x - k + 1$. If $x \in [(k-1)/8, k/8)$ for some $k \in \{1, 2, \ldots, 8\}$, then $f^3(x) = 8x - k + 1$. Note how f is obtained from the identity function in the same way that the successive Rademacher functions were derived from each other as described in Figure 3.6. So, to get f, the identity function is "duplicated" by first scaling its domain from $[0, 1)$ to $[0, 1/2)$ while keeping its range equal to $[0, 1)$, thus obtaining the *left-hand* part of f. Also, the domain is again scaled, this time from $[0, 1)$ to $[1/2, 1)$ again keeping the range equal to $[0, 1)$, thus obtaining the *right-hand* part of f. So the left-hand and right-hand parts of f are each "duplicates" of the identity function. Similarly, the left and right-hand parts of f^2 are each a duplicate of f, and the left and right-hand parts of f^3 are each a duplicate of f^2. In the same way, two duplicates of f^{n-1} produce the function f^n. Compare these comments with those in Section 1.2 of Chapter 1.

At the beginning of this section, it was noted that the identity $d_n = d_1 \circ f^{n-1}$ leads to a dynamical systems formulation of Borel's Theorem, as in (3.48). Here is a formal statement of Borel's Theorem in a dynamical systems form.

Theorem 3.25. *Let* $f : [0, 1) \longrightarrow [0, 1)$ *be the function given by* $f(x) = 2x$ *for* $0 \le x < 1/2$ *and* $f(x) = 2x - 1$ *for* $1/2 \le x < 1$. *Then, there* is a subset $\mathcal{Z}$ of $[0, 1)$ which is of measure zero and is such that for all $x \notin \mathcal{Z}$,

$$\lim_{n \to \infty} \frac{1}{n} \left| \{k : 1 \leq k \leq n \text{ and } f^k(x) \in [0, 1/2)\} \right| = \frac{1}{2},$$

and the same statement is true with $[1/2, 1)$ in place of $[0, 1/2)$. That is, for all $x \in [0, 1)$ except for those x in some set of measure zero, in the limit as $n \longrightarrow \infty$, half of the points in the orbit $\{x, f(x), f^2(x), \ldots\}$ of x lie in the interval $[0, 1/2)$ and a half lie in the other interval $[1/2, 1)$.

Theorem 3.24 shows that outside some set of measure zero, the orbit points of f generally are in a given non-empty subinterval U infinitely often—but this is a "coarse" type of recurrence result because it says nothing about the frequency of orbit points being in U. On the other hand, Theorem 3.25 does give the asymptotic frequency of points being in $[0, 1/2)$ and $[1/2, 1)$, but only for these particular intervals. What we would like is a result like Theorem 3.25, but one which is applicable to a wider class of intervals than just $[0, 1/2)$ and $[1/2, 1)$. Putting this in another way, whereas Borel's Theorem quantifies the Recurrence Theorem, in that it gives the frequency of occurrence of a *single* binary symbol, we seek a result that similarly quantifies the Recurrence Theorem, but for an arbitrary finite sequence of binary symbols. So, the question we now ask is: *given a finite sequence of binary symbols, and given a typical number x in $[0, 1)$, what is the frequency of occurrence of the given finite sequence of symbols in the sequence of binary digits of x?* So, we now set out to investigate the frequency of occurrence in the case of the Recurrence Theorem 3.18. This investigation lies deeper than earlier results.

Now, in Borel's Theorem, we saw that the asymptotic proportion of 1s in the sequence of binary digits of a typical number is equal to $1/2$. However, $1/2$ is also the probability that on an individual toss of an unbiased coin we will get a "head" (that is, a 1). So in a rather vague form we might consider that Borel's Thorem is saying something like

"the average frequency of an event over time = the probability of the event."

Now if we toss a coin s times, the probability of obtaining a given sequence of heads and tails is $1/2^s$. If we now apply these ideas to the

3.10. The dynamical systems approach

case where a finite number $c_1, c_2, \ldots, c_s$ of symbols in $\{0, 1\}$ is given, we might expect that for a typical number x we would have

$$\lim_{n\to\infty} \frac{1}{n} \left| \left\{ j : 1 \le j \le n \text{ and } \right.\right.$$
$$\left.\left. d_j(x) = c_1, d_{j+1}(x) = c_2, \ldots, d_{j+s-1}(x) = c_s \right\} \right| = \frac{1}{2^s}.$$

Here, Borel's Theorem is the case $s = 1$.

Definition A number $x \in [0, 1)$ is called *normal to the base* 2 if, for every finite sequence $c_1, c_2, \ldots, c_s$ of symbols in $\{0, 1\}$,

$$\lim_{n\to\infty} \frac{1}{n} \left| \left\{ j : 1 \le j \le n \text{ and } \right.\right. \qquad (3.60)$$
$$\left.\left. d_j(x) = c_1, d_{j+1}(x) = c_2, \ldots, d_{j+s-1}(x) = c_s \right\} \right| = \frac{1}{2^s}.$$

If a number x has the property that (3.60) holds whenever $s = 1$, then x is called *simply normal to the base* 2. For the time being, the base 2 is understood and we shall refer simply to normal and simply normal numbers as the case may be. If $b \in \mathbb{N}$ and $b \ge 2$, there are corresponding definitions of normal and simply normal numbers to the base b, and there are results that correspond precisely to Borel's Theorem—see [17], for example, Exercise 11, and Chapter 5.

Every normal number is simply normal, but there are simply normal numbers that are not normal. Borel's Theorem says that every number outside of a certain set of measure zero is simply normal. Our aim is to prove the stronger result that every number outside of a certain set of measure zero is normal. The Rademacher functions are well adapted to answering the question of the distribution of the digits of simply normal numbers, but they are too "sparse" in the space of all functions to deal with the question raised by the stronger concept of normal numbers. We need a class of functions adapted to this problem, and this is the class of *Walsh functions*. These are defined in terms of the Rademacher functions, and this is why (3.55) turns out to be useful in considering the prevalence of normal numbers. The Walsh functions are introduced and studied in the next Section.

Proposition 3.23 establishes a very close connection between the dynamical system $([0, 1), f)$ and the binary expansion of numbers, and this connection was used to derive the dynamical systems versions of the Recurrence Theorem 3.18 and Borel's Theorem, as in Theorems 3.24 and 3.25. In fact, Proposition 3.23 can be looked upon as establishing a connection between the dynamical systems $([0, 1), f)$ and (Σ, σ), a connection that leads to the concept of conjugate systems.

Definition Let (S, f) and (T, g) be two dynamical systems. Then the systems are said to be *conjugate* if there is a one-to-one and onto function $h : S \longrightarrow T$ such that

$$h \circ f = g \circ h \text{ or, equivalently, } f = h^{-1} \circ g \circ h.$$

The definition is symmetric in the systems (S, f) and (T, g). The idea behind this definition is that conjugate systems are "equivalent" to each other, and that results in the one system can be converted into corresponding results in the other system. For example, if (S, f) and (T, g) are conjugate and $h : S \longrightarrow T$ is one-to-one and onto and is such that $f = h^{-1} \circ g \circ h$, then if $x \in S$ and $y = h(x) \in T$, we have for $n = 1, 2, \ldots$,

$$\begin{aligned} f^n(x) &= (h^{-1} \circ g \circ h) \circ \cdots \circ (h^{-1} \circ g \circ h)(x) \\ &= h^{-1}(g \circ h \circ h^{-1} \circ g \cdots h \circ h^{-1} \circ g)(h(x)) \\ &= h^{-1} \circ g^n(y). \end{aligned}$$

That is,

$$h(f^n(x)) = g^n(y), \text{ for all } n = 1, 2, 3, \ldots$$

Thus, if we apply h to the orbit $x, f(x), f^2(x), \ldots$ of x in the system (S, f), we obtain the orbit $y, g(y), g^2(y), \ldots$ of y in the system (T, g).

A way of thinking of the meaning of conjugate systems is in terms of the diagram (3.61) below. The top line indicates the action of f in the system (S, f), while the lower line indicates the action of g in the system (T, g). The one-to-one function h maps S onto T, as indicated in the diagram. Conjugacy means that for some h, $h \circ f = g \circ h$. That is,

3.10. The dynamical systems approach

the diagram *commutes* in the sense that if we start with a point in S, and proceed from the top left hand corner to the bottom right hand corner by following the arrows in either of the two possible ways, that is by applying f then h, or h then g, we get the same point of T.

$$\begin{array}{ccc} S & \xrightarrow{f} & S \\ h \downarrow & & h \downarrow \\ T & \xrightarrow{g} & T \end{array} \qquad (3.61)$$

Theorem 3.26. *Let Σ denote the set of all sequences $x = (x_n)$ of zeros and ones such that x does not have the property that for some n_0, $x_n = 1$ for all $n > n_0$. If $x \in \Sigma$ define $\sigma(x) \in \Sigma$ by $\sigma(x)_n = x_{n+1}$ for all $n \in \mathbb{N}$. Also, let $f : [0, 1) \longrightarrow [0, 1)$ be the function given by $f(x) = 2x$ for $0 \leq x < 1/2$ and $f(x) = 2x - 1$ for $1/2 \leq x < 1$. Then the dynamical systems $([0, 1), f)$ and (Σ, σ) are conjugate. In fact, if we let $d : [0, 1) \longrightarrow \Sigma$ be given by saying that for $x \in [0, 1)$, $d(x)$ is the sequence of binary digits in the expansion of x to the base 2, then, d is one-to-one from $[0, 1)$ onto Σ and $d \circ f = \sigma \circ d$.*

Proof. That d is one-to-one and onto is a consequence of Theorem 3.1 and Proposition 3.5. Also, (3.57) shows that $d \circ f = \sigma \circ d$ and conjugacy follows by definition. □

Theorems 3.24 and 3.25 can be regarded as equivalent formulations respectively of the Recurrence Theorem 3.18 and Borel's Theorem 3.22, that arise from the conjugacy of the systems $([0, 1), f)$ and (Σ, σ). The use of systems such as (Σ, σ) are sometimes referred to as *symbolic dynamics* and are used extensively to interpret and understand general dynamical systems, as well as being of interest in their own right (see [10, Chapter 9], for example).

3.11 The Walsh functions

Walsh functions arise by taking products of Rademacher functions. In this way we construct a much larger family of functions on $[0, 1)$ which nevertheless still has many of the properties which made the Rademacher functions so useful in proving Borel's Theorem. In order to define the Walsh functions, we need a result about writing numbers as sums of powers of 2.

Lemma 3.27. *Let $n \in \mathbb{N}$. Then there are unique $n_1, n_2, \ldots, n_k \in \mathbb{Z}_+$ such that $0 \leq n_1 < n_2 < \cdots < n_k$ and*

$$n = 2^{n_1} + 2^{n_2} + \cdots\cdots + 2^{n_k}. \tag{3.62}$$

Also, n is odd if and only if $n_1 = 0$.

Proof. We proceed by induction to prove that each $n \in \mathbb{N}$ can be expressed in the form (3.62). Clearly, this is true for $n = 1$ as $1 = 2^0$. So, assume that $n \in \mathbb{N}$ and that (3.62) holds for all s with $1 \leq s \leq n$. Let q be the element of $\mathbb{N}$ such that

$$2^q \leq n + 1 < 2^{q+1}.$$

If $n + 1 = 2^q$ then (3.62) holds for $n + 1$. Otherwise, $2^q < n + 1$ and we have

$$1 \leq n + 1 - 2^q < 2^{q+1} - 2^q = 2^q \leq n. \tag{3.63}$$

By the inductive assumption, there are $n_1, n_2, \ldots, n_k \in \mathbb{Z}_+$ with $0 \leq n_1 < n_2 < \cdots < n_k$ such that

$$n + 1 - 2^q = 2^{n_1} + 2^{n_2} + \cdots + 2^{n_k}, \tag{3.64}$$

and we deduce that

$$n + 1 = 2^{n_1} + 2^{n_2} + \cdots + 2^{n_k} + 2^q. \tag{3.65}$$

Now, (3.63) and (3.64) give

$$2^{n_k} \leq n + 1 - 2^q < 2^q,$$

3.11. The Walsh functions

so it follows that $n_k < q$. Thus, if we put $n_{k+1} = q$ we have $0 \leq n_1 < n_2 < \cdots < n_{k+1}$ and it follows from (3.65) that (3.62) holds with $n + 1$ in place of n. By induction, every $n \in \mathbb{N}$ may be expressed in the form (3.62).

Concerning uniqueness in (3.62), suppose that we have two distinct representations

$$n = 2^{n_1} + 2^{n_2} + \cdots + 2^{n_k} \quad \text{and} \quad n = 2^{m_1} + 2^{m_2} + \cdots + 2^{m_\ell},$$

where $n_1 < n_2 < \cdots < n_k$ and $m_1 < n_2 < \cdots < m_\ell$. As these representations are distinct, by cancelling out any identical terms and looking at what is left, we may assume that $n_k \neq m_\ell$ and, to be specific, assume that $m_\ell < n_k$. Then,

$$n = 2^{m_1} + 2^{m_2} + \cdots + 2^{m_\ell} \leq 1 + 2 + 2^2 + \cdots + 2^{m_\ell} = 2^{m_\ell+1} - 1$$
$$\leq 2^{n_k} - 1 < 2^{n_k} \leq n.$$

Since we cannot have $n < n$, we have a contradiction, and this establishes the uniqueness of the expression of $n \in \mathbb{N}$ in the form (3.62).

Finally, as 2^k is divisible by 2 for all $k \in \mathbb{N}$, we see that when n is expressed in the form (3.62), n is odd if and only if $n_1 = 0$. □

We are now able to define the Walsh functions[5]. It was Mendès-France [30] who in 1967 first used the Walsh functions in connection with normal numbers. In 1999, by making use of the Walsh functions, Goodman [17] showed how Kac's approach to Borel's Theorem could be extended, to prove a more general result[6]. This result is the Normal Numbers Theorem, discussed in the next section. Our purpose in this section is to derive properties of the Walsh functions for use in the proof of the Normal Numbers Theorem. Recall that r_n denotes the nth Rademacher function, as defined in Section 3.8. The following definitions make use of Lemma 3.27.

Definition Let $n \in \mathbb{N}$ and let $0 < n_1 < n_2 < \cdots < n_k$ be such that

$$2n = 2^{n_1} + 2^{n_2} + \cdots + 2^{n_k}.$$

160   Chapter 3. Probability and Randomness

Then the Walsh function w_n is defined to be

$$w_n = r_{n_1} r_{n_2} \cdots r_{n_k}.$$

The Walsh function w_0 is given by $w_0(x) = 1$ for all $x \in [0, 1)$. In particular, as $2 \cdot 2^{n-1} = 2^n$, note that

$$w_{2^{n-1}} = r_n, \text{ for all } n \in \mathbb{N}.$$

Proposition 3.28. *The following statements hold.*

(1) The Walsh functions are step functions and, for all $n \in \mathbb{Z}_+$, $w_n(x) \in \{-1, 1\}$ for all $x \in [0, 1)$.

(2) If $j, k \in \mathbb{Z}_+$ then,

$$\int_{[0,1)} w_j w_k \, d\mu = \begin{cases} 0, & \text{if } j \neq k; \\ 1, & \text{if } j = k. \end{cases}$$

(3) Let f be the transformation on $[0, 1)$ given by

$$f(x) = \begin{cases} 2x, & \text{if } 0 \leq x < 1/2; \\ 2x - 1, & \text{if } 1/2 \leq x < 1. \end{cases}$$

Then for all $j, k \in \mathbb{Z}_+$,

$$w_j \circ f^k = w_{j2^k}.$$

(4) If $j, n \in \mathbb{N}$,

$$\int_{[0,1)} w_j w_{2j} w_{2^2 j} \cdots w_{2^n j} \, d\mu = 0.$$

Proof. (1) As $w_0(x) = 1$, the result is true for w_0. Now, the Rademacher functions are step functions, as they take values -1 or 1 on each of a finite set of intervals. Also, for $n \geq 1$, by definition the Walsh function w_n is a finite product of Rademacher functions. But by Exercise 8, a product of step functions is a step function. Thus, w_n is a step function.

3.11. The Walsh functions

(2) If $j = k$, $w_j w_k = 1$ so that $\int_{[0,1)} w_j w_k d\mu = 1$. Now, let $j, k \in \mathbb{N}$ with with $j \neq k$. Using Lemma 3.27, let

$$2j = 2^{j_1} + 2^{j_2} + \cdots + 2^{j_q} \text{ and}$$
$$2k = 2^{k_1} + 2^{k_2} + \cdots + 2^{k_s}, \quad (3.66)$$

where $0 < j_1 < j_2 < \cdots < j_q$ and $0 < k_1 < k_2 < \cdots < k_s$. Then,

$$w_j = \prod_{\ell=1}^{q} r_{j_\ell} \text{ and } w_k = \prod_{m=1}^{s} r_{k_m}.$$

Because $j \neq k$, and because Lemma 3.27 gives uniqueness of the representations of $2j, 2k$ in (3.66), we must have: *either* $j_\ell \notin \{k_1, k_2, \ldots, k_s\}$ for some ℓ, *or* $k_m \notin \{j_1, j_2, \ldots, j_q\}$ for some m. Suppose the former occurs, and let $t \in \{1, 2, \ldots, q\}$ be such that $j_t \notin \{k_1, k_2, \ldots, k_s\}$. Then,

$$\int_{[0,1)} w_j w_k \, d\mu = \int_{[0,1)} \left(\prod_{\ell=1}^{q} r_{j_\ell} \right) \left(\prod_{m=1}^{s} r_{k_m} \right) d\mu$$

$$= \int_{[0,1)} r_{j_t} \left(\prod_{\substack{\ell=1 \\ \ell \neq t}}^{q} r_{j_\ell} \right) \left(\prod_{m=1}^{s} r_{k_m} \right) d\mu$$

$$= \int_{[0,1)} r_{j_t} h \, d\mu, \quad (3.67)$$

where h is a product of Rademacher functions, none of which is r_{j_t}. Now the square of a Rademacher function is 1, so h may be written as a product of *distinct* Rademacher functions, none of which is r_{j_t}. Thus, $r_{j_t} h$ is a product of distinct Rademacher functions, so by Proposition 3.20, $\int_{[0,1)} r_{j_t} h \, d\mu = 0$. Then, (3.67) shows that $\int_{[0,1)} w_j w_k d\mu = 0$.

(3) The result is clearly true if $j = 0$ or $k = 0$. So assume that $j, k \in \mathbb{N}$. Let $2j = 2^{j_1} + 2^{j_2} + \cdots + 2^{j_q}$ where $0 < j_1 < j_2 < \cdots < j_q$. Then,

$$2 \cdot 2j = 2^{j_1+1} + 2^{j_2+1} + \cdots + 2^{j_q+1},$$

By the definition of the Walsh function w_{2j} we have

$$w_{2j} = r_{j_1+1} r_{j_2+1} \cdots r_{j_q+1}.$$

Hence, using the fact that $r_j \circ f = r_{j+1}$, which can be checked directly or by appealing to (3.55), we have

$$\begin{aligned} w_j \circ f &= \left(r_{j_1} r_{j_2} \cdots r_{j_q} \right) \circ f \\ &= r_{j_1} \circ f \; r_{j_2} \circ f \cdots r_{j_q} \circ f \\ &= r_{j_1+1} r_{j_2+1} \cdots r_{j_q+1} \\ &= w_{2j}. \end{aligned}$$

So, $w_j \circ f = w_{2j}$. Using this repeatedly gives

$$\begin{aligned} w_j \circ f^k &= \left(w_j \circ f \right) \circ f^{k-1} \\ &= w_{2j} \circ f^{k-1} = \left(w_{2j} \circ f \right) \circ f^{k-2} \\ &= w_{j2^2} \circ f^{k-2} = \cdots \\ &= w_{j2^{k-1}} \circ f \\ &= w_{j2^k}. \end{aligned}$$

(4) Let $j \in \mathbb{N}$ and let $2j = 2^{j_1} + 2^{j_2} + \cdots + 2^{j_q}$ where $0 < j_1 < j_2 < \cdots < j_q$. Then, for $\ell = 0, 1, 2, \ldots, n$,

$$2 \cdot 2^\ell j = 2^{j_1+\ell} + 2^{j_2+\ell} + \cdots + 2^{j_q+\ell},$$

so that

$$w_{2^\ell j} = r_{j_1+\ell} r_{j_2+\ell} \cdots r_{j_q+\ell} = \prod_{k=1}^{q} r_{j_k+\ell}.$$

3.12. Normal numbers and randomness

Using this gives

$$\int_{[0,1)} w_j w_{2j} w_{2^2 j} \cdots w_{2^n j} \, d\mu$$
$$= \int_{[0,1)} \left(\prod_{\ell=0}^{n} w_{2^\ell j} \right) d\mu$$
$$= \int_{[0,1)} \prod_{\ell=0}^{n} \left(\prod_{k=1}^{q} r_{j_k + \ell} \right) d\mu$$
$$= \int_{[0,1)} \left(\prod_{k=1}^{q} r_{j_k} \right) \left(\prod_{\ell=1}^{n} \prod_{k=1}^{q} r_{j_k + \ell} \right) d\mu \quad (3.68)$$
$$= \int_{[0,1)} r_{j_1} \left(\prod_{k=2}^{q} r_{j_k} \right) \left(\prod_{\ell=1}^{n} \prod_{k=1}^{q} r_{j_k + \ell} \right) d\mu$$
$$= \int_{[0,1)} r_{j_1} h \, d\mu,$$

where

$$h = \left(\prod_{k=2}^{q} r_{j_k} \right) \left(\prod_{\ell=1}^{n} \prod_{k=1}^{q} r_{j_k + \ell} \right).$$

But this function h is a product of Rademacher functions r_s where $s > j_1$. Thus, bearing in mind that $r_s^2 = 1$, either $h = 1$ or h is a product of *distinct* Rademacher functions r_s where $s > j_1$. This shows that the integrand in (3.68) is a product of distinct Rademacher functions and so has integral zero by Proposition 3.20. That is, $\int_{[0,1)} w_j w_{2j} w_{2^2 j} \cdots w_{2^n j} \, d\mu = 0$. □

3.12 Normal numbers and randomness

If we toss repeatedly an unbiased coin, eventually we expect to get "heads". Here, "heads" is one of two possible outcomes. We cannot predict exactly when "heads" will occur, but we are confident in predicting that it will occur at some time. What is more, as expressed precisely by Borel's

Theorem, over time we expect to get an asymptotically equal proportion of "heads" and "tails", this being 1/2 because there are only two possible outcomes. In the present case, we consider what happens when we look for a specific *sequence* of outcomes as we toss the coin. We take at random a given finite sequence of "heads" and "tails". If this sequence has a total of r "heads" and "tails", the probability of obtaining this outcome in r consecutive tosses of the coin is 2^{-r}. Now we ask, if we continue to toss the coin again and again, without limit, what will be the eventual proportion of times this given finite sequence occurs in the full sequence of tosses? (Note that the word "occurs" here allows for overlapping in two successive occurrences—see the definition in Section 3.7. If the process is indeed random and unbiased, we would expect that it would occur in the proportion 2^{-r}. When formulated along the lines of Borel's Theorem, we see that this expectation is confirmed: if we take a random finite sequence of 0s and 1s, every number in $[0, 1)$, except for those numbers in some set of measure zero, has the property that the finite sequence will appear in the binary expansion of the number in the predicted proportion. So, although we cannot predict the *positions* of the occurrences of the given finite sequence of digits, we can predict the occurrences in an *averaging* sense. The proof of this result, known as the Normal Numbers Theorem, is based upon a use of the Walsh functions analogous to the use of the Rademacher functions in the proof of Borel's Theorem. Whereas Borel's Theorem was concerned with the proportion of appearances of a *single* symbol, the Normal Numbers Theorem is concerned with the proportion of appearances of a *given finite sequence* of symbols. The approach to the Normal Numbers Theorem that is used here was developed by Gerald Goodman in 1999 [17]. However, the arguments of [17] appeal to measure theory, whereas here we avoid the use of any measure theory, except for the concept of a set of measure zero.

Lemma 3.29. *(The Borel–Cantelli Lemma)* Let (A_n) be a sequence of basic subsets of $[0, 1)$ such that $\sum_{n=1}^{\infty} \mu(A_n) < \infty$. Then, $\bigcap_{n=1}^{\infty} \left(\bigcup_{k=n}^{\infty} A_k \right)$ is a set of measure zero.

Proof. Let $\varepsilon > 0$. Then, because $\sum_{n=1}^{\infty} \mu(A_n) < \infty$, there is $j \in \mathbb{N}$ such that $\sum_{k=j}^{\infty} \mu(A_k) < \varepsilon$. Now each set A_k is basic, and so is the

3.12. Normal numbers and randomness

finite disjoint union of intervals whose total length is $\mu(A_k)$. Hence, the set $\cup_{k=j}^{\infty} A_k$ is the union of a sequence of intervals, the sum of whose lengths is less than ε. So, there is a sequence $J_1, J_2, \ldots$ of intervals such that

$$\bigcup_{k=j}^{\infty} A_k = \bigcup_{k=1}^{\infty} J_k \text{ and } \sum_{k=1}^{\infty} \mu(J_k) < \varepsilon.$$

Hence,

$$\bigcap_{n=1}^{\infty} \left(\bigcup_{k=n}^{\infty} A_k \right) \subseteq \bigcup_{k=j}^{\infty} A_k = \bigcup_{k=1}^{\infty} J_k \text{ and } \sum_{k=1}^{\infty} \mu(J_k) < \varepsilon.$$

This argument may be carried out for every $\varepsilon > 0$, and it follows from the definition that $\bigcap_{n=1}^{\infty} \left(\bigcup_{k=n}^{\infty} A_k \right)$ is a set of measure zero. □

Proposition 3.30. *Let $j \in \mathbb{N}$. Then, there is a subset $\mathcal{Z}$ of $[0, 1)$ which has measure zero and is such that*

$$\lim_{n \to \infty} \frac{1}{n} \left(\sum_{k=0}^{n-1} w_{j2^k}(x) \right) = 0, \text{ for all } x \notin \mathcal{Z}.$$

Proof. Let $j \in \mathbb{N}$ be given. By (1) and (4) of Proposition 3.28, we see that we can apply Proposition 3.21 to deduce that

$$\int_{[0,1)} \frac{1}{n^4} \left(\sum_{k=0}^{n-1} w_{j2^k} \right)^4 d\mu = \frac{3n-2}{n^3}. \tag{3.69}$$

The procedure is now the same as in part of the proof of Borel's Theorem in Section 3.9, although here we have formalized the proof more by using the Borel–Cantelli Lemma. Also, the proof is presented in a more succinct style. Let $\mathcal{Z}$ denote the set of all $x \in [0, 1)$ such that $((\sum_{\ell=0}^{n-1} w_{j2^\ell}(x))/n)$ does *not* converge to 0. Then, for $n = 1, 2, 3, \ldots$, put

$$\mathcal{A}_n = \left\{ x : x \in [0, 1) \text{ and } \frac{1}{n} \left| \sum_{\ell=0}^{n-1} w_{j2^\ell}(x) \right| > \frac{1}{n^{1/8}} \right\}. \tag{3.70}$$

It follows that if $x \in \mathcal{Z}$ and $n \in \mathbb{N}$, there is $k \in \mathbb{N}$ with $k > n$ such that $x \in \mathcal{A}_k$. Thus, when $x \in \mathcal{Z}$, for every n, $x \in \bigcup_{k=n}^{\infty} \mathcal{A}_k$. This gives

$$\mathcal{Z} \subseteq \bigcap_{n=1}^{\infty} \left(\bigcup_{k=n}^{\infty} \mathcal{A}_k \right). \tag{3.71}$$

Now, in view of (3.70), (3.71) and the Borel–Cantelli Lemma 3.29, we see that it suffices to show that $\sum_{n=1}^{\infty} \mu(\mathcal{A}_n) < \infty$. To this end, observe that each function w_k is a step function, so the function $|\sum_{k=0}^{n-1} w_{j2^k}(x)|/n$ is also a step function. Consequently, the set $\mathcal{A}_n$ in (3.70) is a finite union of intervals. Now, (3.70) gives

$$\frac{1}{n^{1/2}} \leq \left(\frac{1}{n} \sum_{k=0}^{n-1} w_{j2^k}(x) \right)^4, \text{ for all } x \in \mathcal{A}_n \tag{3.72}$$

Hence,

$$\mu(\mathcal{A}_n) = n^{1/2} \int_{\mathcal{A}_n} \frac{1}{n^{1/2}} \, d\mu$$

$$\leq n^{1/2} \int_{\mathcal{A}_n} \frac{1}{n^4} \left(\sum_{k=0}^{n-1} w_{j2^k} \right)^4 d\mu, \text{ by (3.72),}$$

$$\leq n^{1/2} \int_{[0,1)} \frac{1}{n^4} \left(\sum_{k=0}^{n-1} w_{j2^k} \right)^4 d\mu$$

$$= n^{1/2} \left(\frac{3n-2}{n^3} \right), \text{ by (3.69),}$$

$$\leq \frac{3}{n^{3/2}}.$$

As $\sum_{n=1}^{\infty} 1/n^{3/2} < \infty$, it follows that $\sum_{n=1}^{\infty} \mu(\mathcal{A}_n) < \infty$. As remarked above, the conclusion follows. □

Lemma 3.31. *The following statements hold.*

(1) If $c_0, c_1, \ldots, c_n \in \mathbb{R}$ are such that $\sum_{j=0}^{n} c_j w_j = 0$, then $c_1 = 0, c_2 = 0, \ldots, c_n = 0$.

3.12. Normal numbers and randomness

(2) Let $n \in \mathbb{N}$ and let J be an interval of the form $J = [(k-1)/2^n, k/2^n)$ for some $k \in \{1, 2, \ldots, 2^n\}$. Then, the Walsh functions $w_0, w_1, w_2, \ldots, w_{2^n-1}$ are constant on J, and there are $c_0, c_1, \ldots, c_{2^n-1}$ such that

$$\sum_{j=0}^{2^n-1} c_j w_j = \chi_J. \tag{3.73}$$

In this case,
$$c_0 = \mu(J) = 2^{-n}.$$

Proof. (1) Let $\sum_{j=0}^n c_j w_j = 0$ and let $k \in \{0, 1, 2, \ldots, n\}$. Then, using (2) of Proposition 3.28, we have

$$\sum_{j=0}^n c_j w_j = 0 \implies \int_{[0,1)} \left(\sum_{j=0}^n c_j w_j \right) w_k \, d\mu = 0$$

$$\implies \int_{[0,1)} \left(\sum_{j=0}^n c_j w_j w_k \right) d\mu = 0$$

$$\implies \sum_{j=0}^n c_j \left(\int_{[0,1)} w_j w_k \, d\mu \right) = 0$$

$$\implies c_k = 0.$$

(2) As $w_0 = 1$, w_0 is constant on J. Let $j \in \{1, 2, \ldots, 2^n - 1\}$ and put $2j = 2^{j_1} + 2^{j_2} + \cdots + 2^{j_k}$ where $0 < j_1 < j_2 < \cdots < j_k$. Then, as $2j \leq 2(2^n - 1) = 2^{n+1} - 2$, we see that $2^{j_k} \leq 2j \leq 2^{n+1} - 2 < 2^{n+1}$. Hence, $j_k \leq n$, and we see that

$$0 < j_1 < j_2 < \cdots < j_k \leq n. \tag{3.74}$$

The definition of the Walsh functions gives

$$w_j = \prod_{\ell=1}^k r_{j_\ell}. \tag{3.75}$$

168   Chapter 3. Probability and Randomness

Now by (3.74), each $j_\ell \leq n$. Then, by the observation just preceding Lemma 3.19, each function r_{j_ℓ} in (3.75) is constant on J. So it follows from (3.75) that w_j is constant on J for all $j = 0, 1, \ldots, 2^n - 1$.

Now, let $K_\ell = [(\ell-1)/2^n, \ell/2^n)$ for each $\ell \in \{1, 2, \ldots, 2^n\}$, and let $0 \leq j \leq 2^n - 1$. We have just shown that this implies that w_j is constant on each interval K_ℓ, so let $w_j(x) = b_{j\ell}$ for $x \in K_\ell$. Then,

$$w_j = \sum_{\ell=1}^{2^n} b_{j\ell} \chi_{K_\ell}. \tag{3.76}$$

(Note that this expresses w_j in an explicit step function form.) Assume that $c_0, c_1, \ldots, c_{2^n-1} \in \mathbb{R}$ are such that

$$\sum_{j=0}^{2^n-1} c_j b_{j\ell} = 0, \text{ for all } \ell = 1, 2, \ldots, 2^n. \tag{3.77}$$

Using (3.76) and (3.77) gives

$$\sum_{j=0}^{2^n-1} c_j w_j = \sum_{j=0}^{2^n-1} c_j \left(\sum_{\ell=1}^{2^n} b_{j\ell} \chi_{K_\ell} \right) = \sum_{\ell=1}^{2^n} \left(\sum_{j=0}^{2^n-1} c_j b_{j\ell} \right) \chi_{K_\ell} = 0.$$

By (1) of the present lemma, already proved, we deduce that $c_0 = 0, c_1 = 0, \ldots, c_{2^n-1} = 0$. This shows that the homogeneous system of 2^n linear equations given in (3.77) has only the zero solution. It is a standard result in linear algebra (see for example [23, p. 22],) that the matrix of such a system is invertible—that is, the $2^n \times 2^n$ matrix $(b_{j\ell})$ is invertible.

Now, J is one of the intervals $K_1, K_2, \ldots, K_{2^n}$. Let $k \in \{1, 2, \ldots, 2^n\}$ be chosen so that $J = K_k$. Because the matrix $(b_{j\ell})$ is invertible, there are $c_0, c_1, \ldots, c_{2^n-1} \in \mathbb{R}$ such that

$$\sum_{j=0}^{2^n-1} c_j b_{j\ell} = \begin{cases} 0, & \text{if } \ell \neq k; \\ 1, & \text{if } \ell = k. \end{cases} \tag{3.78}$$

3.12. Normal numbers and randomness

We claim that $\sum_{j=0}^{2^n-1} c_j w_j = \chi_J$. For, using (3.76) and (3.78),

$$\sum_{j=0}^{2^n-1} c_j w_j = \sum_{j=0}^{2^n-1} c_j \left(\sum_{\ell=1}^{2^n} b_{j\ell} \chi_{K_\ell} \right) = \sum_{\ell=1}^{2^n} \left(\sum_{j=0}^{2^n-1} c_j b_{j\ell} \right) \chi_{K_\ell}$$
$$= \chi_{K_k} = \chi_J,$$

which proves (3.73). Finally, observe that

$$2^{-n} = \mu(J) = \int_{[0,1)} \chi_J \, d\mu$$
$$= \int_{[0,1)} \sum_{j=0}^{2^n-1} c_j w_j \, d\mu$$
$$= \sum_{j=0}^{2^n-1} c_j \int_{[0,1)} w_j \, d\mu$$
$$= c_0 \int_{[0,1)} 1 \, d\mu = c_0,$$

where we have used (2) of Proposition 3.28 with $k = 0$. □

We now state the main result of this section. Recall that if if x is in $[0, 1)$, then $d_1(x), d_2(x), \ldots$ denote the sequence of binary digits of x.

Theorem 3.32. *(The Normal Numbers Theorem) There is a subset $\mathcal{Z}$ of $[0, 1)$ which has measure zero and has the following property: for every $x \in [0, 1)$ with $x \notin \mathcal{Z}$, for every $s \in \mathbb{N}$, and for every choice of symbols $c_1, c_2, \ldots, c_s \in \{0, 1\}$, we have*

$$\lim_{n \to \infty} \frac{1}{n} \left| \left\{ r : 1 \leq r \leq n \text{ and } d_r(x) = c_1, d_{r+1}(x) \right. \right.$$
$$\left. \left. = c_2, \ldots, d_{r+s-1}(x) = c_s \right\} \right| = \frac{1}{2^s}. \quad (3.79)$$

That is, $\mathcal{Z}$ is a set of measure zero and all numbers in $[0, 1)$ that are not in $\mathcal{Z}$ are normal.

Proof. By Proposition 3.30, for each $j \in \mathbb{N}$ there is a subset $\mathcal{Z}_j$ of $[0, 1)$, having measure zero, such that

$$\lim_{n \to \infty} \frac{1}{n} \left(\sum_{k=0}^{n-1} w_{j2^k}(x) \right) = 0, \text{ for all } x \notin \mathcal{Z}_j. \quad (3.80)$$

Put

$$\mathcal{Z} = \bigcup_{j=1}^{\infty} \mathcal{Z}_j. \quad (3.81)$$

Then $\mathcal{Z}$ is a set of measure zero, by (2) of Proposition 3.12.

Now, let $f : [0, 1) \longrightarrow [0, 1)$ be the function given by $f(x) = 2x$ for $0 \le x < 1/2$ and $f(x) = 2x - 1$ for $1/2 \le x < 1$. Let $c_1, c_2, \ldots, c_s \in \{0, 1\}$ be given and let J be the interval given by

$$J = \left[\sum_{j=1}^{s} \frac{c_j}{2^j}, \sum_{j=1}^{s} \frac{c_j}{2^j} + \frac{1}{2^s} \right).$$

In view of (3.59), we have

$d_r(x) = c_1, d_{r+1}(x) = c_2, \ldots, d_{r+s-1}(x) = c_s \iff f^{r-1}(x) \in J.$

Hence,

$1 \le r \le n$ and $d_r(x) = c_1, d_{r+1}(x) = c_2, \ldots, d_{r+s-1}(x) = c_s$
$\iff 1 \le r \le n$ and $f^{r-1}(x) \in J$
$\iff 1 \le r \le n$ and $\chi_J\left(f^{r-1}(x)\right) = 1.$

Hence, to prove (3.79), it suffices to show that if $\mathcal{Z}$ is the set of measure zero given by (3.81), then for all $x \notin \mathcal{Z}$ we have

$$\lim_{n \to \infty} \frac{1}{n} \left(\sum_{r=1}^{n} \chi_J\left(f^{r-1}(x)\right) \right) = \frac{1}{2^s}. \quad (3.82)$$

3.12. Normal numbers and randomness

Now, to prove (3.82), observe that by (2) of Lemma 3.31, there are $b_0, b_1, \ldots, b_{2^s-1}$ such that

$$\sum_{j=0}^{2^s-1} b_j w_j = \chi_J.$$

Hence, as $w_j \circ f^r = w_{j2^r}$ by (3) of Proposition 3.28, we have

$$\chi_J(f^{r-1}(x)) = \sum_{j=0}^{2^s-1} b_j w_j(f^{r-1}(x)) = \sum_{j=0}^{2^s-1} b_j w_{j2^{r-1}}(x). \quad (3.83)$$

Also, if $x \notin \mathcal{Z}$, by (3.81) we must have $x \notin \mathcal{Z}_j$ for all $j \in \mathbb{N}$. Hence, by (3.80) we see that

$$x \notin \mathcal{Z} \implies \lim_{n \to \infty} \frac{1}{n} \left(\sum_{k=0}^{n-1} w_{j2^k}(x) \right) = 0, \quad \text{for all } j = 1, 2, \ldots, 2^s-1. \quad (3.84)$$

Now,

$$\frac{1}{n} \left(\sum_{r=1}^{n} \chi_J(f^{r-1}(x)) \right)$$

$$= \frac{1}{n} \left(\sum_{r=1}^{n} \left(\sum_{j=0}^{2^s-1} b_j w_{j2^{r-1}}(x) \right) \right), \quad \text{by (3.83)},$$

$$= \frac{1}{n} \left(\sum_{j=0}^{2^s-1} \left(\sum_{r=1}^{n} b_j w_{j2^{r-1}}(x) \right) \right)$$

$$= \frac{1}{n} \left(\sum_{r=1}^{n} b_0 w_0(x) + \sum_{j=1}^{2^s-1} b_j \left(\sum_{r=1}^{n} w_{j2^{r-1}}(x) \right) \right)$$

$$= b_0 + \sum_{j=1}^{2^s-1} \frac{b_j}{n} \left(\sum_{r=1}^{n} w_{j2^{r-1}}(x) \right). \quad (3.85)$$

We now have from (3.84) and (3.85) that for every $x \notin \mathcal{Z}$,

$$\lim_{n\to\infty} \frac{1}{n} \left(\sum_{r=1}^{n} \chi_J \left(f^{r-1}(x) \right) \right)$$
$$= b_0 + \left(\sum_{j=1}^{2^s-1} b_j \cdot \lim_{n\to\infty} \frac{1}{n} \left(\sum_{r=1}^{n} w_{j2^{r-1}}(x) \right) \right)$$
$$= b_0 = \frac{1}{2^s},$$

where we have used (2) of Lemma 3.31. Thus, (3.82) has been proved and the result follows. □

The Normal Numbers Theorem 3.32 should be compared with the Recurrence Theorem 3.18. The Recurrence Theorem says that if a finite sequence $c_1, c_2, \ldots, c_s$ of binary digits is given, this sequence occurs infinitely many times in the binary expansion of a typical number in [0, 1). The Normal Numbers Theorem refines this substantially by showing that this finite sequence occurs with a definite asymptotic frequency, and shows that this frequency is $1/2^s$. Like Borel's Theorem as discussed at the end of Section 3.10, the Normal Numbers Theorem may be interpreted as a recurrence result in a dynamical system. This is a subject for investigation at the end of this chapter. A different approach to proving the Normal Numbers Theorem, using trigonometric approximation, may be found in [27, pp.69–70], and another approach using continued fractions may be found in [35].

Finally, at the end of Section 3.10 we commented that Borel's Theorem is saying in its particular context something like

" the average frequency of an event over time equals the probability of the event", (3.86)

and we used this to conjecture that a result like the Normal Numbers Theorem would be true. But, in general, does a statement like (3.86) have substantive content or is it merely a *definition* of the probability of an event? Is it insightful, or is it a tautology? Some ideas related to such questions are considered in the next Section.

3.13 Notions of probability and randomness

This section looks informally at randomness and probability from two different points of view: (1) the axiomatic approach, due to Kolmogorov, where essentially the probabilities are taken as given in advance. This fits in with the preceding sections on recurrence, Borel's Theorem and the Normal Numbers Theorem. (2) The situation where we have data given in advance, and we derive the probabilities from the data—this is the "frequency interpretation" of probability, emphasized by von Mises [31]. This fits in with the ensuing sections on the leading digits of various data.

In Section 3.4 we considered a basic subset of the interval [0, 1) to be a subset of [0, 1) that is a finite union of intervals. Let $\mu(K)$ denote the usual length of an interval K. Then we showed in Proposition 3.10 that this notion of length can be extended to all the basic subsets of [0, 1) so that the natural properties we would expect of length are preserved. This extended notion of length for a basic subset K of [0, 1) was denoted by $\mu(K)$ and was called the *length* of K. We interpreted $\mu(K)$ of a basic set subset K of [0, 1) as the *probability* that the event K will occur. That is, given that x is chosen at random in [0, 1), the probability that $x \in K$ will occur is $\mu(K)$. This line of thought identifies the probability of the event K as being the *proportion* of [0, 1) which is occupied by K. Under this approach, probability is taken as a primitive notion that is given in advance and is assumed to have certain properties. It was A. Kolmogorov who axiomatized this approach in 1935 [26], and it has been almost universally adopted by mathematicians. However, as with other areas of mathematics, there is the problem of how the axioms are related to applications and the world of observed experience. In the case of probability, there has been a school of thought counter to that of Kolmogorov, associated especially with the name of Richard von Mises.

According to von Mises, probability is not a primitive notion, but rather one that is derived from observed data. The primitive notion, according to him, is that of a "collective", by which he means an infinite sequence of observations, or data, satisfying certain conditions. In this approach, probabilities are calculated from the data in a given collective, by identifying probabilities with corresponding limits of frequencies of given attributes of the collective. Thus, according to von Mises, a probability is the probability of an *attribute* for a given collective. His position is crudely summarized in his statement [31, p. 18],

> First the collective—then the probability.

von Mises produced his early ideas on probability in 1919, and modified them over the years in response to criticisms and difficulties. His ideas are readily accessible in [31]. One of the problems faced by von Mises was what he meant by a *random sequence* and by *admissible selection* from the terms of such a sequence, and there is now a considerable literature on this and related issues; in particular there is the notion of *algorithmic complexity*, associated with the names of Chaitin, Kolmogorov, Martin-Löf, and others.[7] Detailed consideration of these matters and the foundations of probability theory lie beyond the scope of this work. However, clear discussion concerning the foundations of probability theory and notions of randomness, and the history of the debates on these issues, may be found in the work of de Finetti [14], Gillies [15], van Lambalgen [28], Li and Vitányi [29], and Volchan [44], where there are further references. The following definitions formally define a notion of probability along the lines of those recommended by von Mises.

Definition Let (c_n) be a sequence of elements of a given set X, and let A be a subset of X. Then, we can consider the "asymptotic proportion" or "asymptotic frequency" of points in the sequence (c_j) that lie within the set A. The frequency of the first n points of the sequence (c_j) that lie in the set A is

$$\frac{1}{n}\left|\left\{j : 1 \leq j \leq n \text{ such that } c_j \in A\right\}\right|.$$

Thus, the "asymptotic frequency" of the points of (c_j) that belong to A is $\nu(A)$ where

$$\nu(A) = \lim_{n\to\infty} \frac{1}{n}\left|\left\{j : 1 \leq j \leq n \text{ such that } c_j \in A\right\}\right|, \qquad (3.87)$$

3.13. Notions of probability and randomness 175

assuming that this limit exists. If the limit $\nu(A)$ exists, as in (3.87), we say that the sequence (c_j) has *frequency behavior* or *statistical behavior relative to* A. In this case, $\nu(A)$ is called the *frequency probability of* A *relative to* (c_j).

Note that in the definition of the frequency probability, the order of the terms of the sequence (c_j) may be important. Also, in the terminology of von Mises, note that the subset A of X corresponds to an attribute for the collective (c_n) or perhaps, more precisely, for the *terms* of the collective (c_n). It is important to note that the frequency probability of a subset A of X may or may not exist. Thus, a question underlying the von Mises approach is: for a given collective, what attributes of the collective have a frequency probability?

Borel's Theorem, appearing in Section 3.10 in the form of Theorem 3.25, in its context can be regarded as reconciling the frequency theory of von Mises with Kolmogorov's axiomatization of probability theory. To see why, recall that if we put $A = [0, 1/2)$, and if $f(x) = \text{frac}(2x)$ for $0 \leq x < 1$, Borel's Theorem says that for all $x \in [0, 1)$, except for those x in some set of measure zero, we have

$$\lim_{n \to \infty} \frac{1}{n} \left| \left\{ j : 1 \leq j \leq n \text{ and } f^{j-1}(x) \in A \right\} \right| = \frac{1}{2} = \mu(A). \quad (3.88)$$

Here, the left-hand side of (3.88) is the frequency probability of $A = [0, 1/2)$ relative to the sequence $x, f(x), f^2(x), \ldots$, while the right-hand side is the "primitive" probability $\mu(A)$ of A. Thus, Borel's Theorem as expressed in (3.88) asserts the equality of the Kolmogorov and the von Mises definitions of probability for the "attribute" or the "event" $[0, 1/2)$. This equivalence of two notions of probabilistic or statistical behavior is present also in Weyl's Theorem 2.20 and it is an idea which appears again in the context of Birkhoff's Individual Ergodic Theorem, discussed in Chapter 5. In the case of Weyl's Theorem, we have α an irrational number in $[0, 1)$ and τ_α the transformation on $[0, 1)$ given by $x \longmapsto \text{frac}(x + \alpha)$. Then, Weyl's result says that if α is irrational, for all $x \in [0, 1)$ and all subintervals J of $[0, 1)$,

$$\lim_{n \to \infty} \frac{1}{n} \left| \left\{ j : 1 \leq j \leq n \text{ and } \tau_\alpha^{j-1}(x) \in J \right\} \right| = \mu(J). \quad (3.89)$$

As in (3.88), the left-hand side of (3.89) is the frequency probability of J relative to the sequence $x, \tau_\alpha(x), \tau_\alpha^2(x), \ldots$, while the right-hand side is the "primitive" probability $\mu(J)$ of J. Hence, once again, the two notions of the probability of an event or an attribute coincide. Thus, (3.88) and (3.89) might be taken to indicate a coincidence in the ideas of Kolmogorov and von Mises.

Now, from the Normal Numbers Theorem 3.32 we know that, in fact, (3.88) holds for every subinterval A of $[0, 1)$ that is of the form $[(k-1)/2^n, k/2^n)$ for some $n \in \mathbb{N}$ and $k \in \{1, 2, \ldots, 2^n\}$. Thus, in both (3.88) and (3.89), the two notions of probability exist and coincide on both "smaller" intervals and "larger" intervals. We might say that the data in either of the dynamical systems in (3.88) and (3.89) has statistical behavior at both "smaller" scales and "larger" scales, or at "coarser" scales and "more refined" scales. However, such behavior does not always occur. The following example shows that a sequence may have statistical behavior for attributes A and B, say, but this does not necessarily mean that the sequence will have statistical behavior for the "finer" attribute $A \cap B$ or for the "coarser" attribute $A \cup B$.

Example Let $X = \{1, 2, 3, 4\}$, let $A = \{1, 2\}$ and $B = \{2, 3\}$. Let $n_0 = 0$ and let $n_1 < n_2 < n_3 < \cdots$ be an increasing sequence in $\mathbb{N}$. Let (c_j) be the sequence in $\mathbb{N}$ given by

$$\underbrace{1, 3, 1, 3, \ldots, 1, 3}, \underbrace{2, 4, 2, 4, \ldots, 2, 4}, \underbrace{1, 3, 1, 3, \ldots, 1, 3}, \underbrace{2, 4, 2, 4, \ldots, 2, 4}, \ldots$$

where the consecutive brackets underneath the terms of the sequence indicate, respectively, $2n_1$ terms, $2n_2$ terms, $2n_3$ terms, and so on, of the sequence. Then it is readily checked that

$$\lim_{\ell \to \infty} \frac{1}{2\ell} \left| \left\{ j : 1 \leq j \leq 2\ell \text{ and } c_j \in A \right\} \right| = \lim_{\ell \to \infty} \frac{\ell}{2\ell} = \frac{1}{2},$$

and that

$$\lim_{\ell \to \infty} \frac{1}{2\ell} \left| \left\{ j : 1 \leq j \leq 2\ell \text{ and } c_j \in B \right\} \right| = \lim_{\ell \to \infty} \frac{\ell}{2\ell} = \frac{1}{2}.$$

It follows that (c_j) has statistical behavior relative to both A and B, with $\nu(A) = \nu(B) = 1/2$. Now, when it comes to considering $A \cap B =$

3.13. Notions of probability and randomness

{2}, observe that for all $k \in \mathbb{N}$ with $k \geq 2$,

$$\frac{\left|\left\{j : 1 \leq j \leq 2n_1 + 2n_2 + \cdots + 2n_{2k} \text{ and } c_j \in A \cap B\right\}\right|}{2n_1 + 2n_2 + \cdots + 2n_{2k}} \quad (3.90)$$
$$= \frac{n_2 + n_4 + \cdots + n_{2k-2} + n_{2k}}{2n_1 + 2n_2 + \cdots + 2n_{2k}},$$

and

$$\frac{\left|\left\{j : 1 \leq j \leq 2n_1 + 2n_2 + \cdots + 2n_{2k-1} \text{ and } c_j \in A \cap B\right\}\right|}{2n_1 + 2n_2 + \cdots + 2n_{2k-1}} \quad (3.91)$$
$$= \frac{n_2 + n_4 + \cdots + n_{2k-4} + n_{2k-2}}{2n_1 + 2n_2 + \cdots + 2n_{2k-1}}.$$

Noting (3.90) and (3.91), we now proceed to choose $n_1, n_2, \ldots$ in a particular way. We arbitrarily select $n_1 \in \mathbb{N}$. Then, choose $n_2 \in \mathbb{N}$ with $n_2 > n_1$ and such that $n_2/(2n_1 + 2n_2) > 1/4$. Then, choose $n_3 \in \mathbb{N}$ with $n_3 > n_2$ and such that $n_2/(2n_1 + 2n_2 + 2n_3) < 1/8$. Now, the terms $n_4, n_5, \ldots$ are defined by an inductive procedure as follows. Assume that $n_1, n_2, \ldots, n_{2k-1}$ have been defined and that

$$\frac{n_2 + n_4 + \cdots + n_{2k-4} + n_{2k-2}}{2n_1 + 2n_2 + \cdots + 2n_{2k-2}} > \frac{1}{4} \text{ and}$$
$$\frac{n_2 + n_4 + \cdots + n_{2k-4} + n_{2k-2}}{2n_1 + 2n_2 + \cdots + 2n_{2k-1}} < \frac{1}{8}. \quad (3.92)$$

Then, let $n_{2k} \in \mathbb{N}$ be such that $n_{2k} > n_{2k-1}$ and

$$\frac{n_2 + n_4 + \cdots + n_{2k-2} + n_{2k}}{2n_1 + 2n_2 + \cdots + 2n_{2k-1} + 2n_{2k}} > \frac{1}{4}. \quad (3.93)$$

Then, let $n_{2k+1} \in \mathbb{N}$ be such that $n_{2k+1} > n_{2k}$ and

$$\frac{n_2 + n_4 + \cdots + n_{2k-2} + n_{2k}}{2n_1 + 2n_2 + \cdots + 2n_{2k} + 2n_{2k+1}} < \frac{1}{8}. \quad (3.94)$$

The Normal Numbers Theorem illustrates that randomness may exist at "all levels of refinement", in the sense that *every* given finite sequence of symbols occurs with the anticipated frequency. However, letting $d_n(x)$ denote the nth binary digit of x, we have

$$\frac{2}{3} = 0.1010101010\ldots, \text{ so that } d_n\left(\frac{2}{3}\right) = \begin{cases} 1, & \text{if } n \text{ is odd,} \\ 0, & \text{if } n \text{ is even.} \end{cases}$$

It follows that at the "first level of refinement", where we just look at the frequency of occurrence of *one* symbol, say 1, we have

$$\lim_{n\to\infty} \frac{1}{n}\left|\left\{j : 1 \leq j \leq n \text{ and } d_j\left(\frac{2}{3}\right) = 1\right\}\right| = \frac{1}{2},$$

and we note that $1/2$ is the "obvious" probability of obtaining 1 so that, so far, all is just as we would expect for a random sequence of 0s and 1s. The same happens for the other single digit, namely 0. Now, if we look at the "second level of refinement", that is if we look at the frequency of occurrence of the possible sequences of *two* symbols, we find that randomness breaks down. There are 4 possible sequences of 2 symbols: 00, 01, 10 and 11. Observe that in the cased of 01 we have $1/2$ as the limiting frequency, for

$$\lim_{n\to\infty} \frac{1}{n}\left|\left\{j : 1 \leq j \leq n \text{ and } d_j\left(\frac{2}{3}\right) = 0 \text{ and } d_{j+1}\left(\frac{2}{3}\right) = 1\right\}\right| = \frac{1}{2},$$

and we can check that the frequencies for 00, 10 and 11 respectively are 0, $1/2$ and 0. Each of these frequencies is inconsistent with randomness of the sequence 10101010... at the second level of refinement, for given such randomness we would expect each limiting frequency to be $1/4$. So, the binary expansion of $2/3$ gives an example of a sequence of 0s and 1s which is random at the "first level of refinement" but is not random at the "second level of refinement". Similarly, we introduce the concept of the "k^{th} level of refinement" by considering the proportion of occurences of all possible k symbols. Then, for example, it can be checked that $226/255$ is random at the first three levels of refinement, but not at the fourth. These ideas suggest that an alternative way of stating Borel's Theorem is: *outside of some set of measure zero, the binary expansion of every point of $[0, 1)$ has random behavior at the first level of refinement*. Similarly, the Normal Numbers Theorem says: *outside of some set of measure zero, the binary expansion of every point of $[0, 1)$ has random behavior at all levels of refinement*.

3.13. Notions of probability and randomness

This defines the sequence $n_1, n_2, \ldots$ by induction, $0 = n_0 < n_1 < n_2 < \cdots$, and we see from (3.92), (3.93) and (3.94) that (3.92) holds for all $k \geq 2$. Thus, if $v(A \cap B)$ exists, it would follow from (3.90) and (3.92) that $v(A \cap B) \geq 1/4$, but it would follow also from (3.91) and (3.92) that $v(A \cap B) \leq 1/8$. Thus, in this case $v(A)$ and $v(B)$ exist but $v(A \cap B)$ does not exist.

Now, we also claim that $v(A \cup B)$ does not exist. For, observe that $A^c \cap B^c = \{4\}$ and consider $v(A^c \cap B^c)$. Because of (3.92), we deduce exactly as above, but with 4 playing the role of 2, that $v(A^c \cap B^c)$ does not exist. However, observe that from (3.87), $v(A)$ exists if and only if $v(A^c)$ exists (in which case $v(A^c) = 1 - v(A)$). Thus we now have

$$v(A^c \cap B^c) \text{ does not exist} \iff v((A \cup B)^c) \text{ does not exist}$$
$$\iff v(A \cup B) \text{ does not exist.}$$

Thus, in this example, $v(A)$ and $v(B)$ exist but neither $v(A \cap B)$ nor $v(A \cup B)$ exists. In a different context, this conclusion may also be found in [14, pp. 74–75].

The example shows that the frequency probabilities may exist for events A and B, but not exist for the events $A \cap B$ or $A \cup B$. That is, whereas a sequence may have statistical behavior for attributes A and B, it need not have statistical behavior for the "finer" attribute $A \cap B$, or for the "coarser" attribute $A \cup B$. However, note that in the above example, the choice of the numbers n_k, on which the example depends, has been carried out by looking at the numbers already constructed—that is, the sequence in the example is *non-random* in the sense that it has been constructed at each stage by looking back and taking account of the preceding terms or "history" of the sequence. On the other hand, as we have seen, the orbits in the dynamical systems of Weyl's Theorem and the Normal Numbers Theorem exhibit statistical behavior at all levels of refinement and all levels of coarseness.

Now, in the case of Borel's Theorem, we saw that the binary expansion of a typical number in $[0, 1)$ produced a "random" sequence of 0s and 1s, in the following sense: the numbers of 0s and 1s in the binary expansions are asymptotically equal, so that the binary expansion shows no preference for either a 0 or a 1. But this observation of randomness is at a coarse level only. The Normal Numbers Theorem shows, in effect, that this phenomenon of randomness in the binary expansion of a typical number persists at ever increasing levels of refinement. For if the binary expansion is indeed random, and if we are given a finite sequence $c_1, c_2, \ldots, c_r$ of 0s and 1s, we would not expect this given sequence to appear in the binary expansion any more than any other given finite sequence of r terms—that is, the given finite sequence should appear in a typical binary expansion at precisely the frequency of the intrinsic probability 2^{-r} of the sequence, and this is precisely what the Normal Numbers Theorem says. So, a typical binary expansion exhibits the *same* statistical behavior at all levels of refinement. Thus, we see that randomness as exhibited in the Normal Numbers Theorem has led to well-defined and even predictable statistical behavior at all levels, whereas the sequence chosen deliberately in the example above apparently has produced *less* statistical behavior.

3.14 The curious phenomenon of the leading significant digit

Suppose we have a collection of positive numbers, perhaps arising from a set of data. Assuming the data is random, we might expect that the leading digits of the numbers in the data would occur with an approximately equal frequency. So, it may come as a very surprising fact that this is often *not* the case. Back in the days when electronic calculators did not exist, arithmetical calculations were carried out using books of logarithmic tables. It seems to have been Simon Newcomb, the professor of mathematics and astronomy at Johns Hopkins University, who observed in 1881 that the pages near the front of books of logarithms were more used than the pages towards the back. In [32] he wrote:

> That the ten digits do not occur with equal frequency must be evident to any one making use of logarithmic tables, and

3.14. The curious phenomenon of the leading significant digit   181

noticing how much faster the first pages wear out than the last ones. The first significant digit is oftener 1 than any other digit, and the frequency diminishes up to 9.

If d denotes a digit between 1 and 9 inclusive, Newcomb stated the values of what he called "the probabilities of occurrence" of d as a first digit and as a second digit, using the base 10. In particular he gave the probability of occurrence of 1 as a first digit to be 0.3010, and the probability of occurrence of 1 as a second digit he gave as 0.1139. He also stated corresponding values for the probability of occurrence as the first digit for the other digits $2, 3, \ldots, 9$, and the probability of occurrence of 0 as a second digit he gave as 0.1197. Newcomb did not support the values of the probabilities by adducing data, rather he supplied a reasoning by which he arrived at these values. Newcomb's reasoning is still worth looking at today, for it contains the germs of most of the subsequent ideas concerning the leading digit problem. Here is his view of how numbers arise in nature:

> As natural numbers occur in nature, they are to be considered as the ratios of quantities. Therefore, instead of selecting a number at random, we must select two numbers, and enquire what is the probability that the first significant digit of their ratio is the digit n ...

He goes on to argue that if these two numbers are written as a number raised to some exponent, taking the ratio of the numbers leads to subtracting the exponents. Also, he argues that we are only interested in the fractional parts of the exponents, because the integer part does not affect the leading digit (in effect he says that $10^{1.25}$ has the same first digit as $10^{.25}$, for example). Letting i denote a number used to represent other numbers as powers of i he says:

> Since these exponents are formed by casting off all the integers from a series of numbers, we may suppose them arranged around a circle according to some law. Then, if we select 2^n exponents at random and call them $s', s'', s''', etc.$, the final ratio, obtained in the manner we have described, will be

$$i^{s'-s''+s'''-s''''+ etc.}$$

The question is, what is the probability that the positive fractional portion of $s' - s'' + s''' - s'''' + etc.$, will be contained between the limits s and $s + ds$. It is evident that, whatever be the original law of arrangement, the fractions will approach to an equal distribution around the circle as n is increased, or the required probability will be equal to ds.

What Newcomb in effect is saying here is that the fractional parts of the exponents are "equally distributed" or "uniformly distributed" around the circle—that is, he is in effect saying that the conclusion of Weyl's Theorem 2.20, concerning the fractional parts of the multiples of an irrational number, should apply to these fractions as they arise in nature. His conclusion?

> The law of probability of the occurrence of numbers is such that all mantissae of their logarithms are equally probable

The point in the mathematical aspect of Newcomb's argument here is his statement that it is evident that the fractional parts of the exponents are distributed uniformly around the circle. Whether one finds this evident or not, Newcomb here puts a finger on a crucial idea—namely, that by taking the fractional part of the logarithms of a geometric sequence, we may obtain a sequence that is "uniformly distributed" on a circle, or an interval. This "uniform distribution" on the circle, as we shall see, corresponds to a *logarithmic* distribution for the original phenomenon. It would then follow from these ideas, although not explicitly stated by Newcomb, that the probability of occurrence of the digit d equals the length of the interval $[\log_{10} d, \log_{10}(d + 1))$, namely $\log_{10}(1 + 1/d)$.

Newcomb's observations do not seem to have made a lasting impression at the time. However, in 1938, the physicist Frank Benford published his paper [5] on "the law of anomalous numbers". Benford does not explicitly mention Newcomb's observation, but he does say:

> There are many instances showing that the geometric series, or the logarithmic law, has long been recognized as a common

3.14. The curious phenomenon of the leading significant digit

phenomenon in factual literature and in the ordinary affairs of life. The wire gauge and drill gauge of the mechanic, the magnitude scale of the astronomer and the sensory response curves of the psychologist are all particular examples of a relationship that seems to extend to all human affairs. The Law of Anomalous Numbers is thus a general probability law of widespread application.

Benford collected data from a wide variety of sources. Some of his data is reproduced in Figure 3.8. Based on this data, Benford concluded that the formula for the frequency of a first digit $d \in \{1, 2, \ldots, 9\}$ is $\log_{10}(1 + 1/d)$, and this is sometimes called *Benford's Law*, although we shall refer to this as *Benford's formulas*. The formulas give the values for the "probabilities of occurrence" as stated by Newcomb.

Newcomb's idea that numbers occurring in nature come from ratios of quantities has been widely used to try and explain the observations of Newcomb and Benford. In this spirit, in 1994 Boyle [8] wrote:

> Frequently, naturally occurring data can be thought of as a result of products or quotients of random variables. For example, the population of a city changes at roughly a rate proportional to the population. If a city has an initial population P_0 and grows by $r_i \%$ in year i, then the population in n years is $P_n = P_0(1 + r_1)(1 + r_2) \cdots (1 + r_n)$, a product of a number of random variables.

In a certain precise sense, Boyle shows that the logarithmic distribution for the significant digit, that is the Benford distribution, is the "limit distribution" obtained under the repeated operations of multiplication, division and taking exponents. He also shows that if we have a random variable—that is, a function—that has a logarithmic distribution, then this distribution will persist under the same operations. In a different context, in 1979 Hamming [18] considered similar questions, arguing that the arithmetic operations of a computer transform various distributions towards a limiting distribution, and the discrete equivalent of this distribution is the logarithmic distribution of Benford and Newcomb.

Table I
Percentages of times the Natural Numbers 1 to 9 are used as first digits in numbers, as determined by 20,229 observations

Group	Title	1	2	3	4	5	6	7	8	9	Count
A	Rivers, Area	31.0	16.4	10.7	11.3	7.2	8.6	5.5	4.2	5.1	335
B	Population	33.9	20.4	14.2	8.1	7.2	6.2	4.1	3.7	2.2	3259
C	Constants	41.3	14.4	4.8	8.6	10.6	5.8	1.0	2.9	10.6	104
D	Newspapers	30.0	18.0	12.0	10.0	8.0	6.0	6.0	5.0	5.0	100
E	Spec. Heat	24.0	18.4	16.2	14.6	10.6	4.1	3.2	4.8	4.1	1389
F	Pressure	29.6	18.3	12.8	9.8	8.3	6.4	5.7	4.4	4.7	703
G	H.P. Lost	30.0	18.4	11.9	10.8	8.1	7.0	5.1	5.1	3.6	690
H	Mol. Wgt.	26.7	25.2	15.4	10.8	6.7	5.1	4.1	2.8	3.2	1800
I	Drainage	27.1	23.9	13.8	12.6	8.2	5.0	5.0	2.5	1.9	159
J	Atomic Wgt.	47.2	18.7	5.5	4.4	6.6	4.4	3.3	4.4	5.5	91
K	$n^{-1}, \sqrt{n}$ ...	25.7	20.3	9.7	6.8	6.6	6.8	7.2	8.0	8.9	5000
L	Design	26.8	14.8	14.3	7.5	8.3	8.4	7.0	7.3	5.6	560
M	*Digest*	33.4	18.5	12.4	7.5	7.1	6.5	5.5	4.9	4.2	308
N	Cost Data	32.4	18.8	10.1	10.1	9.8	5.5	4.7	5.5	3.1	741
O	X-Ray volts	27.9	17.5	14.4	9.0	8.1	7.4	5.1	5.8	4.8	707
P	Am. League	32.7	17.6	12.6	9.8	7.4	6.4	4.9	5.6	3.0	1458
Q	Black body	31.0	17.3	14.1	8.7	6.6	7.0	5.2	4.7	5.4	1165
R	Addresses	28.9	19.2	12.6	8.8	8.5	6.4	5.6	5.0	5.0	342
S	$n^1, n^2, ..., n!$	25.3	16.0	12.0	10.0	8.5	8.8	6.8	7.1	5.5	900
T	Death rate	27.0	18.6	15.7	9.4	6.7	6.5	7.2	4.8	4.1	418
	Average	30.6	18.5	12.4	9.4	8.0	6.4	5.1	4.9	4.7	1011
	Probable Error	±0.8	±0.4	±0.4	±0.3	±0.2	±0.2	±0.2	±0.2	±0.3	—

Figure 3.8. The Figure reproduces a table of data from Benford's paper [5] of 1938. Benford writes: "At the foot of each column of Table I the average percentage is given for each digit, and also the probable error of the average ··· The frequency of first 1's is then seen to be 0.306, which is about equal to the common logarithm of 2. The frequency of first 2's is 0.185, which is slightly greater than the logarithm of 3/2 ··· These resemblances persist throughout, and finally there is 0.047 to be compared with log 10/9, or 0.046."

Raimi [37] has a review of various aspects of the problem. King [25] discusses the problem in relation to Poncelet's Theorem and the Tarski plank problem, the unifying theme being appropriate notions of length that are invariant under the transformations arising in these problems. More recently, Hill [20, 21] has shown that if there is a way of measuring

3.14. The curious phenomenon of the leading significant digit

"length" on a certain family of subsets of $(0, \infty)$, such that this notion of length is "scale invariant" (that is, the length of a set is not changed when the set is multiplied by any $b \in (0, \infty)$), then this notion of "length" is necessarily that of Benford and Newcomb, in which the length of an interval $[d, d + 1]$ is $\log_{10}(d + 1) - \log_{10} d = \log_{10}(1 + 1/d)$. Here, the significance of scale invariance is that the statistical phenomenon in data such as Benford's should persist when different units are used to measure the data.

Now it was noted in Section 3.4 that the length of intervals determines an additive set function μ on $[0, 1)$. Another additive set function on the interval $[1, 10)$ is $\mu_{\log_{10}}$, which is given, as in the comments at the end of Section 3.4, by $\mu_{\log_{10}}([c, d)) = \log_{10} d - \log_{10} c$, when $[c, d) \subseteq [1, 10)$. In discussing Benford's phenomenon and ideas concerning the leading significant digit to the base 10, the above remarks indicate that the logarithmic length $\mu_{\log_{10}}$ plays a role that is analogous to the role played by the usual notion μ of length in discussing Weyl's Theorem in Chapter 2. In fact, as we shall see, Benford's formulas are contained in Weyl's Theorem—they are simply in a different setting, where the usual notion of length is replaced by the notion of "logarithmic length".

The works in [6, 8, 18, 20, 21, 22, 25] deal with various aspects of the significant digit phenomenon, but use more advanced ideas than the ones used here in Chapter 3. The paper of Hill [22] has an interesting discussion of various types of data which give approximate Benford formulas, and discusses reasons for the deviations of these data from Benford's formulas. In Section 3.15, we make use of the basic idea implicit in Newcomb, that taking the fractional part of logarithms of a geometric sequence should give us a "uniformly distributed" sequence, to give an accessible derivation of the formula $\log_{10}(1 + 1/d)$. This depends on Weyl's Theorem discussed in Chapter 2. This approach is also used in Section 3.16, to derive a formula of Diaconis [11], also derived by Hill [20, 21], for the probability of occurrence of a given finite sequence of digits. Final Sections of Chapter 3 give one of various answers to the question as to the reason for Benford's formulas—namely, that if we have a set of data that satisfy Benford's formulas, then those formulas are the only ones possible if we make an additional requirement that the data should satisfy the same formulas under a change in the units used to measure the data. This leads to an alternative approach to some of the results of Hill [20, 21]. Finally, note that by proposing the inverse square law of gravitation, Newton was able to explain Kepler's observation that the planets moved around the sun in ellipses—but what "law" is there, corresponding to the law of gravitation, which gives a causal explanation of a particular observed statistical distribution for a given set of data?

3.15 Leading digits and geometric sequences

Consider the sequence of leading digits of the numbers in the geometric sequence 2, 4, 8, 16, 32, 64, 128, 256, 512, 1024, 2048, 4096, 8192, 16384, ... We obtain a sequence 2, 4, 8, 1, 3, 6, 1, 2, 5, 1, 2, 4, 8, 1, 3 . . . . The problem is: what is the proportion, if there is one, with which a given digit appears in this sequence? We have listed 15 terms of the sequence and we observe that 1 appears 4 times, 2 appears 3 times, 3 appears 2 times, 4 appears 2 times, 5 appears 1 time, 6 appears 1 time, 7 appears 0 times, 8 appears 2 times and 9 appears 0 times. This problem is attributed by A. Avez [3, pp. 63–64], to I. M. Gelfand, and it is also discussed in [25]. A corresponding question also arises for bases other than 10. The answer to these questions is complementary to Newcomb's and Benford's observations about the leading first digits and clarifies what might be happening to explain them. Gelfand proved that for the leading digits of the sequence (2^n) using the base 10, the proportion of terms with leading digit d is $\log_{10}(1 + 1/d)$. The method of proof is, in effect, to make mathematically precise the original idea of Newcomb in 1881—namely, that by taking logarithms and fractional parts we obtain a sequence in $[0, 1)$ that should be uniformly distributed.

Gelfand seems to have posed the above problem in terms of the leading digit relative to base 10, and only for positive integers. We are going to discuss the problem in the context of an arbitrary base and sequences of numbers in $(0, \infty)$. Let $b \in \mathbb{N}$ with $b \geq 2$. We will introduce the

3.15. Leading digits and geometric sequences

concept of the leading digit for an element of $(0, \infty)$. Observe that

$$(0, \infty) = \bigcup_{n=-\infty}^{\infty} \left[b^{n-1}, b^n \right),$$

where the union on the right is one of disjoint intervals. Let $x \in (0, \infty)$ be given. Then, there is a unique $n \in \mathbb{Z}$ such that $x \in [b^{n-1}, b^n)$, so that $x/b^{n-1} \in [1, b)$. Now, $[1, b) = \bigcup_{j=1}^{b-1} [j, j+1)$, so there is a unique $c_1 \in \{1, 2, \ldots, b-1\}$ such that

$$\frac{x}{b^{n-1}} \in [c_1, c_1 + 1).$$

As $x/b^{n-1} \in [c_1, c_1 + 1)$, there is a unique $y \in [0, 1)$ such that

$$\frac{x}{b^{n-1}} = c_1 + y. \tag{3.95}$$

By Theorem 3.4, there are $c_2, c_3, \ldots$ in $\{0, 1, \ldots, b-1\}$ such that the expansion of y in the base b is $y = .c_2 c_3 \ldots = \sum_{j=1}^{\infty} c_{j+1}/b^j$. Thus, we have from (3.95) that

$$x = b^{n-1}(c_1 + y) = b^{n-1}\left(c_1 + \sum_{j=1}^{\infty} \frac{c_{j+1}}{b^j}\right)$$

$$= c_1 b^{n-1} + \sum_{j=1}^{\infty} \frac{c_{j+1}}{b^{j-n+1}}$$

$$= \sum_{j=-n+1}^{\infty} \frac{c_{j+n}}{b^j}. \tag{3.96}$$

Definition The identity (3.96) is called the *expansion of x to the base b*. This extends the notion of expansion to the base b from numbers in $[0, 1)$, as described in Section 3.2, to numbers in $(0, \infty)$.

There are two cases to distinguish in (3.96). If $x \in [1, \infty)$, $n \geq 1$ and (3.96) becomes

$$x = c_1 b^{n-1} + c_2 b^{n-2} + \cdots + c_{n-1} b + c_n + \sum_{j=1}^{\infty} \frac{c_{j+n}}{b^j}, \tag{3.97}$$

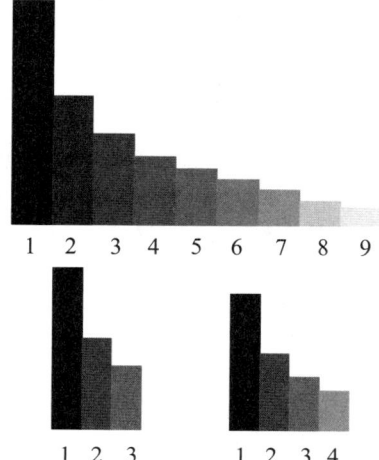

Figure 3.9. This illustrates Benford's formula $\log_b (1 + 1/d)$ for various bases b. The upper picture illustrates the values of $\log_{10} (1 + 1/d)$ for $d = 1, 2, \ldots, 9$. The lower left picture illustrates the values of $\log_4 (1 + 1/d)$ for $d = 1, 2, 3$, and the lower right picture illustrates the values of $\log_5 (1 + 1/d)$ for $d = 1, 2, 3, 4$. Note that if $\log x$ is the logarithm to the natural base e, then

$$\mu_{\log_b}([d, d+1)) = \log_b(d+1) - \log_b d = \log_b\left(1 + \frac{1}{d}\right) = \frac{1}{\log b}\int_d^{d+1} \frac{dx}{x}.$$

That is, Benford's formula to the base b is given by the logarithmic length $\mu_{\log_b}$, whose "density function" is $x \longmapsto 1/x \log b$.

and we write this as

$$x = c_1 c_2 \ldots c_n . c_{n+1} c_{n+2} \ldots . \tag{3.98}$$

On the other hand, if $x \in (0, 1)$, $n \leq 0$ and we write (3.96) as

$$x = .00\ldots 0 c_1 c_2 c_3 \ldots, \tag{3.99}$$

where there are $-n$ zeros which initially appear in the expansion. The number $c_1 \in \{1, 2, \ldots, b-1\}$ appearing in (3.98) or (3.99), as the case may be, is called the *leading significant digit* or *leading digit* of x and is

3.15. Leading digits and geometric sequences

denoted by dig(x) or $\text{dig}_1(x)$. More generally, the numbers $c_1, c_2, c_3, \ldots$ appearing in (3.98) and (3.99) are called the *significant digits of x to the base b* and we write $\text{dig}_j(x) = c_j$ for $j = 1, 2, 3, \ldots$. Then $\text{dig}_j(x)$ is called the jth *significant digit of x relative to the base b*. Note that whereas $\text{dig}_1(x) \in \{1, 2, \ldots, b-1\}$, for $j \geq 2$ we have $\text{dig}_j(x) \in \{0, 1, 2, \ldots, b-1\}$.

Note that if we choose n as in (3.95) and $n \geq 1$, then n is the number of digits in (3.97) that are attached to non-negative powers of b. And if we choose n as in (3.95) and $n \leq 0$, then $-n$ is the number of zeros initially appearing in (3.99) before the first non-zero digit c_1 appears. A special case is when $x \in \mathbb{N}$. Then we would have $c_j = 0$ for all $j \geq n + 1$, and the expansion of x with respect to b would look like $x = c_1 c_2 \ldots c_n.00000$, which may be abbreviated to $x = c_1 c_2 \ldots c_n.0$ or, and this is common usage, $x = c_1 c_2 \ldots c_n$. Thus, the number in base 10 whose expansion is $31415926.00000\ldots$ is commonly written as simply 31415926. In the present context, this latter expression is a bit confusing, because juxtaposition of numbers, especially when they are abstract symbols, generally means multiplication, whereas here it doesn't. In order to avoid this confusion we shall always write in such a situation that $x = c_1 c_2 \ldots c_n.0$. Note that it is not difficult to show that, because of the uniqueness of the digits in the expansion of a number in [0, 1) to the base b, the uniqueness of the digits in the expansion of *any* number in $(0, \infty)$ to the base b follows (see Exercise 18). Also, it follows from (3.97) that when $x \in [1, \infty)$ the integer and fractional parts of x are given by

$$\text{int}(x) = \sum_{j=1}^{n} c_j b^{n-j} = c_1 \ldots c_n.0 \text{ and}$$

$$\text{frac}(x) = \sum_{j=1}^{\infty} \frac{c_{j+n}}{b^j} = .c_{j+n} c_{j+n+1} \ldots.$$

Finally, note that if $x \in (0, 1)$ and $d_j(x)$ is the jth digit of x relative to b as defined in Section 3.2, then $n \leq 0$ in (3.95) and we have $\text{dig}_j(x) = d_{j-n}(x)$ for all $j = 1, 2, \ldots$.

Now, the leading digit of x can be expressed directly in terms of x. To see this, observe that

$$x = b^{\log_b x} = b^{\text{int}(\log_b x) + \text{frac}(\log_b x)} = b^{\text{int}(\log_b x)} b^{\text{frac}(\log_b x)},$$

which gives

$$\frac{x}{b^{\text{int}(\log_b x)}} = b^{\text{frac}(\log_b x)} \in [1, b).$$

By the remarks above on the uniqueness of n and c_1, it follows that

$$n - 1 = \text{int}(\log_b x) \quad \text{and that}$$
$$\text{dig}(x) = c_1 \iff b^{\text{frac}(\log_b x)} \in [c_1, c_1 + 1). \tag{3.100}$$

The latter part of (3.100) can be expressed equivalently by the formula[8]

$$\text{dig}(x) = \text{int}\big(b^{\text{frac}(\log_b x)}\big). \tag{3.101}$$

Gelfand's problem is about the distribution of the leading digits of the numbers in a geometric sequence. It is closely related to Benford's observation concerning the distribution of leading digits in various data and suggests the underlying mathematical phenomenon concealed within Benford's data. This may not *explain* Benford's observations, but it does clarify what is happening. Benford himself was aware of the connection between his observations and geometric sequences and wrote [5]:

> We are so accustomed to labeling things $1, 2, 3, \ldots$ and then saying they are in a natural order that the idea of $1, 2, 4, \ldots$ being a more natural arrangement is not easily accepted. Yet it is in this latter manner that a surprisingly large number of phenomena occur, and the evidence for this is available to everyone.

Let $b \in \mathbb{N}$ with $b \geq 2$ and let $\text{dig}(m)$ denote the leading digit of m relative to the base b. Let $a \in (0, \infty)$ and consider the sequence of numbers $a, a^2, a^3, \ldots$ in $\mathbb{N}$. Then for each $n \in \mathbb{N}$, $\text{dig}(a^n) \in \{1, 2, \ldots, b - 1\}$. Now, let d be some given number in $\{1, 2, \ldots, b - 1\}$. Consider

3.15. Leading digits and geometric sequences

the proportion of elements of $\{a, a^2, a^3, \ldots, a^n\}$ that have leading digit equal to d. This proportion is

$$\frac{1}{n} \left| \left\{ j : 1 \leq j \leq n \text{ and } \mathrm{dig}(a^j) = d \right\} \right|.$$

Consider the existence and value of the limit

$$\lim_{n \to \infty} \frac{1}{n} \left| \left\{ j : 1 \leq j \leq n \text{ and } \mathrm{dig}(a^j) = d \right\} \right|.$$

Newcomb's and Benford's observations would suggest that this limit may exist and, at least in the case where the base is 10, be equal to $\log_{10}(1 + 1/d)$. We now investigate this using Weyl's Theorem 2.20, on the distribution of the fractional parts of the multiples of an irrational number in $[0, 1)$.

Observe that from (3.100),

$$\mathrm{dig}(a^j) = d \iff \mathrm{frac}(\log_b(a^j)) \in \left[\log_b d, \log_b(d+1) \right)$$

$$\iff \mathrm{frac}(j \log_b a) \in \left[\log_b d, \log_b(d+1) \right) \quad (3.102)$$

This shows that if $\log_b a$ is irrational, Weyl's Theorem may be applied to the sequence $\bigl(\mathrm{frac}(j \log_b a)\bigr)$ to deduce a result about the distribution of first digits in the sequence (a^n). Now, $\log_b a$ is rational if and only if there are $p, q \in \mathbb{N}$ such that $\log_b a = p/q$. But

$$\log_b a = p/q \iff a = b^{p/q} \iff a^q = b^p.$$

Thus, if $a^q \neq b^p$ for all $p, q \in \mathbb{N}$, $\log_b a$ is irrational and Weyl's Theorem can be applied to (3.102). We obtain the following result, which in general terms shows that for a given digit, there is an (asymptotic) relative frequency of that digit in the sequence of all leading digits obtained from a geometric sequence of integers, and gives a value for that frequency.

Theorem 3.33. *Let $a \in (0, \infty)$ and let $b \in \mathbb{N}$ with $b \geq 2$, and assume that no positive integer power of a equals a positive integer power of b.*

If $x \in (0, \infty)$ let dig(m) be the leading digit of m relative to the base b, and let $d \in \{1, 2, \ldots, b-1\}$ be given. Then,

$$\lim_{n \to \infty} \frac{1}{n} \left| \{j : 1 \leq j \leq n \text{ and } \mathrm{dig}(a^j) = d\} \right| = \log_b \left(1 + \frac{1}{d}\right).$$

Proof. Because $\log_b a$ is irrational, an application of Weyl's Theorem 2.20 tells us that

$$\lim_{n \to \infty} \frac{1}{n} \left| \{j : 1 \leq j \leq n \text{ and } \mathrm{frac}(j \log_b a) \in [\log_b(d), \log_b(d+1))\} \right|$$
$$= \mu\big([\log_b d, \log_b(d+1))\big).$$

Together with (3.102) this gives

$$\lim_{n \to \infty} \frac{1}{n} \left| \{j : 1 \leq j \leq n \text{ and } \mathrm{dig}(a^j) = d\} \right|$$
$$= \mu\big([\log_b d, \log_b(d+1))\big)$$
$$= \log_b(d+1) - \log_b d$$
$$= \log_b \left(1 + \frac{1}{d}\right).$$

$\square$

Theorem 3.33 should be compared with Borel's Theorem, where we saw that the asymptotic frequency of zeros in the sequence of binary digits of a typical number was $1/2$, the *same* as the asymptotic frequency of ones. Here, the asymptotic frequency of a given digit in the sequence of all leading digits is *not* the same for each digit—we might say that this process is *biased* towards the lower digits, whereas Borel's Theorem is "*unbiased*" in that both 0 and 1 have the same asymptotic frequency. An alternative view might be that the process is not biased, but rather the bias is only *apparent*, and is due to the fact that the terms of a *geometric* sequence are being considered, not a number in $[0, 1)$ that is chosen "at random". A refinement of this view might be that the process is not biased, and we sense this as soon as we use the "appropriate" notion of length, which is $\mu_{\log_b}$, the logarithmic length to the base b. The situation will become clearer in subsequent sections.

3.16 Multiple digits and a result of Diaconis

What can we say about the distribution of digits other than the leading digit? Suppose we have a given base $b \geq 2$ and digits $c_1, c_2, \ldots, c_r \in \{0, 1, 2, \ldots, b-1\}$ with $c_1 \neq 0$. We shall determine when a number $x \in (0, \infty)$ has the property that its first r digits are equal to $c_1, c_2, \ldots, c_r$ respectively. To do this, observe that from (3.100) we deduce

$$\text{dig}(x) = c_1 \iff \text{frac}(\log_b x) \in [\log_b c_1, \log_b(c_1 + 1)).$$

The first step is to extend this formula to multiple digits.

Let $x \in (0, \infty)$ be given and let

$$x = \sum_{j=-n+1}^{\infty} \frac{c_{j+n}}{b^j} \qquad (3.103)$$

be the expansion of x in the base b as in (3.96). Thus, if $r \in \mathbb{Z}$,

$$b^r x = \sum_{j=-n+1}^{\infty} \frac{c_{j+n}}{b^{j-r}} = \sum_{j=-r-n+1}^{\infty} \frac{c_{j+r+n}}{b^j}. \qquad (3.104)$$

A comparison of (3.103) and (3.104) shows that $\text{dig}(x) = \text{dig}(b^r x) = c_1$. Also, the digit in the expansion of $b^r x$ that is attached to the power b^j, namely c_{j+r+n}, is obtained by looking at the digit attached to b^{j+r} in the expansion of x. That is, when we go from the expansion of x to the expansion of $b^r x$ in the base b, the digits stay the same but are shifted r positions to the *left*. We say that multiplication by b^r leads to a *left shift through r positions* in the digits. Similarly, still assuming that $r \in \mathbb{N}$, multiplication by b^{-r} will lead to a *right shift* through r positions in the digits. These facts are, of course, very familiar in the case of base 10, as the examples $100 \times 12.345 = 1234.5$ and $10^{-2} \times .0314159 = .000314159$ show.

Lemma 3.34. *Let* $r, b \in \mathbb{N}$ *with* $b \geq 2$ *and let* $a_1, a_2, \ldots, a_r \in \{0, 1, \ldots, b-1\}$ *with* $a_1 \neq 0$ *be given. Let* $x \in (0, \infty)$ *and let the expansion of* x *to the base* b *be*

$$x = \sum_{j=-n+1}^{\infty} \frac{c_{j+n}}{b^j}, \qquad (3.105)$$

as in (3.103), so that $c_j \in \{0, 1, \ldots, b-1\}$ *for all* j *and* $c_1 \neq 0$. *Then,*

$$c_1 = a_1, \ldots, c_{r-1} = a_{r-1} \text{ and } c_r = a_r$$
$$\iff \operatorname{frac}(\log_b x) + r - 1 \qquad (3.106)$$
$$\in \left[\log_b(a_1 a_2 \ldots a_r.0), \log_b(a_1 a_2 \ldots a_r.0 + 1)\right).$$

Proof. As $n - 1 = \operatorname{int}(\log_b x)$ by (3.100), it follows from (3.105) that

$$b^{\operatorname{frac}(\log_b x)} = \frac{b^{\operatorname{int}(\log_b x) + \operatorname{frac}(\log_b(x))}}{b^{\operatorname{int}(\log_b(x))}} = \frac{x}{b^{n-1}} = c_1 + \sum_{j=1}^{\infty} \frac{c_{j+1}}{b^j}.$$

That is, $b^{\operatorname{frac}(\log_b x)} = c_1.c_2 c_3 \ldots$ So, bearing in mind that multiplication by b^{r-1} will move the digits $r - 1$ positions to the left,

$$b^{\operatorname{frac}(\log_b x) + r - 1} = b^{r-1} b^{\operatorname{frac}(\log_b x)} = c_1 c_2 \ldots c_r.c_{r+1} c_{r+2} \ldots.$$

Thus,

$$b^{frac(\log_b x) + r - 1} \in \left[a_1 a_2 \ldots a_r.0,\ a_1 a_2 \ldots a_r.0 + 1\right)$$
$$\iff c_1 c_2 \ldots c_r.c_{r+1} c_{r+2} \ldots$$
$$\in \left[a_1 a_2 \ldots a_r.0,\ a_1 a_2 \ldots a_r.0 + 1\right)$$
$$\iff \operatorname{int}(c_1 c_2 \ldots c_r.c_{r+1} c_{r+2} \ldots) = a_1 a_2 \ldots a_r.0$$
$$\iff c_1 c_2 \ldots c_r.0 = a_1 a_a \ldots a_r.0$$
$$\iff c_1 = a_1, c_2 = a_2, \ldots, c_r = a_r,$$

the latter coming from the uniqueness of the digits in the expansion to any base. Taking logs now gives

3.16. Multiple digits and a result of Diaconis

$$\text{frac}(\log_b x) + r - 1 \in \left[\log_b(a_1 a_2 \ldots a_r.0), \log_b(a_1 a_2 \ldots a_r.0 + 1)\right)$$
$$\iff c_1 = a_1, \ldots, c_r$$
$$= a_r,$$

as required by (3.106). □

Now, let us apply Lemma 3.34 to calculating the relative frequencies of some preassigned number of digits, thus generalizing Theorem 3.33 which dealt only with one digit. This result, for the base 10, is due to Diaconis [11] in 1978.

Theorem 3.35. *Let* $a \in (0, \infty)$, *let* $b \in \mathbb{N}$ *with* $b \geq 2$, *and let* $\log_b a$ *be irrational. If* $x \in (0, \infty)$ *and* $j \in \mathbb{N}$, *let* $dig_\ell(m)$ *be the* ℓ^{th} *significant digit of x relative to the base b, and let* $c_1, c_2, \ldots, c_r \in \{0, 1, 2, \ldots, b-1\}$ *with* $c_1 \neq 0$ *be given. Then,*

$$\lim_{n \to \infty} \frac{1}{n} \left|\left\{ j : 1 \leq j \leq n \text{ and } dig_\ell(a^j) = c_\ell \text{ for all } \ell = 1, 2, \ldots, r \right\}\right|$$
$$= \log_b\left(1 + \frac{1}{c_1 c_2 \ldots c_r.0}\right). \quad (3.107)$$

Proof. Observe that

$$r - 1 = \log_b(b^{r-1}) \leq \log_b(c_1 c_2 \ldots c_r.0) < \log_b(c_1 c_2 \ldots c_r.0 + 1)$$
$$\leq \log_b\left((b-1)(b^{r-1} + b^{r-2} + \cdots + b + 1) + 1\right)$$
$$= \log_b b^r$$
$$= r.$$

It follows from this that

$$\left[\log_b(c_1 c_2 \ldots c_r.0) - r + 1, \ \log_b(c_1 c_2 \ldots c_r.0 + 1) - r + 1\right) \subseteq [0, 1). \quad (3.108)$$

Using Lemma 3.34 and (3.108) we now have

$\mathrm{dig}_1(a^j) = c_1, \ldots, \mathrm{dig}_{r-1}(a^j) = c_{r-1}$, and $\mathrm{dig}_r(a^j) = c_r$
$\iff \mathrm{frac}(\log_b a^j) + r - 1$
$\in \left[\log_b(c_1c_2\ldots c_r.0), \log_b(c_1c_1\ldots c_r.0+1)\right)$
$\iff \mathrm{frac}(j\log_b a)$
$\in \left[\log_b(c_1c_2\ldots c_r.0) - r + 1, \log_b(c_1c_2\ldots c_r.0+1) - r + 1\right)$
$\subseteq [0, 1).$

As $\log_b a$ is irrational, (3.108) enables us to apply Weyl's Theorem 2.20 to this statement to obtain

$$\lim_{n\to\infty} \frac{1}{n}\left|\left\{j : 1 \le j \le n \text{ and } \mathrm{dig}_\ell(a^j) = c_\ell \text{ for all } \ell = 1, 2, \ldots, r\right\}\right|$$
$$= \log_b(c_1c_2\ldots c_r.0+1) - \log_b(c_1c_2\ldots c_r.0)$$
$$= \log_b\left(1 + \frac{1}{c_1c_2\ldots c_r.0}\right),$$

as required. $\square$

The formula (3.107) is frequently expressed as

$$\lim_{n\to\infty} \frac{1}{n}\left|\left\{j : 1 \le j \le n \text{ and } \mathrm{dig}_\ell(a^j) = c_\ell \text{ for all } \ell = 1, 2, \ldots, r\right\}\right|$$
$$= \log_b\left(1 + \frac{1}{\sum_{j=1}^r c_j b^{r-j}}\right).$$

As well as being proved by Diaconis [11], it is also derived in [20, 21]. In the case where $r = 1$, it is easily checked that this formula reduces to the one for the leading digit, obtained in Section 3.15. In the case $r = 2$

3.16. Multiple digits and a result of Diaconis

we get

$$\lim_{n\to\infty} \frac{1}{n} \left\| \left\{ j : 1 \le j \le n \text{ and } \mathrm{dig}_\ell(a^j) = c_\ell \text{ for } \ell = 1 \text{ and } 2 \right\} \right\|$$
$$= \log_b \left(1 + \frac{1}{c_1 b + c_2}\right).$$

Returning to the general case, (3.107) shows that the greater the number $c_1 c_2, \ldots c_r.0$, the lower is the relative frequency of numbers that have the given digits in the first r positions. In particular, if $c_1 = c_2 = \cdots = c_r = 1$, (3.107) gives the relative frequency as being

$$\log_b \left(1 + \frac{1}{1 + b + b^2 + \cdots + b^{r-1}}\right) = \log_b \left(1 + \frac{b-1}{b^r - 1}\right).$$

More generally, if $c \in \{1, 2, \ldots, b-1\}$ and $c_1 = c_2 = \cdots = c_r = c$, (3.107) becomes

$$\log_b \left(1 + \frac{1}{c(1 + b + b^2 + \cdots + b^{r-1})}\right) = \log_b \left(1 + \frac{b-1}{c(b^r - 1)}\right).$$

The results of this and the preceding section can be used to obtain further results concerning the frequency of digits in geometric sequences. For example, what is the frequency of the appearance of a given digit in the *second* significant digit position for the terms of the sequence (a^j)? If $c \in \{0, 1, 2, \ldots, b-1\}$ is given we have

$$\lim_{n\to\infty} \frac{1}{n} \left\| \left\{ j : 1 \le j \le n \text{ and } \mathrm{dig}_2(a^j) = c \right\} \right\|$$
$$= \sum_{c_1=1}^{b-1} \log_b \left(1 + \frac{1}{c_1 b + c}\right)$$
$$= \log_b \left(\left(1 + \frac{1}{b+c}\right)\left(1 + \frac{1}{2b+c}\right) \cdots \left(1 + \frac{1}{(b-1)b+c}\right)\right).$$

In the case $b = 3$ we have from this that the probability of getting a 2 in the second position is

$$\log_3 \left(\frac{6}{5} \cdot \frac{9}{8}\right) = \log_3 \left(\frac{27}{20}\right).$$

The formulas above also allow us to show that the occurrence of a particular second digit (say) is not generally probabilistically independent of the first digit, as noted by Hill [20].

3.17 Dynamical systems and changes of scale

Benford's Law says that the leading significant digit in certain data sets satisfies a logarithmic distribution. Now, as in Figure 3.8, these data sets may arise from the use of a certain unit of measurement, different for the different data sets. However, in the presence of a genuine statistical phenomenon, we expect that the observed distribution of a data set will be *independent* of the particular unit of measurement we use—for it would seem that a genuine statistical phenomenon should not depend upon the units of measurement. So, if Benford's formulas apply to the lengths of rivers when the length is measured in miles, we expect that they will also apply if we choose to measure the lengths of the rivers in kilometers instead. Now, if a change in units means a change in *scale*, we can say that we expect that a data set that satisfies Benford's formulas should also satisfy Benford's formulas upon a change of scale—that is, Benford's formulas should be *scale invariant*. This seems to have been realized by Newcomb in 1881 [32] when he wrote "As natural numbers occur in nature, they are to be considered as the *ratios* of quantities" (my emphasis) and, of course, if it is only the ratios that are significant, the units are not significant.

Now a change in scale corresponds to multiplying the original units by a positive constant. Let a be such a constant. Then the corresponding change of units or scale is given by the function $M_a : (0, \infty) \longrightarrow (0, \infty)$, where $M_a(x) = ax$ for all $x \in (0, \infty)$. Thus, the dynamical system $((0, \infty), M_a)$ seems to be a system that we should consider in relation to Benford's formulas. However, a problem is that $(0, \infty)$ is unbounded. The solution is to replace the system $([0, \infty), M_a)$ by an equivalent system that is bounded. To see how to proceed, suppose that $b \in \mathbb{N}$ with $b \geq 2$. The crucial observations are that if $x \in (0, \infty)$, there is a unique $n \in \mathbb{Z}$ such that $x/b^n \in [1, b)$ and that $\text{dig}(x/b^n) = \text{dig}(x)$.

3.17. Dynamical systems and changes of scale

Thus, the value of the leading significant digit of x to the base b is reduced to what it is for the number x/b^n, and this number is in the *bounded* interval $[1, b)$. The effect is that if we consider the distribution of the leading significant digits of a sequence of points in the unbounded interval $(0, \infty)$, it is equivalent to consider the sequence of corresponding points in the bounded interval $[1, b)$. For example, to the base 10, the number .000000314159265358... in $(0, \infty)$ produces the corresponding number 3.14159265358... in $[1, 10)$, both having a leading significant digit of 3.

Definition Let $b \in (1, \infty)$ be given. Although b need not be in $\mathbb{N}$, we continue to call b the *base*. It follows, as in deriving (3.95), that if $x \in (0, \infty)$ there is a unique $z \in [1, b)$ and a unique $n \in \mathbb{Z}$ such that

$$x = b^{n-1} z. \tag{3.109}$$

The number z in (3.109) is called the *mantissa of* x and is denoted by mant(x). Thus, if mant denotes the function $x \mapsto \text{mant}(x)$, we see that

$$\text{mant} : (0, \infty) \longrightarrow [1, b). \tag{3.110}$$

Note that the definition of the mantissa is *relative* to the given base, but as long as only one base is being considered, no confusion arises. Also, note that whereas this definition of the mantissa is the one used by Boyle [8] and Hill [20], in older usage the mantissa is the fractional part of the logarithm. That is, traditionally, the mantissa is frac($\log_b x$). Here, it is convenient to follow Boyle and Hill in the definition of the mantissa for, as we shall see, this definition of the mantissa of x is strictly analogous to the fractional part of x; it is simply relative to multiplication instead of addition. When $b \in \mathbb{N}$, as far as Benford-type data are concerned, in (3.109) it suffices to consider z instead of x, as x and z have the same sequence of significant digits. Note that in (3.110), mant(x) = x for all $x \in [1, b)$.

Now, let us "change the units" in $(0, \infty)$, by multiplying by some given number in $(0, \infty)$, a say. Then, we replace each $x \in (0, \infty)$ by ax. That is, the change of units corresponds to applying the function

$M_a : (0, \infty) \longrightarrow (0, \infty)$ given by $M_a(x) = ax$. Thus, to the extent that we are interested in Benford-type phenomena, we examine mant(x) before the change of units and mant(ax) after the change of units. The situation is illustrated in the following diagram.

$$\begin{array}{ccc} (0, \infty) & \xrightarrow{M_a} & (0, \infty) \\ \text{mant} \downarrow & & \text{mant} \downarrow \\ [1, b) & \xrightarrow{T_a} & [1, b) \end{array} \qquad (3.111)$$

The picture suggests how we might interpret the change of units in $(0, \infty)$ as a transformation on $[1, b)$. We seek a transformation $T_a : [1, b) \longrightarrow [1, b)$ such that the diagram "commutes"—that is for all $x \in (0, \infty)$, $T_a(\text{mant}(x)) = \text{mant}(M_a(x))$, which is to say

$$T_a \circ \text{mant} = \text{mant} \circ M_a,$$

as indicated by the arrows in the diagram. In particular this gives, for all $x \in [1, b)$,

$$T_a(x) = T_a(\text{mant}(x)) = \text{mant}(ax). \qquad (3.112)$$

We now look at some properties of the mantissa and the transformation T_a, showing in particular that T_a depends only upon the mantissa of a, and that, if T_a is given by (3.112), then (3.111) commutes.

Proposition 3.36. *Let $b \in (1, \infty)$ be a given base, let $a, a', x \in (0, \infty)$ and let $T_a : (0, \infty) \longrightarrow [1, b)$ be given by $T_a(x) = \text{mant}(ax)$. Then the following hold.*

(1) $\text{mant}(x) \in [1, b)$, $\text{mant}(x) = b^{\text{frac}(\log_b(x))}$ and, for $b \in \mathbb{N}$ with $b \geq 2$, $\text{dig}(x) = \text{int}(\text{mant}(x))$.

(2) $\text{mant}(ax) = \text{mant}(a \, \text{mant}(x))$, and the diagram in (3.111) commutes.

(3) $T_a = T_{\text{mant}(a)}$.

3.17. Dynamical systems and changes of scale

(4) $T_a \circ T_{a'} = T_{aa'}$.
(5) $T_a^n = T_{a^n}$ for all $n \in \mathbb{N}$.

Proof. Clearly, $\text{mant}(x) \in [1, b)$, by the definition. If we take logarithms in (3.109) we obtain $\log_b x = n - 1 + \log_b z = n - 1 + \log_b(\text{mant}(x))$. As $\text{mant}(x) \in [1, b)$, $\text{frac}(\log_b(x)) = \log_b(\text{mant}(x))$, and it follows that $\text{mant}(x) = b^{\text{frac}(\log_b(x))}$. That $\text{dig}(x) = \text{int}(\text{mant}(x))$ comes from (3.101). Hence, (1) holds.

Now, recall that $\text{frac}(x + y) = \text{frac}(x + \text{frac}(y))$. Using this and (1) we have, for $a, x \in (0, \infty)$,

$$\begin{aligned}
\text{mant}(a \, \text{mant}(x)) &= b^{\text{frac}(\log_b(a \, \text{mant}(x)))} \\
&= b^{\text{frac}(\log_b a + \log_b(\text{mant}(x)))} \\
&= b^{\text{frac}(\log_b a + \text{frac}(\log_b x))} \\
&= b^{\text{frac}(\log_b a + \log_b x)} \\
&= b^{\text{frac}(\log_b ax)} \\
&= \text{mant}(ax).
\end{aligned}$$

(This identity also can be proved directly from the definition of the mantissa.)

The equation $\text{mant}(a \, \text{mant}(x)) = \text{mant}(ax)$ implies that $T_a \circ \text{mant} = \text{mant} \circ M_a$, so that (3.111) commutes. This shows that (2) holds.

Using (2), we now have for $x \in [1, b)$, $T_a(x) = \text{mant}(ax) = \text{mant}(\text{mant}(a)x) = T_{\text{mant}(a)}(x)$, and so (3) holds. Using (2) again we have, for $x \in [1, b)$, $(T_a \circ T_{a'})(x) = \text{mant}(a \, \text{mant}(a'x)) = \text{mant}(aa'x) = T_{aa'}(x)$, and (4) follows. Finally, (5) is immediate from (4). □

Part (3) of Proposition 3.36 shows that in considering the transformation T_a, we may as well assume that $a \in [1, b)$, because $\text{mant}(a) \in [1, b)$ and $T_a = T_{\text{mant}(a)}$. Now as $T_a(x) = \text{mant}(ax)$ for all $x \in [1, b)$, we see that when $a \in [1, b)$ we can write $T_a(x)$ down explicitly as

$$T_a(x) = \begin{cases} ax, & \text{if } x \in \left[1, \frac{b}{a}\right); \\ \frac{ax}{b}, & \text{if } x \in \left[\frac{b}{a}, b\right). \end{cases} \quad (3.113)$$

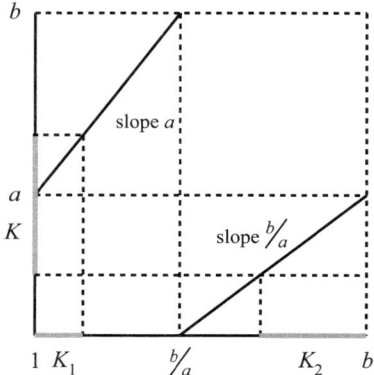

Figure 3.10. Let $b \in (1, \infty)$ and let $1 < a < b$. The figure illustrates the graph of T_a, which lies within the square $[1, b) \times [1, b)$. On the interval $[1, b/a)$, $T_a(x) = ax$ and the graph has slope a; while on the interval $[b/a, b)$, $T_a(x) = ax/b$ and the graph has slope a/b. Despite the differing slopes of the graph on the two different parts of $[1, b)$, the inverse image under T_a of a subinterval K has the same total logarithmic length as K. In the figure, the inverse image of the interval K is the union of the intervals K_1 and K_2. Now, the logarithmic length to the base b of an interval $J = [c, d)$ is $\mu_{\log_b}(J) = \log_b d - \log_b c$, as described in Section 3.4. In the figure, the logarithmic length of K equals the sum of the logarithmic lengths of K_1 and K_2. The calculations are in (3.115), (3.116) and (3.117) below.

In summary, we have seen that, in so far as we are interested in Benford-type data relative to a given base $b \in \mathbb{N}$ with $b \geq 2$, instead of looking at $x \in (0, \infty)$ we need only look at mant$(x) \in [1, b)$. In the case when the base b is in $(1, \infty)$, a change of scale in $(0, \infty)$ corresponds to multiplication by some number $a \in (0, \infty)$, and we have seen that it is enough to assume that $a \in [1, b)$. Then, the calculation that led to (3.113) shows that if $a \in [1, b)$ and T_a is given by (3.113), the dynamical system $([1, b), T_a)$, illustrated in Figure 3.10 corresponds to the change of scale given by multiplication by a. A dynamical system $([1, b), T_a)$ for some $a \in [1, b)$ is called a *Benford system*.

We conclude this section by showing that if (c_n) is a sequence of data points satisfying certain Benford's formulas, then under every change of

3.17. Dynamical systems and changes of scale

scale, the new sequence of data points obtained will also satisfy Benford's formulas. The technical formulation of this idea takes place in the dynamical system $([1, b), T_a)$, and culminates in Theorem 3.39 below. There is a slight but significant difference in our theorem, in that whereas Benford's formulas involved the occurrence only of the discrete digits $1, 2, \ldots, 9$, here the formulation is what can naturally be called *continuous* (see later comments).

Definition Let $b \in (1, \infty)$, and for a subinterval K of $[1, b)$ with end points c, d with $c \leq d$, put

$$\lambda(K) = \log_b d - \log_b c.$$

Then, in the notation introduced at the end of Section 3.4, $\lambda(K)$ is $\mu_{\log_b}(K)$, the logarithmic length of K to the base b. A sequence (c_n) of points in $[1, b)$ *obeys Benford's law to the base b* if, for all subintervals K of $[1, b)$,

$$\lim_{n \to \infty} \frac{1}{n} \left| \left\{ j : 1 \leq j \leq n \text{ and } c_j \in K \right\} \right| = \lambda(K). \tag{3.114}$$

Now, a basic set is a finite disjoint union of intervals, by Proposition 3.7. Consequently, if a sequence (c_n) obeys Benford's Law to the base b, using (3.114) it is easy to check that for every basic subset K of $[1, b)$,

$$\lim_{n \to \infty} \frac{1}{n} \left| \left\{ j : 1 \leq j \leq n \text{ and } c_j \in K \right\} \right| = \lambda(K).$$

That is, if (3.114) holds for subintervals of $[1, b)$, it holds also for basic subsets of $[1, b)$. Note that from Proposition 3.11, λ is an additive set function on the basic subsets of $[1, b)$. These facts are used in proving the following result.

Proposition 3.37. *Let $b \in (1, \infty)$ and let λ denote the logarithmic length relative to the base b. Let $a \in [1, b)$ and let $T_a : [1, b) \longrightarrow [1, b)$ be given by $T_a(x) = \text{mant}(ax)$. Let (c_n) be a sequence of points in $[1, b)$ that obeys Benford's law to the base b. Then the sequence $(T_a(c_n))$ also obeys Benford's law to the base b.*

Proof. Let $K = [c, d)$ be a subinterval of $[1, b)$. If $K \subseteq [1, a)$, it follows from (3.113) that $T_a^{-1}(K) = [bc/a, bd/a)$ and so

$$\lambda(T_a^{-1}(K)) = \log_b\left(\frac{bd}{a}\right) - \log_b\left(\frac{bc}{a}\right) = \log_b d - \log_b c = \lambda(K). \tag{3.115}$$

Similarly, if $K \subseteq [a, b)$, $T_a^{-1}(K) = [c/a, d/a)$ and so

$$\lambda(T_a^{-1}(K)) = \log_b\left(\frac{d}{a}\right) - \log_b\left(\frac{c}{a}\right) = \log_b d - \log_b c = \lambda(K). \tag{3.116}$$

In general, $T_a^{-1}(K)$ is not an interval, so we put

$$K_1 = K_1 \cap [1, a) \text{ and } K_2 = K \cap [a, b). \tag{3.117}$$

Then (3.115), (3.116) and (3.117) show that $K = K_1 \cup K_2$ and that this is a disjoint union of intervals. We see that $T_a^{-1}(K) = T_a^{-1}(K_1) \cup T_a^{-1}(K_2)$ and that this also is a disjoint union of intervals. Thus, $T_a^{-1}(K)$ is a basic set. Using the fact that λ is an additive set function, and using (3.115) and (3.116) again, we have

$$\begin{aligned}
\lambda(T_a^{-1}(K)) &= \lambda(T_a^{-1}(K_1 \cup K_2)) \\
&= \lambda(T_a^{-1}(K_1) \cup T_a^{-1}(K_2)) \\
&= \lambda(T_a^{-1}(K_1)) + \lambda(T_a^{-1}(K_2)) \\
&= \lambda(K_1) + \lambda(K_2) \\
&= \lambda(K).
\end{aligned} \tag{3.118}$$

Now, (c_n) satisfies Benford's law to the base b. So, using (3.114) and (3.118) we now have

$$\begin{aligned}
\lim_{n \to \infty} \frac{1}{n}\left|\left\{ j : 1 \leq j \leq n \text{ and } T_a(c_j) \in K \right\}\right| \\
= \lim_{n \to \infty} \frac{1}{n}\left|\left\{ j : 1 \leq j \leq n \text{ and } c_j \in T_a^{-1}(K) \right\}\right| \\
= \lambda(T_a^{-1}(K)) \\
= \lambda(K).
\end{aligned}$$

As this holds for every subinterval K of $[1, b)$, we see that $(T_a(c_n))$ also satisfies Benford's law to the base b. □

3.18 The equivalence of Kronecker and Benford systems

A Kronecker dynamical system is a system of the form $([0, 1), \tau_\alpha)$, where α is in $[0, 1)$ and $\tau_\alpha : [0, 1) \longrightarrow [0, 1)$ is the transformation given by $\tau_\alpha(x) = \text{frac}(x - \alpha)$. Such systems were extensively discussed in Chapter 3, and a transformation τ_α was called a *translation*. Kronecker's Theorem 2.5 on the density, and Weyl's Theorem 2.20 on the uniform distribution, of the sequence of fractional parts of the multiples of an irrational number are results obtained for systems of this type. Now, there is a connection between the Kronecker system $([0, 1), \tau_\alpha)$, and the Benford system $([1, b), T_a)$—for in the system $([1, b), T_a)$, the mantissa plays the role which was played by the fractional part in the system $([0, 1), \tau_\alpha)$. Whereas it seems that Kronecker systems are those that are appropriate to study in relation to addition, Benford systems are appropriate to study in relation to multiplication. In this section we see that, in a precise sense, the two types of system are equivalent.

Let $a, b \in (0, \infty)$ with $b > 1$. In view of Proposition 3.36, if we put $\alpha = \log_b a$ and use the definition of τ_α we have, for all $x \in [1, b)$,

$$T_a(x) = \text{mant}(ax) = b^{\text{frac}(\log_b(ax))} = b^{\text{frac}(\log_b a + \log_b x)} = b^{\tau_\alpha(\log_b x)}.$$
(3.119)

Let $\theta : [1, b) \longrightarrow [0, 1)$ be given by $\theta(x) = \log_b x$. Then θ is one-to-one and onto and its inverse $\theta^{-1} : [0, 1) \longrightarrow [1, b)$ is given by $\theta^{-1}(y) = b^y$. Then, (3.119) can be written as

$$T_a(x) = b^{\tau_\alpha(\log_b x)} = \theta^{-1}(\tau_\alpha(\theta(x))) = \left(\theta^{-1} \circ \tau_\alpha \circ \theta\right)(x).$$

That is, if $\alpha = \log_b a$,

$$T_a = \theta^{-1} \circ \tau_\alpha \circ \theta,$$
(3.120)

or, alternatively,

$$\theta \circ T_a = \tau_\alpha \circ \theta.$$
(3.121)

The situation expressed by (3.120) and (3.121) can be illustrated in the diagram (3.122) below, and (3.121) expresses the fact that the diagram commutes.

206   Chapter 3. Probability and Randomness

$$\begin{array}{ccc} [1,b) & \xrightarrow{T_a} & [1,b) \\ \theta \downarrow & & \theta \downarrow \\ [0,1) & \xrightarrow{\tau_\alpha} & [0,1) \end{array} \quad (3.122)$$

The relationships in (3.120) and (3.121), illustrated by the commutative diagram (3.122), in effect say that the dynamical systems $([0, 1), \tau_\alpha)$ and $([1, b), T_a)$ are "equivalent", in the sense that whatever one knows about one of the systems can be interpreted equally as something we know about the other. (In fact, (3.121) and (3.122) illustrate that the systems $([0, 1), \tau_\alpha)$ and $([1, b), T_a)$ are conjugate in the sense of Section 3.10.) As an illustration of this idea, we look at how the orbit of a point in one system is related to the orbit of the corresponding point in the other system. Let J be a subinterval of $[1, b)$, let $x \in J$ and let $n \in \mathbb{N}$. Then, as $T_a = \theta^{-1} \circ \tau_\alpha \circ \theta$ by (3.120) we have

$$\begin{aligned} T_a^n(x) &= \left(\theta^{-1} \circ \tau_\alpha \circ \theta\right)^n(x) \\ &= \left(\theta^{-1} \circ \tau_\alpha \circ \theta \circ \theta^{-1} \circ \cdots \theta^{-1} \circ \tau_\alpha \circ \theta\right)(x) \\ &= \left(\theta^{-1} \circ \tau_\alpha^n \circ \theta\right)(x). \end{aligned}$$

Thus,

$$T_a^n(x) \in J \iff \left(\theta^{-1} \circ \tau_\alpha^n \circ \theta\right)(x) \in J \iff \tau_\alpha^n(\theta(x)) \in \theta(J). \quad (3.123)$$

This shows a point in the T_a-orbit of x is in J if and only if the corresponding point in the τ_α-orbit of $\theta(x)$ is in $\theta(J)$. Denoting the subinterval J of $[1, b)$ by $[c, d)$, and using the fact that $T_a^n(x) = T_{a^n}(x) =$ mant$(a^n x)$ as found in (5) of Proposition 3.36, and taking the case $x = 1$ in (3.123), we have

$$\begin{aligned} \text{mant}(a^n) \in [c, d) &\iff T_a^n(1) \in J \iff \tau_\alpha^n(\theta(1)) \in \theta(J) \\ &\iff \tau_\alpha^n(0) \in \theta(J) \iff \text{frac}(n\alpha) \in [\log_b c, \log_b d). \end{aligned} \quad (3.124)$$

Now by Weyl's Theorem 2.20, if $\alpha = \log_b a$ is irrational,

$$\lim_{n \to \infty} \frac{1}{n} \left|\left\{j : 1 \leq j \leq n \text{ and frac}(j\alpha) \in J\right\}\right| = \mu(J).$$

3.18. The equivalence of Kronecker and Benford systems

Thus, when $\log_b a$ is irrational, it follows from (3.124) that

$$\lim_{n\to\infty} \frac{1}{n} \left| \left\{ j : 1 \leq j \leq n \text{ and } \text{mant}(a^j) \in [c, d) \right\} \right|$$
$$= \lim_{n\to\infty} \frac{1}{n} \left| \left\{ j : 1 \leq j \leq n \text{ and } \text{frac}(j\alpha) \in [\log_b c, \log_b d) \right\} \right| \quad (3.125)$$
$$= \mu\big([\log_b c, \log_b d)\big)$$
$$= \log_b d - \log_b c$$
$$= \lambda([c, d)),$$

where λ is the logarithmic length to the base b.

Weyl's Theorem 2.20 describes the distribution properties of the sequence (frac($n\alpha$)), and the frac function arises from the operation of addition on $\mathbb{R}$. The following result is immediate from (3.125), and is a strict analogue of Weyl's Theorem, obtained when we replace the operation of addition by multiplication, when we replace the fractional part by the mantissa, and when we replace the normal notion μ for the length of subintervals of $[0, 1)$ by the logarithmic length λ for subintervals of $[1, b)$.

Theorem 3.38. *Let $a, b \in (0, \infty)$ with $b > 1$ and $\log_b a$ irrational. Then, for each subinterval J of $[1, b)$,*

$$\lim_{n\to\infty} \frac{1}{n} \left| \left\{ j : 1 \leq j \leq n \text{ and } \text{mant}(a^j) \in J \right\} \right| = \lambda(J).$$

A special case of Theorem 3.38 immediately gives the result of Gelfand, obtained earlier as Theorem 3.33.

Theorem 3.39. *Let b be a positive integer with $b \geq 2$, let $a \in (0, \infty)$ and assume that $\log_b a$ is irrational. Let $\text{dig}(x)$ denote the leading significant digit with respect to the base b, and let $d \in \{1, 2, \ldots, b-1\}$ be given. Then,*

$$\lim_{n\to\infty} \frac{1}{n} \left| \left\{ j : 1 \leq j \leq n \text{ and } \text{dig}(a^j) = d \right\} \right| = \log_b \left(1 + \frac{1}{d}\right).$$

Proof. It follows from (3.101) or (1) of Proposition 3.36 that $\text{dig}(x) = \text{int}(\text{mant}(x))$. Thus, if $d \in \{1, 2, \ldots, b-1\}$, $\text{dig}(a^j) = d$ if and only if $\text{mant}(a^j) \in [d, d+1)$. The statement now follows from Theorem 3.38 upon noting that $\lambda([d, d+1)) = \log_b(d+1) - \log_b(d) = \log_b(1 + 1/d)$. □

The results of Theorems 3.38 and 3.39 have essentially been obtained from the fact that the systems $([0, 1), \tau_\alpha)$ and $([1, b), T_a)$ are equivalent, or conjugate in the sense of Section 3.10, as exemplified by the commutative diagram (3.122). All we have then done is to transfer the information we have about the system $([0, 1), \tau_\alpha)$ with the usual notion of length, as obtained by Weyl's Theorem in Chapter 2, to the "equivalent" system $([1, b), T_a)$ with the logarithmic notion of length. In doing this, we have given a dynamical systems interpretation of Newcomb's original insight in his 1881 paper, which he expressed by saying

> The law of probability of the occurrence of numbers is such that all mantissae [the fractional parts] of their logarithms are equally probable.

(The comment in brackets is to indicate Newcomb's original meaning of the mantissa, as distinct from the definition used here and in [8] and [20].)

3.19 Scale invariance and the necessity of Benford's law

One of the questions raised by Benford's data in Figure 3.8 is the following: why should the leading digits in such data obey Benford's logarithmic formula $\log_b(1 + 1/d)$, for $d = 1, 2, \ldots b - 1$? In this section we endeavor to answer this question by showing that if the data are such that the distribution of the leading digits remains the same under every change of scale, that is under every change of units, the distribution must necessarily be given by Benford's formula. Such a result would suggest that Benford's formula is simply a logical necessity for any data

3.19. Scale invariance and the necessity of Benford's law

whose statistical properties remain the same under every change of units or scale. In fact, we show the necessity of Benford's formula even if the data have the same statistical behavior under a *single* change of scale, in certain circumstances. The results in this section should be compared with the recent works of Hill [20, 21, 22] and that of earlier researchers (see especially [21, 37] for references).

The method for tackling this matter is in terms of the concept of an additive set function, introduced in Section 3.4. Each additive set function determines a "statistical law", which a given sequence of data may or may not obey. The general approach in this section is to show that if a sequence of data obeys a law ν, and if under a change of scale the sequence obeys the same law ν, then ν must be the law determined by the logarithmic length. That is, if there is a law that a sequence of data obeys and continues to obey under a change of scale, then that law is unique and must be an appropriate logarithmic law (under certain conditions anyway).

Definition Let S be a given interval and let $\mathcal{B}$ denote the family of basic subsets of S. Let $\nu : \mathcal{B} \longrightarrow [0, \infty]$ be a given additive set function, as defined in Section 3.4, such that $\nu(S) = 1$. That is, $\nu(S) = 1$ and if $A_1, A_2, \ldots, A_n$ are disjoint sets in $\mathcal{B}$ then

$$\nu\left(\bigcup_{j=1}^{n} A_j\right) = \sum_{j=1}^{n} \nu(A_j).$$

Then, a sequence $c_1, c_2, c_3, \ldots$ of elements of S *obeys the law* ν if for all $K \in \mathcal{B}$,

$$\lim_{n \to \infty} \frac{1}{n} \left|\left\{j : 1 \leq j \leq n \text{ and } c_j \in K\right\}\right| = \nu(K). \quad (3.126)$$

Note that the restriction that $\nu(S) = 1$ is necessary, for if (3.126) holds it must hold for $K = S$, so as to ensure that both sides of (3.126) equal 1. The function μ which assigns to each basic set its usual length is an additive set function, as proved in Proposition 3.11. Also, if $b > 1$ is given, the logarithmic length to the base b is denoted by λ and it also is an additive set function on $[1, b)$, by Proposition 3.11. Note that λ is given, as before, by

$$\lambda([c, d)) = \log_b d - \log_b c, \text{ for } [c, d) \subseteq [1, b).$$

Thus, in the case when $b \in (1, \infty)$ is given and $S = [1, b)$, and when the set function ν in (3.126) is taken to be λ, the definition in (3.126) is saying that (c_n) obeys Benford's law relative to the base b, in accordance with the earlier definition as expressed by (3.114).

If ν is an additive set function on the basic subsets $\mathcal{B}$ of an interval S, then $A, B \in \mathcal{B}$ and $A \subseteq B$ imply that $\nu(A) \leq \nu(B)$. For, if $A \subseteq B$, we have $\nu(A) \leq \nu(A) + \nu(B \cap A^c) = \nu(B)$. This fact will be used repeatedly in this section.

Let $f : S \longrightarrow S$ be a transformation such that $f^{-1}(L) \in \mathcal{B}$ for all $L \in \mathcal{B}$. The additive set function ν is called f-*invariant* if

$$\nu(f^{-1}(K)) = \nu(K), \text{ for all } K \in \mathcal{B}. \tag{3.127}$$

The additive set function ν is f-invariant if and only if $\nu(f^{-1}(K)) = \nu(K)$ for all subintervals K of J. (Proving this is Exercise 19.)

Now, the idea of (3.126) is that it expresses a statistical law satisfied by the data $c_1, c_2, \ldots$. The left-hand side of (3.126) says there is always a limiting frequency in the behavior of the data, namely "how often" we have $c_j \in K$, while the right-hand side is the "probability" $\nu(K)$ that a point of J belongs to K. The relationship between (3.126) and (3.127) is given below in Proposition 3.40. The point of this result is that it changes the question about the invariance of the behavior of the data under a transformation of the data to a question of the invariance of the law which the data satisfies. Note that if f is one-to-one and onto and such that both f and f^{-1} map sets in $\mathcal{B}$ to sets in $\mathcal{B}$, then ν is invariant if and only if $\nu(f(A)) = \nu(A)$ for all $A \in \mathcal{B}$.

Proposition 3.40. *Let S be a given interval, let $\mathcal{B}$ denote the family of basic subsets of S, and let ν be an additive set function on $\mathcal{B}$ with $\nu(S) = 1$. Let $f : S \longrightarrow S$ be a transformation such that $f^{-1}(L) \in \mathcal{B}$ for all $L \in \mathcal{B}$. Let (c_n) be a sequence in S that obeys the law ν. Then*

$$(f(c_n)) \text{ obeys the law } \nu \iff \nu \text{ is } f\text{-invariant}.$$

3.19. Scale invariance and the necessity of Benford's law

Proof. As (c_n) obeys the law ν, if $K \in \mathcal{B}$ we have

$$\nu(f^{-1}(K)) = \lim_{n \to \infty} \frac{1}{n} \left| \left\{ j : 1 \leq j \leq n \text{ and } c_j \in f^{-1}(K) \right\} \right|$$

$$= \lim_{n \to \infty} \frac{1}{n} \left| \left\{ j : 1 \leq j \leq n \text{ and } f(c_j) \in K \right\} \right|.$$

It follows that $\nu(f^{-1}(K)) = \nu(K)$ for all $K \in \mathcal{B}$ if and only if

$$\lim_{n \to \infty} \frac{1}{n} \left| \left\{ j : 1 \leq j \leq n \text{ and } f(c_j) \in K \right\} \right| = \nu(K)$$

for all $K \in \mathcal{B}$, which is to say that $(f(c_n))$ obeys the law ν if and only if ν is f-invariant. $\square$

Proposition 3.40 shows that if a sequence obeys the same law under an application of a transformation f, then this means that the law, expressed by a certain additive set function, is f-invariant. So, the following question arises: how many f-invariant additive set functions are there? In certain situations, there is essentially only one f-invariant additive set function, which means that there is a *unique* law for data that satisfies a law and whose transformed data also satisfies the law. We proceed to show that Benford's Law is a law of this type.

The main technique in this section is to show that if ρ is an irrational translation on $[0, 1)$, and if ν is an additive set function on the basic subsets of $[0, 1)$ that is ρ-invariant, then ν must be a constant multiple of the set function μ. That is, following multiplication of ν by some constant, we have

$$\nu([c, d)) = \mu([c, d)) = d - c, \text{ for all subintervals } [c, d) \text{ of } [0, 1).$$

This result in $[0, 1)$ can then be interpreted as a result in $[1, b)$, using the ideas in Section 3.18. We then see that Benford's formulas are a necessary consequence of the following more general observation: if a set of data satisfies the same statistical law under even a *single* change of scale, then that law must be logarithmic. Note that for full validity, this comment must be formulated more precisely.

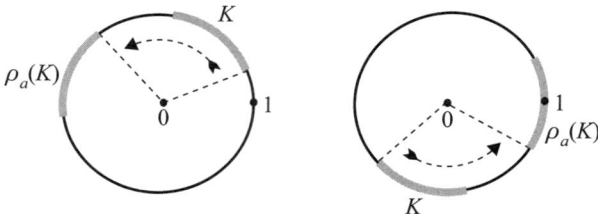

Figure 3.11. This illustrates the circle interpretation of Lemma 3.41. On the unit circle $\mathbb{T}$, ρ_α denotes rotation anti-clockwise through an angle $2\pi\alpha$. On the left and on the right, K denotes an arc and $\rho_\alpha(K)$ is the rotated arc. In each case, the arcs K and $\rho_\alpha(K)$ clearly have the same length, which expresses the fact that arc length on the circle is ρ_α-invariant. On the left, *none* of K pass through 1 as the rotation occurs. But on the right, *part* of K goes through 1. A third case, not illustrated, is where *all* of K goes through 1. In all cases, the rotation does not alter the length of the arc—the arc length is invariant under the rotation. On the circle, the three cases are all the same, in effect. But for $[0, 1)$ the three cases are illustrated respectively below, in the interval interpretation of the circle and the translation interpretation of the rotation.

Lemma 3.41. *Let $\alpha \in [0, 1)$, let τ_α denote the translation on $[0, 1)$ given by $x \longmapsto \mathrm{frac}(\alpha + x)$, let $\mathcal{B}$ denote the family of basic subsets of $[0, 1)$ and let $\mu(B)$ denote the usual length of a basic set $B \in \mathcal{B}$. Then, μ is τ_α-invariant.*

Proof. Here is a formal proof for the interval interpretation, and Figure 3.12 illustrates the unit circle interpretation.

It suffices to prove that if $K = [a, b)$ is a subinterval of $[0, 1)$, then $\mu(\tau_\alpha(K)) = \mu(K)$. Now we have

$$\tau_\alpha([a, b)) = \begin{cases} [a + \alpha, b + \alpha), & \text{if } b + \alpha \leq 1; \\ [a + \alpha, 1) \cup [0, b + \alpha - 1), & \text{if } a + \alpha \leq 1 \text{ and } b + \alpha > 1; \\ [a + \alpha - 1, b + \alpha - 1), & \text{if } a + \alpha > 1. \end{cases}$$

3.19. Scale invariance and the necessity of Benford's law

Thus,

$$\mu(\tau_\alpha(K)) = \begin{cases} b + \alpha - a - \alpha, & \text{if } b + \alpha \le 1; \\ 1 - a - \alpha + b + \alpha - 1, & \text{if } a + \alpha \le 1 \text{ and } b + \alpha > 1; \\ b + \alpha - 1 - a - \alpha + 1), & \text{if } a + \alpha > 1 \end{cases}$$
$$= b - a$$
$$= \mu([a, b))$$
$$= \mu(K).$$

It follows that μ is τ_α-invariant. □

Theorem 3.42. *Let $\alpha \in [0, 1)$ be an irrational number, let τ_α denote the translation on $[0, 1)$ given by $x \mapsto \text{frac}(\alpha + x)$, let $\mathcal{B}$ denote the family of basic subsets of $[0, 1)$, and let μ be the additive set function on $\mathcal{B}$ that assigns to each basic set B its usual length. Then, if ν is a τ_α-invariant additive set function on $\mathcal{B}$, there is $c \in [0, \infty)$ such that $\nu = c\mu$.*

Ideas in the Proof We first note that we can assume without loss of generality that $\nu([0, 1)) = 1$. Then, we have to show that for each subinterval K of $[0, 1)$, $\mu(K) = \nu(K)$. From the irrationality of α and the τ_α-invariance of ν, it is easy to show that for every $x \in [0, 1)$, $\nu(\{x\}) = 0$. This means that $\nu(K)$ is the same regardless of whether the endpoints of K belong to K. So, we may as well assume that K is a subinterval of $[0, 1)$ that is of the form $[a, b)$. The main idea in proving that $\mu(K) = \nu(K)$ is to approximate K by a finite number of intervals, $J_1, J_2, \ldots, J_q$ as described below, such that $\mu(J_j)$ has the same value for each j, $\nu(J_j)$ has the same value for each j, and $\mu(J_j)$ approximately equals $\nu(J_j)$ for each j. This is possible because of the τ_α-invariance of μ and the assumed τ_α-invariance of ν. This means that $\mu(K)$ is "approximately equal" to $\nu(K)$. By increasing the number of intervals $J_1, J_2, \ldots, J_q$ and by estimating the accuracy of the approximations, we find that the approximations become more accurate the more intervals we take, and we can conclude that $\mu(K) = \nu(K)$. Figures 3.12 and 3.13 illustrate the ideas in more detail.

```
┌─╳─╳─╳─)      ┌─╳─╳─⟩
0  J₁ J₂ ------------------ J_q J_{q+1}  1
```

Figure 3.12. This illustrates ideas in part of the argument used to prove that the only τ_α- invariant additive set function ν on $[0, 1)$ such that $\nu([0, 1)) = 1$ is μ. Let $k \in \mathbb{N}$ and put $\beta = \text{frac}(k\alpha)$. Let $q \in \mathbb{N}$ be chosen so that $q\beta \leq 1 < (q + 1)\beta$. Thus,

$$\frac{1}{q+1} < \beta \leq \frac{1}{q}.$$

For $j = 1, 2, \ldots, q$ put $J_j = [(j-1)\beta, j\beta)$. Each interval J_j has μ-length β, a length that can be made as small as we wish by Kronecker's Theorem. Also, as ν is τ_β-invariant and $\tau_\beta(J_{j-1}) = J_j$, there is θ such that $\nu(J_1) = \nu(J_2) = \cdots = \nu(J_q) = \theta$. Put $J_{q+1} = [q\beta, 1)$ and note that $\mu(J_{q+1}) \leq \beta$. As $J_{q+1} \subseteq \tau_\beta(J_q)$, $\nu(J_{q+1}) \leq \nu(\tau_\beta(J_q)) = \theta$. Then, as $\cup_{j=1}^q J_j \subseteq [0, 1) = \cup_{j=1}^{q+1} J_j$, we have $q\theta \leq 1 \leq (q+1)\theta$. Thus,

$$\frac{1}{q+1} \leq \theta \leq \frac{1}{q}.$$

The above displayed inequalities give

$$|\beta - \theta| \leq \frac{1}{q} - \frac{1}{q+1} = \frac{1}{q(q+1)}. \tag{$*$}$$

If β is small, q is large, so we see that for small β, β and θ are "approximately" equal, as expressed precisely by $(*)$.

Proof. We may assume that $\nu \neq 0$. In this case, we may also assume that $\nu([0, 1)) = 1$. For, let $\nu' = \nu([0, 1))^{-1}\nu$. Then ν' is a τ_α-invariant additive set function with $\nu'([0, 1)) = 1$. So, if the result is true for additive set functions ν with $\nu([0, 1)) = 1$, we deduce that $\nu([0, 1))^{-1}\nu = \nu' = \mu$ and the result $\nu = c\mu$ is true with $c = \nu([0, 1))$.

Now, let $\alpha \in (0, 1)$ be irrational and let ν denote an additive set function on $\mathcal{B}$ that is τ_α-invariant and is such that $\nu([0, 1)) = 1$. First, we show that if $x \in [0, 1)$, $\nu(\{x\}) = 0$. Note from Proposition 2.9 that $\tau_\alpha^n = \tau_{n\alpha}$ so that $\text{frac}(x + n\alpha) = \tau_\alpha^n(x)$ for all $x \in [0, 1)$. Thus, as ν is τ_α-invariant,

$$\nu(\{\text{frac}(x + n\alpha)\}) = \nu(\{\tau_\alpha^n(x)\}) = \nu(\{x\}), \text{ for all } n \in \mathbb{N}.$$

3.19. Scale invariance and the necessity of Benford's law 215

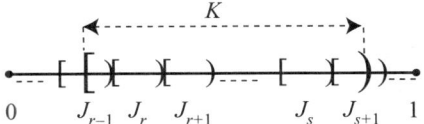

Figure 3.13. This illustrates a further part of the argument used in proving Theorem 3.42. The interval K is "approximately" equal to the union of the $s-r+1$ disjoint intervals $J_r, J_{r+1}, \ldots, J_s$, and there are at most q of these intervals. Consequently, $\mu(K)$ approximately equals $(s-r+1)\beta$ and $v(K)$ approximately equals $(s-r+1)\theta$, and $s-r+1 \leq q$. But now the inequality in $(*)$ above shows that these approximations differ by at most $q|\beta - \theta| \leq 1/(q+1)$. By making β smaller and smaller, q becomes larger and larger and we can deduce that these approximations get closer and closer to each other, and that simultaneously they approximate more and more closely the true values of $\mu(K)$ and $v(K)$. This enables us to deduce that $\mu(K) = v(K)$.

Then, as all the points x, $\text{frac}(x+\alpha)$, $\text{frac}(x+2\alpha)$, ... are distinct,

$$\sum_{n=1}^{\infty} v(\{x\}) = \lim_{r \to \infty} v(\{\text{frac}(x+\alpha), \text{frac}(x+2\alpha), \ldots, \text{frac}(x+r\alpha)\})$$
$$\leq \lim_{r \to \infty} v([0,1)) = 1 < \infty.$$

If $v(\{x\}) > 0$, this would give a contradiction. So, we have

$$v(\{x\}) = 0 \text{ for all } x \in [0, 1). \quad (3.128)$$

Now, let K be a subinterval of $[0, 1)$. Because of (3.128), the value of $v(K)$ is not affected by whether the end points of K are in K. Hence, without loss of generality we may take K to be of the form $[a, b)$ and take $\mu(K) > 0$. We now prove that $v(K) = \mu(K)$.

As α is irrational, by Kronecker's Theorem 2.5 there is a positive integer k such that $\text{frac}(k\alpha) < \mu(K)/3$. Put $\beta = \text{frac}(k\alpha)$, and observe that $\beta < 1/3$ and that, in fact, β can be made as small as we wish. As $\tau_\beta = \tau_\alpha^k$ (see the remarks following Proposition 2.9, every τ_α-invariant finitely additive set function is also τ_β-invariant. Let q be the greatest number in $\mathbb{N}$ such that $q\beta \leq 1$, which implies that $q\beta \leq 1 < (q+1)\beta$ and that $q > 2$. Put $J_0 = \emptyset$ and then, for $j = 1, 2, \ldots, q$, put $J_j =$

$[(j-1)\beta, j\beta)$. Also, put $J_{q+1} = [q\beta, 1)$. Then, $\tau_\beta(J_{j-1}) = J_j$, for $j = 2, \ldots, q$ and $J_{q+1} \subseteq \tau_\beta(J_q)$. As ν is τ_β-invariant,

$$\nu(J_1) = \nu(J_2) = \cdots = \nu(J_q) \text{ and } \nu(J_{q+1}) \leq \nu(J_q).$$

Put $\theta = \nu(J_1) = \cdots = \nu(J_q)$, and note that $\nu(J_{q+1}) \leq \theta$ (in fact, $\nu(J_{q+1}) < \theta$ as β is irrational). Now, $[0, 1)$ is the disjoint union of $J_1, J_2, \ldots, J_q, J_{q+1}$, so we have

$$q\theta = \sum_{j=1}^{q} \nu(J_j) \leq 1 = \sum_{j=1}^{q+1} \nu(J_j) \leq q\theta + \theta = (q+1)\theta.$$

However, we also have $q\beta \leq 1 < (q+1)\beta$, by the definition of q. Thus,

$$\frac{1}{q+1} \leq \theta \leq \frac{1}{q} \text{ and } \frac{1}{q+1} < \beta \leq \frac{1}{q}. \tag{3.129}$$

Hence,

$$|\theta - \beta| \leq \frac{1}{q} - \frac{1}{q+1} = \frac{1}{q(q+1)}. \tag{3.130}$$

This argument is illustrated in Figure 3.12.

Now, for $0 \leq r \leq s \leq q+1$, put $J(r, s) = \bigcup_{j=r}^{s} J_j$. As $\beta < \mu(K)/3$, there are unique r, s with $1 \leq r \leq s \leq q$ such that

$$J(r, s) \subseteq K \subseteq J(r-1, s+1). \tag{3.131}$$

(A situation where neither 0 nor 1 is an endpoint of K is illustrated in Figure 3.13.) Using (3.129) and (3.130) we have

$$|\mu(J(r,s)) - \nu(J(r,s))| = \left|\sum_{j=r}^{s} \mu(J_j) - \nu(J_j)\right| \leq q|\theta - \beta| \leq \frac{1}{q+1} < \beta. \tag{3.132}$$

Using (3.131) and (3.132) we get

$$\nu(K) \leq \nu(J(r,s)) + \nu(J_{r-1} \cup J_{s+1}) < \mu(J(r,s)) + \beta + 2\theta \leq \mu(K) + \beta + 2\theta. \tag{3.133}$$

3.19. Scale invariance and the necessity of Benford's law

Similarly, we also have

$$\mu(K) \leq \mu(J(r,s)) + \mu(J_{r-1} \cup J_{s+1}) < \nu(J(r,s)) + \beta + 2\beta \leq \nu(K) + 3\beta.$$
(3.134)

As $q > 2$, $2/q < 1$. Thus, from (3.129), (3.130), (3.133) and (3.134), we have

$$|\mu(K) - \nu(K)| \leq \max\{\beta + 2\theta, 3\beta\} \leq \beta + 2\beta + \frac{2}{q(q+1)} < 4\beta.$$
(3.135)

But β was a positive number that could be chosen to be as small as we wish, while keeping K unchanged. Thus, (3.135) gives $\mu(K) = \nu(K)$, as required. □

Note that the conclusion in Theorem 3.42 fails when the translation τ_α arises from a rational number α (see Exercise 20).

In Section 3.18 we commented that the commutative diagram (3.122) says that the dynamical systems $([0, 1), \tau_\alpha)$ and $([1, b), T_a)$ are "equivalent" in the sense that whatever one knows about one of the systems can be interpreted equally as something we know about the other. Now, Theorem 3.42 says something about the dynamical system $([0, 1), \tau_\alpha)$. So we expect that if $\alpha = \log_b a$, something equivalent may be said about the system $([1, b), T_a)$. This is expressed by Theorem 3.43 below, essentially proved by Hill [20, Theorem 3.8]. It asserts that every T_a-invariant additive set function on an interval $[1, b)$ is a multiple of the logarithmic length $\mu_{\log_b}$ to the base b, as described in Proposition 3.11.

Theorem 3.43.[9] *Let $b > 1$ be a given real number and let $a \in [1, b)$ be such that $\log_b a$ is irrational. Let $T_a : [1, b) \longrightarrow [1, b)$ denote the function given by*

$$T_a(x) = \mathrm{mant}(ax) = \begin{cases} ax, & \text{if } x \in \left[1, \frac{b}{a}\right); \\ \dfrac{ax}{b}, & \text{if } x \in \left[\dfrac{b}{a}, b\right). \end{cases}$$

Let C denote the family of basic subsets of $[1, b)$, let λ be the additive set function on C that assigns to each basic set $C = [c, d)$ the value

$\lambda(C) = \lambda([c, d)) = \log_b d - \log_b c$. That is, λ is the logarithmic length $\mu_{\log_b}$ to the base b. Then, if η is a T_a-invariant additive set function on C there is $c \in [0, \infty)$ such that $\eta = c\lambda$.

Proof. Let $\mathcal{B}$ denote the family of basic subsets of $[0, 1)$, let $\theta : [1, b) \longrightarrow [0, 1)$ be the one-to-one and onto function given by $\theta(x) = \log_b x$, and let $\alpha = \log_b a$. Observe that α is irrational and that $\lambda = \mu \circ \theta$. Then, if η is a T_a-invariant additive set function on C, $\eta \circ \theta^{-1}$ is a τ_α-invariant additive set function on $\mathcal{B}$. For, as the properties of the log function ensure that a subset A of $[0, 1)$ is in $\mathcal{B}$ if and only if $\theta^{-1}(A)$ is in C, and as $\theta^{-1} \circ \tau_\alpha = T_a \circ \theta^{-1}$ by (3.121),

$$\begin{aligned}(\eta \circ \theta^{-1})(\tau_\alpha(A)) &= \eta((\theta^{-1} \circ \tau_\alpha)(A)) \\ &= \eta((T_a \circ \theta^{-1})(A)) = \eta(T_a(\theta^{-1}(A))) \\ &= \eta(\theta^{-1}(A)) = (\eta \circ \theta^{-1})(A),\end{aligned}$$

which shows that $\eta \circ \theta^{-1}$ is τ_α-invariant. As α is irrational, Theorem 3.43 applies and it follows that there is $c \in (0, \infty)$ such that $\eta \circ \theta^{-1} = c\mu$. But then, $\eta = c\mu \circ \theta = c\lambda$. □

Note that just as Theorem 3.42 fails when α is rational (see Exercise 20), so too does Theorem 3.43 fail when $\log_b a$ is rational. Now, we saw in Proposition 3.40 that if a sequence of points obeys a law, a transformed sequence of these points obeys the same law if and only if the law is invariant under the transformation. Thus, we can interpret Theorem 3.43 as saying that if we have a sequence of points in $[1, b)$ that obeys a law ν, and if this sequence obeys the same law under a change of units, then ν must be the logarithmic length to the base b. This demonstrates the necessity of the logarithmic law as the only possible law for scale-invariant data. Consequently, Benford's Law is a law of necessity. Here is formal statement of this result.

Theorem 3.44. *Let $b > 1$ be a given real number and let $a \in [1, b)$ be such that $\log_b a$ is irrational. Let $T_a : [1, b) \longrightarrow [1, b)$ denote the*

3.19. Scale invariance and the necessity of Benford's law

function given by

$$T_a(x) = \text{mant}(ax) = \begin{cases} ax, & \text{if } x \in \left[1, \dfrac{b}{a}\right); \\ \dfrac{ax}{b}, & \text{if } x \in \left[\dfrac{b}{a}, b\right). \end{cases}$$

Let C denote the family of basic subsets of $[1, b)$ and let ν be an additive set function on C such that $\nu([1, b)) = 1$. Let (c_n) be a sequence of points in $[1, b)$ that obeys the law ν and such that $(T_a(c_n))$ also obeys the law ν. Then, $\nu = \mu_{\log_b}$, the logarithmic length to the base b.

Proof. This follows from Proposition 3.40 and Theorem 3.43. □

Finally, note that in this chapter we have distinguished between *Benford's Law* and *Benford's formulas*. We might say that if $b \in \mathbb{N}$ with $b \geq 2$, a sequence (x_n) in $(0, \infty)$ satisfies Benford's formulas to the base b if

$$\lim_{n \to \infty} \frac{1}{n} |\{j : 1 \leq j \leq n \text{ and } \text{dig}(x_j) = d\}|$$
$$= \log_b\left(1 + \frac{1}{d}\right), \text{ for all } d = 1, 2, \ldots, b - 1.$$

This is equivalent to having

$$\lim_{n \to \infty} \frac{1}{n} |\{j : 1 \leq j \leq n \text{ and } \text{mant}(x_j) \in J\}| = \mu_{\log_b}(J), \quad (3.136)$$

for all intervals $J = [d, d + 1)$ with $d = 1, 2, \ldots, b - 1$. The point is that in (3.136), equality is only required for the finite number of intervals $[1, 2), [2, 3), \ldots, [b - 1, b)$ whereas, in our definition of Benford's Law, equality in (3.136) is required to obtain for *all* subintervals J of $[1, b)$. We might say that our version of Benford's Law is *continuous*, whereas Benford's formulas are a less restrictive *discrete* law. So, the question remains: is the logarithmic law the only possible one in the discrete situation of Benford's formulas? Any answer to this depends upon a more precise formulation of the question, but the problem will not be addressed further here. However, see the recent work of Fewster [13] and also [34]. Also, note that Ross [42] obtains accessible results concerning probabilistic aspects of Benford's law.

Exercises

1. Let Σ denote the set of all sequences of 0s and 1s. That is, $(x_n) \in \Sigma$ means that $x_n \in \{0, 1\}$ for all $n \in \mathbb{N}$. If $x \in [0, 1)$ let $d_n(x)$ denote the nth binary digit of x, and let $\phi : [0, 1) \longrightarrow \Sigma$ be the map $x \longmapsto (d_n(x))$. Prove that ϕ is one-to-one and prove that the range of ϕ has a complement in Σ that is countable. Using Exercises 1.6 and 1.7, deduce that the range of ϕ is uncountable and that $[0, 1)$ is uncountable. (Note that as ϕ is one-to-one, $[0, 1)$ may be identified with the range of ϕ in Σ.)

2. Let $b \in \mathbb{N}$ with $b \geq 2$. If $x \in [0, 1)$, let $d_n(x)$ denote the nth digit of x in its expansion to the base b. Now, for each $n \in \mathbb{N}$ and $k \in \{0, 1, 2, \ldots, b^n - 1\}$, put $J_{k,n} = [k/b^n, (k+1)/b^n)$. Note that the intervals $J_{k,n}$ are pairwise disjoint and that $\cup_{k=0}^{b^n-1} J_{k,n} = [0, 1)$. Thus, given $x \in [0, 1)$ and given n, there is a unique $k \in \{0, 1, 2, \ldots, b^n - 1\}$ such that $x \in J_{k,n}$. Then, given $c \in \{0, 1, \ldots, b-1\}$, prove that $d_n(x) = c$ if and only if the remainder when k is divided by b is c. Deduce that if $c \in \{0, 1, \ldots, b-1\}$,

$$d_n(x) = c \iff x \in \bigcup_{j=0}^{b^{n-1}-1} \left[\frac{jb+c}{b^n}, \frac{jb+c+1}{b^n}\right).$$

In the special case when $b = 2$, deduce that

$$d_n(x) = \begin{cases} 0, & \text{if } x \in \left[\frac{j-1}{2^n}, \frac{j}{2^n}\right) \text{ and } j \text{ is odd;} \\ 1, & \text{if } x \in \left[\frac{j-1}{2^n}, \frac{j}{2^n}\right) \text{ and } j \text{ is even.} \end{cases}$$

3. Let $b \in \mathbb{N}$ with $b \geq 2$. If $k \in \{1, 2, \ldots, b^n\}$, prove that there are $c_1, c_2, \ldots, c_n \in \{0, 1, \ldots, b-1\}$ such that

$$\left[\frac{k-1}{b^n}, \frac{k}{b^n}\right) = \left[\sum_{j=1}^{n} \frac{c_j}{b^j}, \sum_{j=1}^{n} \frac{c_j}{b^j} + \frac{1}{b^n}\right).$$

4. Let S be an interval and let $x \in S$. If A is a basic subset of S, put

$$\delta_x(A) = \begin{cases} 1, & \text{if } x \in A; \\ 0, & \text{if } x \notin A. \end{cases}$$

Exercises

Prove that δ_x is an additive set function on S. Note that δ_x is variously called the *Dirac function* at x or the *Dirac measure at x*.

5. Let S be an interval and let $\mathcal{B}$ be its family of basic subsets. Let ν be an additive set function on $\mathcal{B}$ such that $\nu(S) < \infty$.
   (i) If $\nu(\{x\}) = 0$ for all $x \in S$, prove that there is a non-decreasing function $\upsilon : S \longrightarrow \mathbb{R}$ such that $\nu = \mu_\upsilon$.
   (ii) Put $Z = \{z : z \in S \text{ and } \nu(\{z\}) > 0\}$. Prove either that Z is finite or that the elements of Z may be written as a sequence $z_1, z_2, \ldots$ of points in S.
   (iii) Let $\nu(\{x\}) > 0$ for some $x \in S$ and let Z be as in (ii). Prove that there is a non-decreasing function $\upsilon : S \longrightarrow \mathbb{R}$ such that
   $$\nu = \mu_\upsilon + \sum_{z \in Z} \nu(\{z\}) \delta_z,$$
   where δ_z is the Dirac measure at z as defined in Exercise 4 above. Note that it may be necessary to explain the interpretation of $\sum_{z \in Z} \nu(\{z\}) \delta_z$ as an additive set function on S.

6. Let S be an interval of positive length.
   (i) If $J_1, J_2, \ldots, J_r$ are intervals such that $S \subseteq \cup_{j=1}^{r} J_j$, prove that $\mu(S) \leq \sum_{j=1}^{n} \mu(J_j)$.
   (ii) Prove that S is not a set of measure zero. [Hint: use the Heine-Borel Theorem, mentioned in Section 1.3 of Chapter 1 and discussed in [4, pp. 320-321] and in [40, pp. 65-66].

7. Prove that the Cantor set is uncountable.

8. Let S be an interval and let f, g be step functions on S. Prove that $|f|$, $\max\{f, g\}$, $\min\{f, g\}$ and fg are step functions.

9. Let (ϕ_n) be a non-decreasing sequence of non-negative step functions on a bounded interval S, and suppose that there is $M \in (0, \infty)$ such that $\int_S \phi_n(x) \, dx \leq M$ for all $n \in \mathbb{N}$. Prove that there is a subset $\mathcal{Z}$ of S which has measure zero and is such that if $x \notin \mathcal{Z}$ then $(\phi_n(x))$ is a convergent sequence. As a start on this problem, take $\mathcal{Z}$ to be the set $\{z : z \in S \text{ and } (\phi_n(x)) \text{ does } not \text{ converge}\}$. Also, for each $\varepsilon > 0$ and $n \in \mathbb{N}$, let
$$A_{n,\varepsilon} = \left\{ x : x \in S \text{ and } \phi_n(x) > \frac{M}{\varepsilon} \right\}.$$
As each ϕ_n is a step function, the set $A_{n,\varepsilon}$ is a finite, disjoint union of intervals. Then,
$$\frac{M}{\varepsilon} \chi_{A_{n,\varepsilon}} \leq \phi_n, \text{ and } \frac{M}{\varepsilon} \mu(A_{n,\varepsilon}) \leq \int_S \phi_n(x) \, dx \leq M.$$

It follows that $\mu(A_{n,\varepsilon}) \leq \varepsilon$, for all $n = 1, 2, 3, \ldots$. Now, complete the proof by proving that $A_{n,\varepsilon} \subseteq A_{n+1,\varepsilon}$, that $\mathcal{Z} \subseteq \cup_{n=1}^{\infty} A_{n,\varepsilon}$, and that $\mathcal{Z}$ is contained in a union of a sequence of intervals the sum of whose lengths is at most ε.

10. Let $b \in \mathbb{N}$ with $b \geq 2$. If $x \in [0, 1)$, let $d_n(x)$ denote the nth digit of x in its expansion to the base b. Let $f : [0, 1) \longrightarrow [0, 1)$ be given by

$$f(x) = bx - kx, \text{ for } \frac{k}{b} \leq x < \frac{k+1}{b}, \text{ where } k = 0, 1, 2, \ldots, b-1.$$

   (i) Show that for all $x \in [0, 1)$ and $n \in \mathbb{N}$, $f^n(x) = \text{frac}(b^n x)$.
   (ii) If $x \in [0, 1)$ and $n \in \mathbb{N}$, prove that $d_n(f(x)) = d_{n+1}(x)$.
   (iii) If $x \in [0, 1)$ and $c \in \{0, 1, \ldots, b-1\}$, prove that

   $$d_n(x) = c \iff f^{n-1}(x) \in \left[\frac{c}{b^n}, \frac{c+1}{b^n}\right).$$

11. If $x \in [0, 1)$ let $\sum_{j=1}^{\infty} d_j(x)/2^j$ be the binary expansion of x. Let $r \in \mathbb{N}$ and put $b = 2^r$.

   (i) Prove that

   $$x = \sum_{j=1}^{\infty} \frac{d_j(x)}{2^j} = \sum_{j=1}^{\infty} \frac{1}{b^j} \left(\sum_{s=0}^{r-1} 2^s d_{jr-s}(x)\right).$$

   (ii) For $x \in [0, 1)$ let $g_n(x)$ denote the nth digit in the expansion of x to the base b. Prove that, for all $j \in \mathbb{N}$,

   $$g_j(x) = \sum_{s=0}^{r-1} 2^s d_{jr-s}(x).$$

   (iii) Prove the following form of Borel's Theorem for numbers to the base $b = 2^r$. There is a subset $\mathcal{Z}$ of $[0, 1)$ that has measure zero and is such that if $x \in [0, 1)$ and $x \notin \mathcal{Z}$, then for every $h_1, h_2, \ldots, h_q \in \{0, 1, \ldots, 2^r - 1\}$,

   $$\lim_{n \to \infty} \frac{1}{n} \left|\left\{j : 1 \leq j \leq n \text{ and }\right.\right.$$
   $$\left.\left. g_j(x) = h_1, g_{j+1}(x) = h_2, \ldots, g_{j+q-1}(x) = h_q\right\}\right| = \frac{1}{b^q}.$$

Exercises

12. According to Borel's Theorem 3.22, if $x \in [0, 1)$ and $d_j(x)$ denotes the jth digit of x to the base 2, there is a subset $\mathcal{Z}$ of $[0, 1)$ that has measure zero and is such that if $x \notin \mathcal{Z}$,

$$\lim_{n \to \infty} \frac{d_1(x) + d_2(x) + \cdots + d_n(x)}{n} = \frac{1}{2}.$$

    Prove that such a subset $\mathcal{Z}$ is uncountable.

13. We say that a sequence (u_n) of numbers in $[0, 1)$ is *uniformly distributed* in $[0, 1)$ if, for each subinterval J of $[0, 1)$,

$$\lim_{n \to \infty} \frac{1}{n} \left| \left\{ j : 1 \leq j \leq n \text{ and } u_j \in J \right\} \right| = \mu(J).$$

    Let f be the transformation on $[0, 1)$ given by

$$f(x) = \begin{cases} 2x, & \text{if } 0 \leq x < 1/2; \\ 2x - 1, & \text{if } 1/2 \leq x < 1. \end{cases}$$

    Prove that if $x \in [0, 1)$, then x is normal to the base 2 if and only if $(f^n(x))$ is uniformly distributed in $[0, 1)$.

14. Let $b \in \mathbb{N}$ and let $f : [0, 1) \longrightarrow [0, 1)$ be given by

$$f(x) = \text{frac}(bx).$$

    Prove that the dynamical system $([0, 1), f)$ is transitive and sensitive to initial conditions, where these terms are as defined in Chapter 2, Section 2.9. Prove also that this system has a dense set of periodic points (a point x is *periodic* under the transformation f if $f^n(x) = x$ for some $n \in \mathbb{N}$). Deduce that $([0, 1), f)$ is a chaotic dynamical system, in the sense that it is transitive, sensitive to initial conditions, and has a dense set of periodic points. (This is the definition of a chaotic system in [10, p. 119].)

15. Consider the number x that is given in terms of its binary expansion by

$$x = .00011011110001101011000101100110\cdots,$$

    where the dots indicate that the first 32 digits are repeated in that order. Then x is rational as its binary expansion is periodic. Prove that $x = 233003187/2147483648$, and check that the binary expansion of x shows random behavior at the first and second levels of refinement, but not at the third level of refinement.

16. Let $b \in \mathbb{N}$ with $b \geq 2$, and let $d \in \{1, 2, \ldots, b-1\}$. The leading digit of x with respect to b is denoted by $\text{dig}(x)$. Show that the set
$$\{x : x \in (0, 1) \text{ and } \text{dig}(1/x) = d\}$$
is a union of a sequence of disjoint intervals, and show that the sum of the lengths of the intervals in this sequence is $b/(b-1)d(d+1)$. Prove that $\sum_{d=1}^{b-1} b/[(b-1)d(d+1)] = 1$, and explain why this was to be expected.

17. Let $b \in \mathbb{N}$ with $b \geq 2$, and let $x \in [1, \infty)$. The leading digit of x with respect to b is denoted by $\text{dig}(x)$. A sequence (t_n) of positive numbers is called a *Benford sequence relative to the base* b if, for all $d \in \{1, 2, 3 \ldots, b-1\}$,
$$\lim_{n \to \infty} \frac{1}{n} \left| \{j : 1 \leq j \leq n \text{ and } \text{dig}(t_j) = d\} \right| = \log_b \left(1 + \frac{1}{d}\right).$$
Now, if a sequence (u_n) of positive numbers is such that $(\text{frac}(u_n))$ is uniformly distributed in $[0, 1)$, prove that the sequence (b^{u_n}) is a Benford sequence relative to the base b.

18. Let $b \in \mathbb{N}$ with $b \geq 2$. Then, if $n \in \mathbb{N}$, prove that there are uniquely determined $a_1, a_2, \ldots, a_k \in \{0, 1, \ldots, b-1\}$ and $n_1, n_2, \ldots, n_k \in \mathbb{Z}_+$ with $0 \leq n_1 < n_2 < \cdots < n_k$ such that
$$n = a_1 b^{n_1} + a_2 b^{n_2} + \cdots + a_n b^{n_k}.$$
Hence, or otherwise, deduce that if $x \in (0, \infty)$, the expansion of x to the base b, as described in Section 3.15, uniquely determines the sequence of digits in the expansion.

19. Let J be an interval and let $\mathcal{B}$ be the family of all basic subsets of J. Let $f : J \longrightarrow J$ be a transformation such that $f^{-1}(L) \in \mathcal{B}$ for all $L \in \mathcal{B}$. If ν is a finitely additive set function on $\mathcal{B}$ such that $\nu(f^{-1}(K)) = \nu(K)$ for all subintervals K of J, prove that ν is f-invariant. That is, prove that $\nu(f^{-1}(L)) = \nu(L)$ for all basic subsets L of J.

20. Let $\alpha = p/q$ where $p, q \in \mathbb{N}$ and p, q have no common factors. Let $\tau_\alpha : [0, 1) \longrightarrow [0, 1)$ be given by $\tau_\alpha(x) = \text{frac}(x + \alpha)$. Let μ be the finitely additive set function on the basic sets that assigns to an interval its ordinary length. Prove the following.
   (i) $\{\tau_\alpha^n(0) : n \in \mathbb{N}\} = \{0, 1/q, 2/q, \ldots, (q-1)/q\}$.
   (ii) There is a non-empty basic set A such that $0 < \mu(A) < 1$ and $\tau_\alpha(A) = A$.

(iii) There is a finitely additive set function ν on the basic sets of $[0, 1)$ such that ν is τ_α-invariant and $\nu([0, 1)) = 1$, but $\nu \neq \mu$.

21. Let $\xi : [0, 1] \longrightarrow \mathbb{R}$ be given by $\xi(0) = 0$ and

$$\xi(x) = 1 - x \operatorname{int}\left(\frac{1}{x}\right), \text{ for } x \in (0, 1].$$

Sketch the graph of ξ and then prove the following.

(i) For $x \in (0, 1]$, $\xi(x) = x \operatorname{frac}(1/x)$ and ξ maps $[0, 1]$ into $[0, 1]$.
(ii) Observe that $(0, 1] = \cup_{n=1}^\infty (1/(n+1), 1/n]$ and that this union is disjoint. Thus, for $x \in (0, 1]$, there is a unique $n \in \mathbb{N}$ such that $x \in (1/(n+1), 1/n]$. Prove that $\operatorname{int}(1/x) = n$ and that $\xi(x) = 1 - nx$.
(iii) Prove that $\xi(x) < x$ for all $x \in (0, 1]$.
(iv) Prove that $\lim_{n \to \infty} \xi^n(x) = 0$ for all $x \in [0, 1]$.
(v) Prove that if $x \in [0, 1]$, then $\xi^n(x) = 0$ for some $n \in \mathbb{N}$ if and only if x is rational.

22. Let $(K, *)$ be a group. That is, there is a function $(x, y) \longmapsto x * y$ from $K \times K$ into K, an element $e \in K$, and a function $x \longmapsto x^{-1}$ from K into K such that for all $x, y, z \in K$ we have: $(x * y) * z = x * (y * z)$, $x * e = e * x = x$ and $x * x^{-1} = x^{-1} * x = e$. The element e is unique and is called the *identity* of the group, and x^{-1} is unique and is called the *inverse* of the element $x \in K$. Now let X be a set and let $\theta : K \longrightarrow X$ be one-to-one and onto. For $x, y \in X$, make the definition that

$$x \bullet y = \theta\left(\theta^{-1}(x) * \theta^{-1}(y)\right).$$

Prove that $(X, \bullet)$ is a group, identifying the identity of this group and the inverse of an element of this group.

Investigations

1. Towards the end of Section 3.10, Borel's Theorem 3.22 was interpreted in the form of Theorem 3.32, so as to reveal a "coarse" type of recurrence result in a dynamical system. Now, the Normal Numbers Theorem 3.32 is an advance on Borel's Theorem. So, investigate in what way the Normal Numbers Theorem implies a result about recurrence in an appropriate dynamical system.

2. In the country of Laputa, the government runs lotteries to raise funds for mathematical research. Each time there is a lottery, the government sells n tickets, where n is known in advance for each lottery but the value of n may change from one lottery to another. In each lottery, the tickets are numbered $1, 2, \ldots, n$, and the numbers are written on the tickets to the base 10 in the usual way. When the lottery is drawn, it is by randomly drawing one of these tickets, where each ticket is equally likely to be drawn. If one is trying to buy a winning ticket, is it a reasonable strategy to insist on buying tickets whose leading digit is a 1? If not, is there some sort of reasonable strategy for buying a winning ticket based on the leading digits of tickets? How do the suggested strategies, if any, vary with n, the total number of tickets in the lottery?

3. Let $a, b \in \mathbb{N}$ with $b \geq 2$ and let $c \in (0, \infty)$ all be given. Investigate whether a Benford type of phenomenon occurs for the sequence (ca^n). That is, for $d \in \{1, 2, \ldots, b-1\}$, investigate whether

$$\lim_{n \to \infty} \frac{1}{n} \left| \left\{ j : 1 \leq j \leq n \text{ and } \mathrm{dig}(ca^j) = d \right\} \right| = \log_b \left(1 + \frac{1}{d} \right).$$

Then, investigate the corresponding situation for multiple digits, along the lines of the results of Diaconis as presented in Theorem 3.35.

4. (Some prior knowledge of elementary group theory would assist in this investigation. Also, see Exercise 22) Let $(G, *)$ be a group whose identity is e and in which the inverse of an element x is denoted by x^{-1}. A *subgroup* of a group $(G, *)$ is a group $(H, *)$, where H is a subset of G and the restriction of the function $(x, y) \longmapsto x * y$ on $G \times G$ to $H \times H$ is also denoted by $*$. A group $(G, *)$ is *abelian* if $x * y = y * x$ for all $x, y \in G$. If $(G, *)$ is an abelian group and $(H, *)$ is a subgroup of G, let $G/H = \{xH : x \in G\}$. Each element of G/H is a subset of G called an H-*coset*, and any two H-cosets are either identical or disjoint. Then $(G/H, \star)$ is a group if we define $(xH) \star (yH) = xyH$. This definition of $\star$ is unambiguous, the identity element of G/H is H, and the inverse of xH in G/H is $x^{-1}H$. We call G/H the *quotient group* of G with H. Two groups $(G, *)$ and $(H, \bullet)$ are *isomorphic* if there is a one-to-one and onto function $\theta : G \longrightarrow H$ such that $\theta(x * y) = \theta(x) \bullet \theta(y)$ for all $x, y \in G$, in which case θ is called an *isomorphism*. If the groups G, H are isomorphic, so too are H, G.

   (i) Prove that $(\mathbb{R}, +)$ is a group and that $(\mathbb{Z}, +)$ is a subgroup.
   (ii) Prove that $([0, 1), \oplus)$ is a group if $\oplus$ is given by $x \oplus y = \mathrm{frac}(x + y)$. Identify the identity element of this group, and the inverse of an element of this group. Prove that $([0, 1), \oplus)$ is isomorphic to the quotient group of $(\mathbb{R}, +)$ with $(\mathbb{Z}, +)$, and identify the isomorphism.

(iii) Prove that $(0, \infty)$ is a group under multiplication. Identify the identity element, and the inverse elements of elements of $(0, \infty)$. If $b > 1$ prove that $G_b = \{b^n : n \in \mathbb{Z}\}$ is a subgroup of $(0, \infty)$. Identify the G_b-cosets in $(0, \infty)/G_b$.

(iv) Let $b > 1$ be given, and for $x, y \in [1, b)$ let $x \otimes y = \text{mant}(xy)$. Prove that $([1, b), \otimes)$ is a group, and prove that it is isomorphic to the quotient group $(0, \infty)/G_b$, identifying this isomorphism.

(v) Prove that the groups $([0, 1), \oplus)$ and $([1, b), \otimes)$ are isomorphic and identify the isomorphism.

The above are technical preliminaries. Let $b > 1$ be a given base. Recall that the function denoted by dig assigns to each element $x \in (0, \infty)$ its leading significant digit. Given the remarks at the beginning of Section 3.16, it seems likely that dig is constant on each G_b-coset. Investigate this specific possibility and then systematically look at leading significant digit phenomena from the group theory point of view.

Notes

1. [Page 117] Sets of measure zero play a fundamental role in analysis, for important results such as Borel's Theorem and Birkhoff's Ergodic Theorem are stated in terms of an "exceptional" set of measure zero. Sets of measure zero correspond to events of "probability zero". It is usual to find that results stated using the notion of a set of measure zero are proved using the techniques of Lebesgue integration or general measure theory. It is one purpose of this work to examine to what extent these general theories of integration are essential for discussing some results that traditionally have used them—the answer seems to be that they are unnecessary for a discussion of Weyl's Theorem, a restricted form of Poincaré recurrence, Borel's Theorem, and a restricted version of Kac's formula for recurrence times, but they do seem much more difficult to avoid in proving Birkhoff's Individual Ergodic Theorem. The paper of F. Riesz [38] has a general discussion and survey of classic results in analysis involving sets of measure zero. His comments there on Borel's Theorem on normal numbers are of most specific relevance for this work.

2. [Page 130] The first type of recurrence result was proved by the French mathematician Henri Poincaré. Such results now are usually considered within a measure theory framework.

3. [Page 136] These functions are named after the German born analyst and number theorist Hans Rademacher (1892–1969).

4. [Page 144] Borel presented an argument for this in [7], although there have been questions over the argument's validity. The proof of Borel's Theorem given here has been chosen so as to minimize the number of preliminaries needed to carry out the proof. However, hidden away in the proof is a more general result concerning the integration of step functions: if (ϕ_n) is a nondecreasing sequence of step functions on the bounded interval S such that there is $K \in [0, \infty)$ with the property that $\int_S \phi_n(x)\, dx \leq K$ for all $n = 1, 2, \ldots$, then there is a subset $\mathcal{Z}$ of S such that $\mathcal{Z}$ has measure zero and for all $x \notin \mathcal{Z}$, $(\phi_n(x))$ converges (to a finite limit) (see Exercise 9 above).This provides an approach to a more general integration theory that is mentioned by F. Riesz [38], and is developed in more detail by B. Sz.-Nagy and F. Riesz [39, Chapter II], and A. J. Weir [46]. It's not altogether clear to whom the proof of Borel's Theorem by means of the Rademacher functions is due. This proof is the one used, for example, by M. Kac in [24], and he deduces that

$$\sum_{n=1}^{\infty} \int_0^1 \left(\frac{1}{n} \sum_{j=1}^n r_j(x) \right)^4 dx < \infty.$$

Kac then invokes measure theory to deduce that $\sum_{n=1}^{\infty} \left(\frac{1}{n} \sum_{j=1}^n r_j(x) \right)^4$ is finite, for almost all x. He deduces that for almost all x it must happen that $\lim_{n \to \infty} \frac{1}{n} \left(\sum_{j=1}^n r_j(x) \right) = 0 \ldots$. However, the ideas of using the Rademacher functions and integrating only step functions are mentioned by F. Riesz in [38, p. 221], where he attributes the idea to A. Khintchine. Riesz can avoid using measure theory, but on this point his discussion is allusive rather than complete. The approach in [33] follows the Rademacher function approach and avoids measure theory.

5. [Page 159] Introduced by J. L. Walsh in 1923.

6. [Page 159] In [17], Goodman proved for a general base the results presented in this section for the base 2. To do this he uses "Rademacher functions to the base b"—whereas the Rademacher functions to the base 2 take on the values -1 and 1, which are the second roots of unity, these functions to a general base are complex valued and take their values in the set of bth roots of unity. There are extra technicalities arising from all this, and that is why the approach here is restricted to the base 2. Also, Goodman assumes a knowledge of measure theory, whereas here we don't.

7. [Page 174] Algorithmic complexity requires the mathematical notion of a *Turing machine*. A Turing machine T takes as input a finite sequence v of symbols and either "halts" after a finite number of "computations" and

produces an output consisting of (another) finite sequence $w = T(v)$ of symbols; or it doesn't halt, and computes forever. The notation $T(v) = w$ indicates that a Turing machine is a (special type of) function. Note that T is not one-to-one. We let $|z|$ denote the number of symbols in a finite sequence z of symbols. The finite sequence w of symbols has an *algorithmic complexity* $K_T(w)$ defined by

$$K_T(w) = \begin{cases} \infty, & \text{if } T(v) = w \text{ for no } v; \\ \min\{|v| : T(v) = w\}, & \text{otherwise.} \end{cases}$$

This is sometimes called *Kolmogorov complexity*. The idea is that the "complexity" or "degree of randomness" of w can be estimated by comparing $|w|$ with $K_T(w)$—if $K_T(w)$ is small, w has low complexity, while if $K_T(w)$ is larger, w has higher complexity. This comes from the equation $T(v) = w$, by thinking of v as an (encoded) computer program or algorithm that describes w, the idea being that the more complex or random w is, the longer a program or algorithm must be to describe w. Thus, an infinite sequence $z = (z_n)$ of symbols might be called *random* if there is $c > 0$ such that $K_T(z_1 z_2 \ldots z_n) \geq n - c$, for all $n \in \mathbb{N}$. In fact, it is often assumed that the Turing machine T is *prefix-free*, which means that if T halts for each of the inputs v and v', then neither v nor v' is an initial segment of the other. This leads to the notion of the *Chaitin complexity* of a finite sequence of symbols, with a corresponding definition of a random sequence. Now, imagine we have a partition of $[0, 1)$ into an infinite family $(J_j)_{j=1}^{\infty}$ of disjoint intervals, each of which is of the form $J_j = [(k_j - 1)/2^{n_j}, k_j/2^{n_j})$, for some $n_j \in \mathbb{N}$ and $k_j \in \{1, 2, \ldots, 2^{n_j}\}$. By Theorem 3.4 with $b = 2$, with each J_j is associated a finite sequence $v_j = c_{j1} c_{j2} \ldots c_{jn_j}$ of binary symbols determined by the property that $J_j = \left[\sum_{k=1}^{n_j} c_{jk}/2^{n_j}, \sum_{k=1}^{n_j} c_{jk}/2^{n_j} + 1/2^{n_j}\right)$. Because the intervals J_j form a partition, given two such finite sequences, neither is an initial segment of the other, and we have also

$$\sum_{j=1}^{\infty} 2^{-|v_j|} = \sum_{j=1}^{\infty} 2^{-n_j} = \sum_{j=1}^{\infty} \mu(J_j) = 1.$$

This observation is related to the *Kraft inequality* and also to the question of the probability of when a randomly chosen program—that is, a finite sequence of symbols—causes the machine to halt within a finite number of steps when it is made an input into the machine. If $\mathcal{H}$ denotes the subset of all the possible inputs into the machine that cause the machine to halt after a finite number of steps, the probability that a randomly chosen program is in $\mathcal{H}$ is $\sum_{v \in \mathcal{H}} 2^{-|v|}$, which is an example of a *Chaitin constant*. Discussion of the above ideas may be found in [9, 29, 44], where there are further references, and with [44] recommended as a wide-ranging but succinct introduction.

8. [Page 190] This formula was pointed out to me by Keith Tognetti.

9. [Page 217] A *measure* is a set function μ that is *countably* additive rather than merely finitely additive, in the sense that if (A_n) is a sequence of disjoint sets that have a measure, then $\cup_{n=1}^{\infty} A_n$ has a measure and

$$\mu(\cup_{n=1}^{\infty} A_n) = \sum_{n=1}^{\infty} \mu(A_n).$$

Theorems 3.42 and 3.43 are closely related to a classical result in abstract harmonic analysis, namely that a Haar measure on a compact group—that is, a measure that is invariant under all translations in the group—is essentially unique (see [19, pp.185–186 and 193–195]). The result of Theorem 3.43 reveals the uniqueness of an invariant finitely additive measure assuming Benford-type data under a *single* change of scale. The main results in this section fall into the general category of results about "uniquely ergodic" transformations, which are transformations for which there is a single invariant measure (see for example the discussion by Parry [36, Theorem 2, pp. 14–15], which uses techniques of functional analysis). In [20, Theorem 3.8], Hill uses measure theory and appeals to the fact that an irrational translation is uniquely ergodic to deduce that the logarithmic distribution λ is uniquely determined by Benford-type data. In effect, Theorem 3.42 shows that an irrational translation is uniquely ergodic on $[0, 1)$ and Theorem 3.43 shows that a transformation T_a such that no power of a is a power of b is uniquely ergodic on $[1, b)$.

Bibliography

[1] T. M. Apostol, *Calculus, Volume I*, Blaisdell, New York, 1962.

[2] J. Arledge and S. Tekanski, *A new property of repeating decimals*, The College Math. Journal, **39** (2008), 107–111.

[3] A. Avez, *Ergodic Theory of Dynamical Systems, Volume 1*, University of Minnesota, School of Mathematics, Minneapolis Minnesota, 1966.

[4] R. G. Bartle and D. R. Sherbert, *Introduction to Real Analysis*, third edition, Wiley, New York, 2000.

Bibliography

[5] F. Benford, *The law of anomalous numbers*, Proc. Amer. Phil. Soc., **78** (1938), 551–572.

[6] A. Berger and T. P. Hill, *Newton's method obeys Benford's Law*, Amer. Math. Monthly, **114** (2007), 588–601.

[7] E. Borel, *Les probabilités dénombrables et leurs applications arithmétiques*, Rendiconti del Circ. Math. Palermo, **27** (1909), 247–271.

[8] J. Boyle, *An application of Fourier series to the most significant digit problem*, Amer. Math. Monthly, **102** (1994), 879–886.

[9] C. S. Calude and G. J. Chaitin, *What is... a halting probability?*, Notices Amer. Math. Soc. **57** (2010), 236–237.

[10] R. Devaney, *A First Course in Chaotic Dynamical Systems*, Addison Wesley, Massachusetts, 1992.

[11] P. Diaconis, *The distribution of leading digits and uniform distribution* mod 1, Proc. Amer. Phil. Soc., **78** (1978), 551–572.

[12] W. Feller, *An Introduction to Probability Theory and its Applications*, Third Edition, volume I, Wiley, New York 1968.

[13] R. M. Fewster, *A simple explanation of Benford's Law*, The American Statistician, **63** (2009), 26–32.

[14] B. de Finetti, *Probability, Induction and Statistics: the art of guessing*, Wiley, Aberdeen, 1972.

[15] D. A. Gillies, *An Objective Theory of Probability*, Methuen, London 1973.

[16] B. D. Ginsberg, *Midy's (nearly) secret theorem—an extension after 165 years*, The College Math. Journal, **35** (2004), 26–30.

[17] G. Goodman, *Statistical independence and normal numbers: an aftermath to Marc Kac's Carus Monograph*, American Mathematical Monthly, **106** (1999), 112–126.

[18] R. W. Hamming, *On the distribution of numbers*, The Bell System Technical Journal, **49** (1970), 1609–1625.

[19] E. Hewitt and K. A. Ross, *Abstract Harmonic Analysis, volume I*, Springer, Berlin, 1963.

[20] T. P. Hill, *Base invariance implies Benford's Law*, Proc. Amer. Math. Soc., **123** (1995), 887–895.

[21] T. P. Hill, *The significant-digit phenomenon*, Amer. Math. Monthly, **102** (1995), 322–327.

[22] T. P. Hill, *A statistical derivation of the significant-digit law*, Statistical Science, **10** (1995), 354–363.

[23] K. Hoffman, *Linear Algebra*, Prentice-Hall, Englewood Cliffs, 1971.

[24] M. Kac, *Statistical Independence in Probability, Analysis and Number Theory*, Carus Mathematical Monographs, no. 12, The Mathematical Association of America, Washington DC 1959.

[25] J. L. King, *Three problems in search of a measure*, American Mathematical Monthly, **101** (1994), 609–628.

[26] A. N. Kolmogorov, *Foundations of the Theory of Probability*, translated by N. Morrison, Chelsea, New York 1956.

[27] L. Kuipers and H. Niederreiter, *Uniform Distribution of Sequences*, Pure and Applied Maths no. 29, Wiley, New York, 1974.

[28] M. van Lambalgen, *Random Sequences*, Doctoral thesis, University of Amsterdam, 1987.

[29] M. Li and P. Vitányi, *An Introduction to Kolmogorov Complexity and its Applications*, Springer, New York, 1993.

[30] M. Mendès-France, *Nombres Normaux. Applications aux fonctions pseudo-aléatoires*, Journal d'Analyse Mathématique., **20** (1967), 1–56.

[31] R. von Mises, *Probability, Statistics and Truth* (English translation of the German, originally published 1928), Dover, New York, 1981.

[32] S. Newcomb, *Note on the frequency of use of the different digits in natural numbers*, Amer. Journal Math., **4** (1881), 39–40.

[33] R. Nillsen, *Normal numbers without measure theory*, Amer. Math. Monthly, **107** (2000), 639–644.

[34] R. Nillsen, *Conditions for the necessity of Benford's logarithmic laws*, preprint.

[35] I. Niven, *Irrational Numbers*, Carus Mathematical Monographs No.11, Mathematical Association of America, Washington DC, 1967.

[36] W. Parry, *Topics in Ergodic Theory*, Cambridge University Press, Cambridge, 1981.

[37] R. Raimi, *The first digit problem*, American Mathematical Monthly, **83** (1976), 521–538.

[38] F. Riesz, *Les ensembles de mesure nulle et leur rôle dans l'analyse*, Oeuvres Complètes, I, Académie des Sciences de Hongrie, Budapest, 1960, 353–372.

Bibliography 233

[39] F. Riesz and B. Sz.-Nagy, *Functional Analysis*, Ungar, New York, 1965

[40] K. A. Ross, *Elementary Analysis: the theory of calculus*, Springer, New York, 1980.

[41] K. A. Ross, *Repeating decimals: a period piece*, Math. Magazine, **83** (2010), 33–45.

[42] K. A. Ross, *Benford's Law, a growth industry*, Amer. Math. Monthly, (to appear).

[43] V. Subramanyam, *Cyclic decimal expansions*, Resonance, (2003), 75–80.

[44] S. B. Volchan, *What is a random sequence?*, Amer. Math. Monthly, **109** (2002), 46–63.

[45] D. D. Wall, *On Normal Numbers*, Ph.D. thesis, Univ. of California, Berkeley, 1949.

[46] A. J. Weir, *Lebesgue Integration and Measure*, Cambridge University Press, Cambridge, 1973.

Chapter 4 Flow Chart

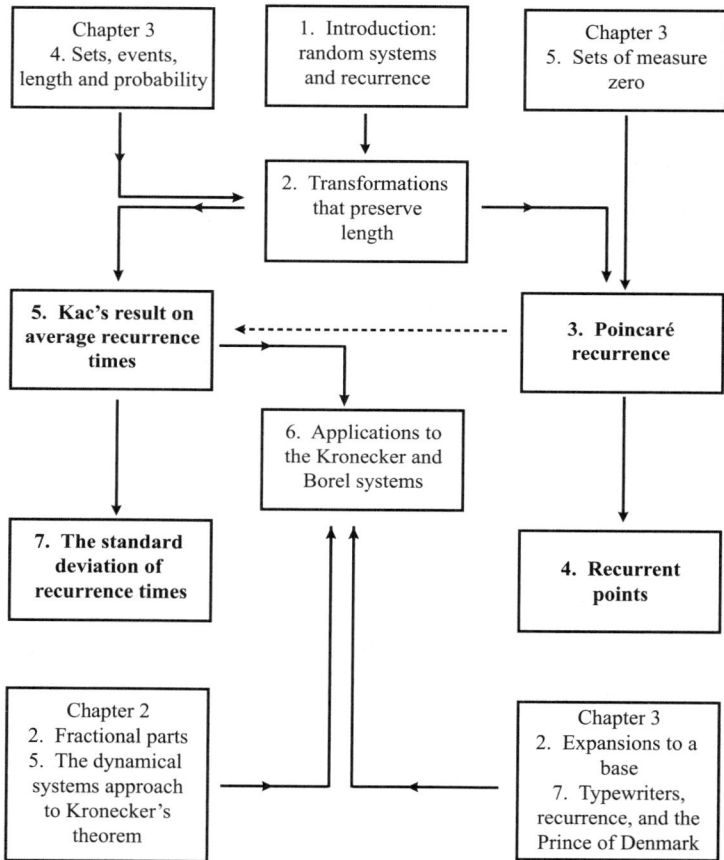

In the diagram, the sections in bold are those with the main results. The unbroken arrows indicate the sequences of sections necessary to get to the results of subsequent sections. The dotted arrows indicate that a preceding section is complementary to the following section, or that it may be useful to the following section without it being essential.

Chapter 4

Recurrence

> The sum total of the energy in the universe is determinate, it is not infinite. Consequently the number of positions, changes, combinations of this energy, although tremendously large and practically "innumerable", is nevertheless determinate and not infinite. But time, in which the universe exercises its energy, is infinite ... Consequently the development at this moment must be a repetition, so too that which it produces and that from which it arises, and so forwards and backwards ... the total arrangement of all forms of energy ever recurs.
>
> *Friedrich Nietzsche* (1844–1900)

4.1 Introduction: random systems and recurrence

Imagine an experiment where an unbiased coin is tossed at random over and over again. Then, recurrence in this process is illustrated by the fact that, sooner or later, we expect to get the outcome "heads". What is more, if we continue tossing, sooner or later we expect to get the result "heads" again—the outcome "heads" will then have *recurred*.

More generally, imagine a process developing over time. One state of the process may be observed and then, at a later time, the same state, or one close to it, may be observed again. We might say that the original state of the process has recurred or has recurred in some approximate sense. Such recurrence phenomena are common in nature. For example, the seasons recur at fixed times of the year, the sun rises once every twenty four hours and many astronomical phenomena recur in a like manner. However, recurrence may not be strictly periodic. During a drought, rain is anticipated to occur at some future time, but there may be little or no knowledge as to when this will happen. Or a gambler on a "losing streak" will anticipate that he must have a big win sooner or later, but he has no idea when it will be, and he may have lost all his money in the meantime. There are different beliefs about human life—orthodox Christianity holds that death is a conclusive event; whereas Buddhists believe that on death one will generally be reincarnated, and this can be thought of as belief in a form of recurrence for individual human life. The German philosopher Friedrich Nietzsche seems to have held that history is recurrent, in that historical events will eventually repeat themselves.

Recurrence may be regular or even predictable, as in the case of the sun and the tides; or it may be unpredictable as in the case of rain. We might think of the former type of recurrence as "deterministic" and the latter type as "random", although it might be argued that what appears to us to be random may be due to a lack of knowledge or understanding. In Chapter 4 we look at the question of recurrence from a mathematical viewpoint, emphasizing the case of recurrent phenomena in dynamical systems on a bounded interval of numbers.

Recall that if S is a set, a function $f : S \longrightarrow S$ is called a *transformation* on S and, in this case, the pair (S, f) is called an (abstract) *dynamical system*. The points of S describe the possible "states" of the system, and the transformation f describes how the system evolves over time. If the system happens to be in state x (say), then $f(x)$ is the state of the system after the elapse of one time unit. If $n \in \mathbb{N}$, then $f^n(x)$ can be interpreted as the state the system is in after n time units, given that the system started in state x. That is, if the system started in state x, the sequence $x, f(x), f^2(x), f^3(x), \ldots$ represents the sequence of successive states of the system as it evolves, and this sequence is called the *orbit* of x.

Insofar as coin tossing is concerned, the preceding ideas may be developed in the context of a specific dynamical system. We let

$$f : [0, 1) \longrightarrow [0, 1)$$

4.1. Introduction: random systems and recurrence

be the binary function given by

$$f(x) = \begin{cases} 2x, & \text{if } 0 \leq x < 1/2; \\ 2x - 1, & \text{if } 1/2 \leq x < 1, \end{cases}$$

and consider the dynamical system $([0, 1), f)$ as a model for repeated coin tossing. Then, if $x \in [0, 1)$ is given, we can interpret x as an individual coin tossing experiment by saying:

"if $f^{n-1}(x) \in [0, 1/2)$, the experiment has resulted in heads on the nth toss", and

"if $f^{n-1}(x) \in [1/2, 1)$, the experiment has resulted in tails on the nth toss".

Now, it may not be clear, at this stage, why these interpretations are suggested as being reasonable; rather, this is something to think about. However, if we accept these interpretations, at least provisionally, the statement

"the 'experiment' x will eventually produce the result 'heads'",

can be expressed mathematically by saying:

"there is a value $n \in \mathbb{N}$ such that $f^n(x) \in [0, 1/2)$." (4.1)

Statement (4.1) expresses in a precise mathematical way the idea of "recurrence" as mentioned in the opening paragraph.

In the case of a dynamical system (S, f), where S is an interval, notions of recurrence arise when questions such as the following are considered: (i) if J is a subinterval of S having positive length and if x is in J, is there $n \in \mathbb{N}$ such that $f^n(x)$ is in J? (ii) if J is a subinterval of S having positive length and if x is in J, is $f^n(x)$ in J for infinitely many n? If either of these questions has an answer of "yes" for all or even some subintervals of S, we could say that the system has a form of recurrence. If the system is "random", we would possibly expect this to occur. For, in this case, if the system starts off in a state x, the points $f(x), f^2(x), f^3(x), \ldots$ should be scattered randomly throughout the system, and so some of them should be in any part of the system that is specified in advance.

One of the main results in Chapter 4 is a form of the Recurrence Theorem proved by Henri Poincaré in 1904 which says, in the present context: if S is a bounded interval and (S, f) is a dynamical system in which the probability of an event is independent of the past history of the system, then for every subinterval J and for almost all points x in J, $f^n(x)$ is in J for some $n \in \mathbb{N}$—that is, under iteration by f, almost all points of J return to J at some future time. Note that in a system as given by the identity transformation $x \longmapsto x$ on an interval, all points in a subinterval will return to the subinterval after a single iteration, even though the system would not be considered as random in any sense. So, recurrence phenomena may be linked to systems which are not random. However, Poincaré's result poses another question: given that x is in J, what is the value of the *recurrence time*, which we take to be the *least* value of $n \in \mathbb{N}$ such that $f^n(x)$ is in J? If the question of the recurrence time is too difficult, we can ask instead: what is the expected or *average* value of the recurrence time as x varies over J? This value would seem to be smaller when J is larger and, under certain conditions, it is proportional to the inverse of the length of J. In a "random" dynamical system (S, f), the average recurrence time for the event J is $\mu(S)/\mu(J)$, where $\mu(J)$ denotes the length, or probability, of the interval J, a result obtained by Marc Kac in 1947 [8].

These ideas lead to the question: how do we formulate precisely the concept of a system in which the probability of an event is independent of the past history of the system? One answer, whose implications are explored here, is that such systems have a transformation that preserves basic sets and their lengths. Such transformations are discussed in Section 2. Whereas the usual treatments of recurrence phenomena in dynamical systems assume a knowledge of measure theory, the aim here is to avoid any such assumptions. The discussion in Chapter 4 is restricted essentially to dynamical systems where the set of states is a bounded interval. When it comes to considering Kac's result on average recurrence times, we need an additional assumption, which is in effect that almost all orbits are spread throughout the system. That is, for Kac's result we need a "random" system, as intuitively described in Section 1.2 of Chapter 1, and elsewhere.

4.2 Transformations that preserve length

When S is an interval and f is a transformation on S, the probabilistic independence from the past of the behavior of the dynamical system (S, f) may be expressed by a "length-preserving" property of the transformation f. Recall from Section 3.4 of Chapter 3 that a *basic set* is defined as a set that is a finite union of intervals, and that the *length* of a basic set is the sum of the lengths of the intervals in a finite family of pairwise disjoint intervals whose union is the given set. If A is a basic set, we denoted the length of A by $\mu(A)$. Thus, when A is an interval, $\mu(A)$ denotes the usual length of the interval. The concept of a *length-preserving* transformation might be understood in different ways. However, the following is the definition we use.

Definition Let f be a transformation on an interval S. Then f is called *length-preserving* if for every basic subset B of S, $f^{-1}(B)$ is basic and $\mu(f^{-1}(B)) = \mu(B)$.[1] Thus, f is length-preserving if $f^{-1}(B)$ is basic whenever B is a basic set and these two sets always have the same length.

Let us consider the possible meanings of length-preserving transformations in terms of the more intuitive notion of probability. For this purpose, it is convenient to assume that $\mu(S) = 1$. Let us think of S as the set of all possible states of the dynamical system or process (S, f). Then, as $f^{-1}(B) = \{x : f(x) \in B\}$, $\mu(f^{-1}(B))$ may be thought of as the probability that $f(x)$ is in B. Or, $\mu(f^{-1}(B))$ may be thought of as the probability that the system finds itself in a state in B after the elapse of one time unit. The point is that because the transformation is length-preserving, $\mu(f^{-1}(B))$ depends only upon $\mu(B)$, not upon the history of the system, nor upon the position of B within S.

Now, if f is length-preserving, if B is a basic set, and if $n \in \mathbb{N}$,

$$\mu(f^{-n}(B)) = \mu(f^{-1}(f^{-(n-1)}(B)))$$
$$= \mu(f^{-(n-1)}(B)) = \cdots$$
$$= \mu(f^{-1}(B)) = \mu(B).$$

The equality $\mu(f^{-n}(B)) = \mu(B)$ may be thought of as saying that after the elapse of n time units, the probability that the state of the system is in B is equal to $\mu(B)$, the probability that the system was initially in a state in B. That is, as time progresses, the probability that the system is in a certain set of states does not change. For this reason, such processes or systems are sometimes called *stationary*.

Note that it can be shown that if f is contracting or expanding, as n increases towards infinity, $\mu(f^{-n}(B))$ increases if f is contracting, and decreases if f is expanding. These ideas are illustrated in Figure 4.1. Length-preserving transformations may be seen as being associated with systems that are precisely "balanced" between expansion and contraction. Such systems may exhibit behavior that is either static or periodic or, at the other extreme, random and chaotic.

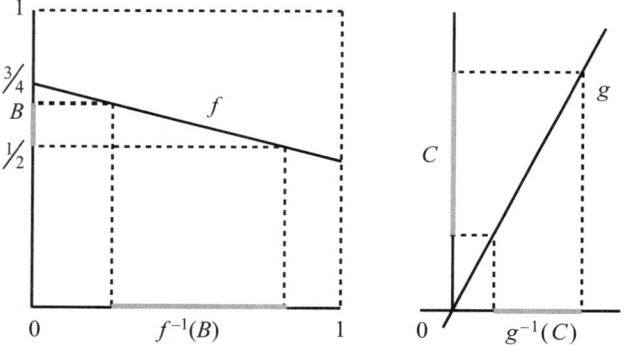

Figure 4.1. The left picture illustrates the transformation $f : [0, 1] \longrightarrow [0, 1]$ given by $f(x) = -x/4 + 3/4$. Then, f is contracting in the sense that $|f(x) - f(y)| = |x - y|/4$, so that the distance between $f(x), f(y)$ is $1/4$ of the distance between x, y. In this case we see that, for the indicated set B, $\mu(f^{-1}(B)) = 4\mu(B)$, illustrating that when f is contracting, the sequence $(\mu(f^{-n}(B)))$ is increasing, a situation corresponding to *information gain* (compare this with Figure 1.4). Note that $\mu(f^{-1}([1/2, 3/4]) = 1$. On the right, $g : \mathbb{R} \longrightarrow \mathbb{R}$ is given by $g(x) = ax$, for some given $a > 1$. We see in this case that g is expanding in the sense that $|f(x) - f(y)| = a|x - y|$ for all x, y, and that for the set C, the sequence $(\mu(g^{-n}(C)))$ is decreasing, a situation corresponding to *information loss* (compare this with Figure 1.5 in Chapter 1).

4.2. Transformations that preserve length

Now consider the case of the *binary transformation*. This is the transformation on $[0, 1)$ given by

$$f(x) = \begin{cases} 2x, & \text{for } 0 \le x < 1/2; \\ 2x - 1, & \text{for } 1/2 \le x < 1, \end{cases}$$

and the dynamical system $([0, 1), f)$ is called the *Borel* system. Given a subinterval $[a, b)$ of $[0, 1)$,

$$f^{-1}([a, b)]) = \left[\frac{a}{2}, \frac{b}{2}\right) \cup \left[\frac{a+1}{2}, \frac{b+1}{2}\right),$$

so that $f^{-1}([a, b))$ is the union of two disjoint intervals and so is a basic set. The disjointness gives

$$\mu(f^{-1}([a, b))) = \mu\left(\left[\frac{a}{2}, \frac{b}{2}\right)\right) + \mu\left(\left[\frac{a+1}{2}, \frac{b+1}{2}\right)\right)$$
$$= \frac{b}{2} - \frac{a}{2} + \frac{b+1}{2} - \frac{a+1}{2}$$
$$= b - a$$
$$= \mu([a, b)).$$

Thus, the basic sets $[a, b)$ and $f^{-1}([a, b))$ have the same length for all a, b. It is easy to see from this (the details in a more general case are considered below), that for every basic subset B of $[0, 1)$, $f^{-1}(B)$ is basic and $\mu(f^{-1}(B)) = \mu(B)$. This shows that f is a length-preserving transformation in the sense of the definition. Figure 4.2 illustrates the fact that the binary transformation is length-preserving.

The following Lemma describes an often useful way of checking whether a transformation is length- preserving.

Lemma 4.1. *Let f be a transformation on an interval S. Assume that f has the properties that whenever J is a subinterval of S, $f^{-1}(J)$ is a basic subset of S and $\mu(f^{-1}(J)) = \mu(J)$. Then f is a length-preserving transformation.*

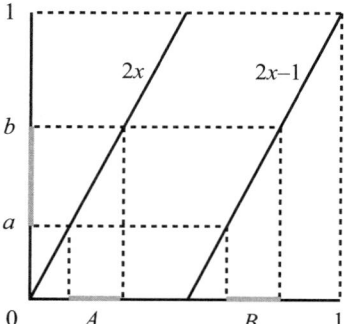

Figure 4.2. The graph is of the transformation f on $[0, 1)$, given by $f(x) = 2x$ for $0 \leq x < 1/2$ and $f(x) = 2x - 1$ for $1/2 \leq x < 1$. The inverse image of $[a, b)$ under f is $A \cup B$, where $A = [a/2, b/2)$ and $B = [(a+1)/2, (b+1)/2)$, as indicated. We have

$$\mu(f^{-1}([a, b))) = (b-a)/2 + (b-a)/2 + (b-a)/2 = b - a = \mu([a, b)),$$

which expresses the fact that f is length-preserving.

Proof. If B is a basic subset of S, there are disjoint subintervals $J_1, J_2, \ldots, J_k$ of S such that $B = \bigcup_{j=1}^{k} J_j$. Then,

$$f^{-1}(B) = f^{-1}\left(\bigcup_{j=1}^{k} J_j\right)$$

$$= \bigcup_{j=1}^{k} f^{-1}(J_j),$$

and it follows that $f^{-1}(B)$ is basic because it is the union of a finite number of basic sets. Also, because the intervals $J_1, J_2, \ldots, J_k$ are disjoint, the basic sets $f^{-1}(J_1), \ldots, f^{-1}(J_k)$ are disjoint. Hence, using (2) of Proposition 3.10,

4.2. Transformations that preserve length

$$\mu\left(f^{-1}(B)\right) = \mu\left(\bigcup_{j=1}^{k} f^{-1}(J_j)\right)$$
$$= \sum_{j=1}^{k} \mu\left(f^{-1}(J_j)\right)$$
$$= \sum_{j=1}^{k} \mu(J_j)$$
$$= \mu\left(\bigcup_{j=1}^{k} J_j\right) = \mu(B).$$

Thus, f satisfies the definition for being a length-preserving transformation. □

We now introduce a general class of length-preserving transformations. To this end, let S be an interval and let $f : S \longrightarrow \mathbb{R}$ be a function. If there are $s, t \in \mathbb{R}$ such that $f(x) = sx + t$, for all $x \in S$, then f is called a *linear* function on S and s is called the *slope*. The range of a linear function is an interval, and if the slope is non-zero, a linear function is one-to-one. The graph of a linear function is a straight line segment. If f is a linear function with slope $s \neq 0$, and if U is a subinterval of the range of f, observe that $|s| = \mu(U)/\mu(f^{-1}(U))$, so that

$$\mu(f^{-1}(U)) = \frac{\mu(U)}{|s|}. \tag{4.2}$$

It is necessary to consider functions that are not necessarily linear over the whole interval S, but that are linear on each of a finite number of subintervals of S.

Definition A real-valued function f on an interval S is said to be *piecewise linear* or to be *piecewise linear on S* if there is a partition of S into a finite number of disjoint intervals such that, on each of these

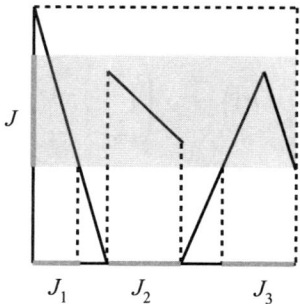

Figure 4.3. The figure depicts the graph of a piecewise linear function on an interval. The function has four linear components. In the figure, the inverse image of the interval J may be found by looking at the parts of the graph of the function that lie within the grey horizontal strip obtained from J. By doing this we identify the intervals J_1, J_2 and J_3, and we see that if we denote the function by f,

$$f^{-1}(J) = J_1 \cup J_2 \cup J_3.$$

Thus, $f^{-1}(J)$ is a finite union of intervals and so is a basic set. This type of argument works for every piecewise linear function, not just for the particular one in the figure.

intervals, f is a linear function. That is, f is piecewise linear on S means that there are subintervals $S_1, S_2, \ldots, S_r$ of S and real numbers $s_1, t_1, s_2, t_2, \ldots, s_r, t_r$ such that

(1) $S = \bigcup_{j=1}^{r} S_j$, where this union is disjoint, and

(2) for each $j = 1, 2, \ldots, r$, $f(x) = s_j x + t_j$ for all $x \in S_j$.

The case of f being linear on S corresponds to the case $r = 1$ in the definition of a piecewise linear function.

Lemma 4.2. *Let f be a piecewise linear function on the interval S. Then for every interval J, $f^{-1}(J)$ is a basic subset of S.*

4.2. Transformations that preserve length

Proof. Let f be piecewise linear on S, and let f be given as in (1) and (2) above, and let J be an interval. Now,

$$f^{-1}(J) = \left(\bigcup_{j=1}^{r} S_j\right) \cap f^{-1}(J) = \bigcup_{j=1}^{r} S_j \cap f^{-1}(J).$$

But, as f is piecewise linear as given by (2) above, each set $S_j \cap f^{-1}(J)$ is an interval. Hence, $f^{-1}(J)$ is a finite union of intervals and so is basic. □

Figure 4.3 illustrates the argument used in proving Lemma 4.2.

The following result gives conditions on a piecewise linear transformation that ensure that it is length-preserving.

Theorem 4.3. *Let S be a bounded interval with positive length, and let f be a piecewise linear transformation on S. Let $S_1, S_2, \ldots, S_r$ be a partition of S into subintervals such that f is linear on each interval S_j. Assume further that when f is restricted to any one of the intervals $S_1, S_2, \ldots, S_r$, the range of f is either S, or S less one or two endpoints. Then f is a length-preserving transformation.*

Proof. The argument is illustrated in Figure 4.4. Let U be a subinterval of S of positive length, and observe that as f has slope s_j on S_j, and as $f(S_j) = S$ except maybe for one or two endpoints, we have $s_j = \mu(U)/\mu(f^{-1}(U) \cap S_j)$, for $j = 1, 2, \ldots, r$. Hence,

$$\mu(f^{-1}(U)) = \mu\left(\bigcup_{j=1}^{r} f^{-1}(U) \cap S_j\right)$$

$$= \sum_{j=1}^{r} \mu(f^{-1}(U) \cap S_j)$$

$$= \mu(U) \sum_{j=1}^{r} |s_j|^{-1}. \tag{4.3}$$

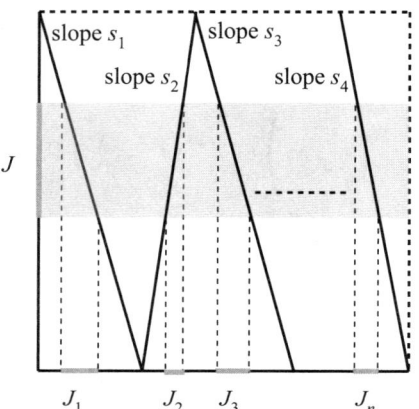

Figure 4.4. The figure depicts a piecewise linear transformation on $[0, 1]$. The transformation has n linear components, each of which has range equal to $[0, 1]$, or $[0, 1]$ less one or two endpoints. The linear components have slopes $s_1, s_2, \ldots, s_n$, as indicated. In the figure, J denotes a subinterval of $[0, 1]$ and the inverse image of J under the transformation is the disjoint union of $J_1, J_2, \ldots, J_n$, as indicated. Now, for each $j = 1, 2, \ldots, n$, the definition of slope means that for $j = 1, 2, \ldots, n$, $|s_j| = \mu(J)/\mu(J_j)$. Thus, denoting the transformation by f we have

$$\mu(f^{-1}(J)) = \mu\left(\bigcup_{j=1}^{n} J_j\right) = \sum_{j=1}^{n} \mu(J_j) = \sum_{j=1}^{n} \frac{\mu(J)}{|s_j|} = \mu(J) \sum_{j=1}^{n} \frac{1}{|s_j|}.$$

Taking $J = [0, 1)$ in this equation, we have $1 = \mu(f^{-1}([0, 1))) = \mu([0, 1))$. We deduce that

$$\sum_{j=1}^{n} \frac{1}{|s_j|} = 1 \text{ so that } \mu(f^{-1}(J)) = \mu(J),$$

and we see that f is length-preserving.

By taking $U = S$ we deduce that $\sum_{j=1}^{r} |s_j|^{-1} = 1$, and then it follows from (4.3) that $\mu(f^{-1}(U)) = \mu(U)$. It follows from Lemma 4.1 and Lemma 4.2 that f is length-preserving. □

4.2. Transformations that preserve length

Note that the binary transformation on $[0, 1)$ is piecewise linear, as depicted in Figure 4.2. Theorem 4.3 implies in particular that the binary transformation is length-preserving.

There is a different situation from that of Theorem 4.3 that still ensures that a piecewise linear transformation is length-preserving.

Theorem 4.4. *Let S be a bounded interval, and let f be a piecewise linear transformation on S. Let $S_1, S_2, \ldots, S_r$ be a partition of S into subintervals such that f is linear on each interval S_j. Assume that on each of the intervals S_j, the slope of f is either 1 or -1, and that if $j \neq k$ then $f(S_j) \cap f(S_k)$ is either empty or a single point. Then f is a length-preserving transformation.*

Proof. For each $j \in \{1, 2, \ldots, r\}$, $f(S_j)$ is an interval, and because f has slope 1 or -1 on each interval, S_j, $\mu(f(S_j)) = \mu(S_j)$. Then, because any two of the sets $f(S_1), f(S_2), \ldots, f(S_r)$ have at most one point in common, we have

$$\mu(S) \geq \mu\left(\bigcup_{j=1}^{r} f(S_j)\right) = \sum_{j=1}^{r} \mu(f(S_j))$$

$$= \sum_{j=1}^{r} \mu(S_j)$$

$$= \mu\left(\bigcup_{j=1}^{r} S_j\right)$$

$$= \mu(S).$$

Since the range $f(S)$ of f equals $\bigcup_{j=1}^{r} f(S_j)$, it follows that the range of f is a basic subset of S which has length equal to the length of S. It follows that $\mu(S \cap f(S)^c) = \mu(S) - \mu(f(S)) = 0$. So, $f(S)$ is a basic subset of S whose complement in S is finite. Hence, if B is a basic subset of S,

$$\mu(B \cap f(S)) = \mu(B). \tag{4.4}$$

Thus, if U is a subinterval of S,

$$\mu\left(f^{-1}(U)\right) = \sum_{j=1}^{r} \mu\left(f^{-1}(U) \cap S_j\right)$$
$$= \sum_{j=1}^{r} \mu\left(f^{-1}(U \cap f(S_j))\right)$$
$$= \sum_{j=1}^{r} \mu(U \cap f(S_j)), \text{ by (4.2)},$$
$$= \mu\left(U \cap f(S)\right)$$
$$= \mu(U), \text{ by (4.4)}.$$

That f is length-preserving now follows from Lemma 4.1. □

Theorems 4.3 and 4.4 describe the main types of length-preserving transformations that we use in this work. The binary transformation is a simple and important example of a length-preserving transformation on $[0, 1)$ which is covered by Theorem 4.3. Perhaps the simplest type of length-preserving transformations covered by Theorem 4.4 are the translations on $[0, 1)$. If $\alpha \in [0, 1)$, the translation corresponding to α is the function f_α, where $f_\alpha(x) = \text{frac}(x + \alpha)$ for all $x \in [0, 1)$. Alternatively,

$$f_\alpha(x) = \begin{cases} x + \alpha, & \text{if } 0 \leq x < 1 - \alpha; \\ x + \alpha - 1, & \text{if } 1 - \alpha \leq x < 1, \end{cases}$$

an expression which makes it clear that f_α must be length-preserving because of Theorem 4.4. Some comments on other types of length-preserving transformations, not covered by Theorems 4.3 or 4.4, are depicted in Figure 4.5.

4.3 Poincaré recurrence

Henri Poincaré's[2] result on recurrence is concerned with when a point in a given set will return to the set at some future time, under iteration by

4.3. Poincaré recurrence

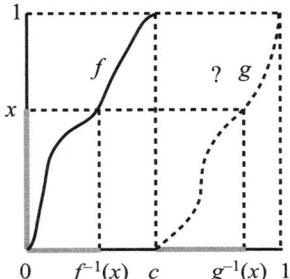

Figure 4.5. In the figure, c is a given number between 0 and 1, and f is a given continuous and strictly increasing function on $[0, c)$ such that $f(0) = 0$ and the range of f on $[0, c)$ is $[0, 1)$. Now, if g is a function mapping $[c, 1)$ into $[0, 1)$, we can define a transformation h on $[0, 1)$ by

$$h(u) = \begin{cases} f(u), & \text{for } 0 \leq u < c; \\ g(u), & \text{for } c \leq u < 1. \end{cases}$$

Here the question is: can we choose the function g so that the transformation h becomes length-preserving? If the function $x \mapsto x - f^{-1}(x) + c$ is strictly increasing and maps from $[0, 1)$ onto $[c, 1)$, we can take the function g to be the one determined by its inverse function, given by $g^{-1}(x) = x - f^{-1}(x) + c$. In this way we see that g may sometimes be chosen so that h is a length-preserving transformation.

a given transformation. Intuitively, if the system is "random", we would expect such recurrence to be the norm—because as we take successive points $x, f(x), f^2(x), \ldots$ of the orbit of a point x in a set U, we would expect these points to be scattered randomly throughout the system, and so some of these points should be in U, demonstrating recurrence. In fact, recurrence phenomena occur when the transformation is length-preserving. Recall from Section 3.5 of Chapter 3 that a subset $\mathcal{Z}$ of $\mathbb{R}$ is a set of measure zero if, for each $\varepsilon > 0$, there is a sequence (A_n) of intervals such that $\mathcal{Z} \subseteq \bigcup_{n=1}^{\infty} A_n$ and $\sum_{n=1}^{\infty} \mu(A_n) < \varepsilon$. Also, recall that we say a statement holds for *almost all* x in a set U if the statement holds for all values of x in U that are outside some set of measure zero. As presented in this section, Poincaré's result is that in a dynamical system (S, f), where S is a bounded interval and f is a length-preserving transformation, for every basic set U of positive length, almost every point of U will, under iteration by f, eventually appear again in U. Poincaré's result originally appeared in [12] and in his *Les Méthodes Nouvelles de la Mécanique Céleste* (the latter appears in English in [13]). Approaches to Poincaré's Theorem based upon measure theory may be found in [4, 11]. The present approach avoids measure theory, and uses only the notions of a basic set on an interval and the length of a basic set. Note that Theorem 3.18 in Chapter 3 presented a recurrence result, saying that for almost all points x of $[0, 1)$, every given finite sequence of binary digits will appear infinitely many times in the binary expansion of x. This is a recurrence result closely related to Poincaré's Theorem. We continue to let $\mu(A)$ denote the length of a basic set A.

A basic observation concerning recurrence is the following. Let S be an interval, let f map S into S and let V be a subset of S. Then for $x \in S$ and $n \in \mathbb{N}$ we have

$$x \in V \text{ and } f^n(x) \in V \iff x \in V \cap f^{-n}(V).$$

Thus, for a given $n \in \mathbb{N}$, V contains a point x such that $f^n(x) \in V$ if and only if $V \cap f^{-n}(V) \neq \emptyset$.

The following is a simple recurrence result, and gives conditions that ensure that there is a point in a set U that will return to U or "recur" in U within a finite number of iterations of the transformation.

Proposition 4.5. *Let S be a bounded interval, let U be a basic subset of S having positive length, and let $f : S \longrightarrow S$ be a length-preserving transformation. Then, there is $x \in U$ and $n \in \mathbb{N}$ such that $f^n(x) \in U$.*

Proof. Suppose that the sets $U, f^{-1}(U), f^{-2}(U) \ldots$ are pairwise disjoint. Then, for all $n \in \mathbb{N}$,

$$n\mu(U) = \sum_{j=1}^{n} \mu(f^{-j}(U)) = \mu\left(\bigcup_{j=1}^{n} f^{-j}(U)\right) \leq \mu(S) < \infty.$$

As this holds for all $n \in \mathbb{N}$ we must have $\mu(U) = 0$. Thus, if $\mu(U) > 0$, there are $j, k \in \mathbb{Z}_+$ with $j > k$ such that $f^{-k}(U) \cap f^{-j}(U) \neq \emptyset$.

4.3. Poincaré recurrence

Hence,

$$f^{-k}(U \cap f^{-(j-k)}(U)) = f^{-k}(U) \cap f^{-j}(U) \neq \emptyset$$
$$\implies U \cap f^{-(j-k)}(U) \neq \emptyset.$$

But then, if $x \in U \cap f^{-(j-k)}(U)$, put $n = j - k$ and we have $x \in U$ and $f^n(x) \in U$. □

Note that if U is an open interval and if $x \in U$ is such that $f^j(x) \in U$ for some $j \in \mathbb{N}$, this does not mean necessarily that $f^n(x) \in U$ for infinitely many values of $n \in \mathbb{N}$ (see Figure 4.6 and Exercise 5). We now proceed to show that under the conditions of Proposition 4.5, almost all points of S will recur in U.

Lemma 4.6. *Let S be a bounded interval and let $f : S \longrightarrow S$ be a length-preserving transformation. Let $n \in \mathbb{N}$ with $n > 1$, let A be a basic subset of S, and let*

$$B = A \cap \left(\bigcap_{j=1}^{n-1} f^{-j}(A^c) \right).$$

Then B is a basic set and

$$\mu(B) \leq \frac{\mu(S)}{n}.$$

Proof. As A is basic, A^c is basic. Also, as f is length-preserving, $f^{-j}(A^c)$ is basic for $j \in \mathbb{N}$. Thus, B is a finite intersection of basic sets and so is basic. Now observe that

$$x \in B \implies x \in A \cap \left(\bigcap_{j=1}^{n-1} f^{-j}(A^c) \right)$$
$$\implies f^j(x) \notin A \text{ for all } j = 1, 2, \ldots, n - 1$$
$$\implies f^j(x) \notin B \text{ for all } j = 1, 2, \ldots, n - 1, \text{ as } B \subseteq A,$$
$$\implies x \notin f^{-j}(B) \text{ for all } j = 1, 2, \ldots, n - 1.$$

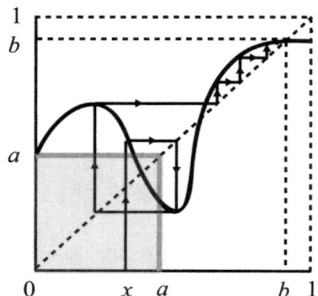

Figure 4.6. In the figure, the graph of a transformation on [0, 1] is pictured. The number a is given between 0 and 1, and x is the point indicated in $[0, a]$. If the graph of the transformation on $[0, a]$ were to lie entirely within the grey-colored square $[0, a] \times [0, a]$, then the orbit of x would lie entirely within $[0, a]$. As it is, we can follow the arrows and we see that $x \in [0, a]$, $f(x) \notin [0, a]$, $f^2(x) \in [0, a]$, $f^3(x) \notin [0, a]$, $f^4(x) \notin [0, a], \ldots, f^n(x) \notin [0, a], \ldots$. In fact, we see that $\lim_{n \to \infty} f^n(x) = b$, where b is the indicated fixed point of the transformation. Thus, the point x is in $[0, a]$ to start with, it leaves $[0, a]$ on the first iteration, returns to $[0, a]$ on the second iteration, and after that it leaves $[0, a]$ permanently and converges to a point in another part of the system. So, x returns *once* to $[0, a]$ but never returns to $[0, a]$ after that. The above picture also can produce examples of points in $[0, a]$ that return to $[0, a]$ an arbitrarily large number of times before leaving $[0, a]$ permanently and then converging to the fixed point b. Of course, the transformation in this figure is not length-preserving.

Thus, $B \cap f^{-j}(B) = \emptyset$ for all $j = 1, 2, \ldots, n-1$. Hence if $j, k \in \{1, 2, \ldots, n\}$ and $j < k$, $k - j \in \{1, 2, \ldots, n-1\}$ and

$$f^{-j}(B) \cap f^{-k}(B) = f^{-j}(B \cap f^{-(k-j)}(B)) = f^{-j}(\emptyset) = \emptyset.$$

Thus, the sets $f^{-j}(B)$ for $j = 1, 2, \ldots, n$ are pairwise disjoint and we have

$$n\mu(B) = \sum_{j=1}^{n} \mu(B) = \sum_{j=1}^{n} \mu(f^{-j}(B)) = \mu\left(\bigcup_{j=1}^{n} f^{-j}(B)\right) \leq \mu(S).$$

This gives $\mu(B) \leq \mu(S)/n$, as required. □

4.4. Recurrent points

Theorem 4.7. *(Poincaré's Recurrence Theorem, 1890).*[3] *Let S be a bounded interval, let U be a basic subset of S of positive length, and let f be a length-preserving transformation from S to S. Then, for almost all $x \in U$, there is $n \in \mathbb{N}$ such that $f^n(x) \in U$.*

Proof. For $n = 1, 2, 3, \ldots$ let

$$U_n = U \cap \left(\bigcap_{j=1}^{n-1} f^{-j}(U^c) \right),$$

and let

$$\mathcal{Z} = \bigcap_{n=1}^{\infty} U_n = U \cap \left(\bigcap_{j=1}^{\infty} f^{-j}(U^c) \right).$$

Note that by Lemma 4.6, each set U_n is basic and $\mu(U_n) \leq \mu(S)/n$, so that $\lim_{n \to \infty} \mu(U_n) = 0$. As $\mathcal{Z} = \bigcap_{n=1}^{\infty} U_n$, $\mathcal{Z}$ is a set of measure zero. Now, if $x \in U$ and $x \notin \mathcal{Z}$, we see from the definition of $\mathcal{Z}$ that there is $n \in \mathbb{N}$ such that $x \notin f^{-n}(U^c)$. That is, $x \in U$ but $f^n(x) \notin U^c$, which is to say that $x \in U$ and $f^n(x) \in U$, as required. □

4.4 Recurrent points

Poincaré's Recurrence Theorem asserts the existence almost everywhere of points that return to a given set, but it does not tell us whether a point will return to the set *close to the original point*. In this section the result of Poincaré is refined by introducing the notion of a *recurrent point*. This notion is intrinsic to the point itself. The main result is essentially a strengthening of Poincaré's Theorem and says that, under given conditions, almost all points in the space are recurrent.

Definition Let S be a bounded interval, and let f be a transformation on S. Then a point $x \in S$ is said to be *periodic for f*, or to be simply *periodic*, if there is $n \in \mathbb{N}$ such that $f^n(x) = x$. In this case, there is a least element $r \in \mathbb{N}$ such that $f^r(x) = x$ and this number r is called the *period* of x. A periodic point has period 1 if and only if $f(x) = x$, in which case x is called a *fixed point* of f. The point x in S is called *recurrent* if for each open interval V with $x \in V$, there is some $n \in \mathbb{N}$ such that $f^n(x) \in V$.

Note that if x is a periodic point, then it is recurrent. For, in this case there is $r \in \mathbb{N}$ such that $f^r(x) = x$. Hence, if V is an interval with $x \in V$, $f^r(x) = x \in V$. It then follows that x is a recurrent point, by the definition. Note also that in this case, for all $n \in \mathbb{N}$,

$$f^{rn}(x) = (f^r)^n(x) = f^r(f^r(\cdots f^r(x)\cdots)) = x \in V. \quad (4.5)$$

Now let x be a recurrent point that is *not* periodic, and let V be an open interval that contains x. It is convenient to denote V by V_1. Then because x is recurrent, there is a least value $n_1 \in \mathbb{N}$ such that $f^{n_1}(x) \in V_1$, and $f^{n_1}(x) \neq x$ because x is not periodic. Let V_2 be an open interval such that $x \in V_2 \subseteq V_1$ and $f^{n_1}(x) \notin V_2$. As x is recurrent, there is a least value $n_2 \in \mathbb{N}$ such that $f^{n_2}(x) \in V_2$ and $f^{n_2}(x) \neq x$ because x is not periodic. Also, necessarily, $n_2 > n_1$. Let V_3 be an open interval such that $x \in V_3 \subseteq V_2$ and $f^{n_2}(x) \notin V_3$. As x is recurrent, there is a least value $n_3 \in \mathbb{N}$ such that $f^{n_3}(x) \in V_3$ and $f^{n_3}(x) \neq x$ because x is not periodic. Necessarily, $n_3 > n_2$. Continuing in this way, we see that there is an infinite non-increasing sequence (V_n) of subintervals of S, starting with the given interval V, and there is a strictly increasing sequence $n_1, n_2, n_3, \ldots$ in $\mathbb{N}$ such that

$$x \in V_j \subseteq V \text{ and } f^{n_j}(x) \in V_j \subseteq V, \text{ for all } j \in \mathbb{N}. \quad (4.6)$$

With a little reflection it is clear that the intervals $V_2, V_3, V_4, \ldots$ above may be chosen so that their lengths are as small as we wish. Then, by considering the cases of periodic and non-periodic points separately, the following result is immediate from (4.5) and (4.6).

Proposition 4.8. *Let S be an interval, let $f : S \longrightarrow S$ be a transformation on S, and let x be a point in S that is a recurrent point of f. Let V be an open subinterval of S with $x \in V$. Then, there is an infinite sequence $n_1, n_2, n_3, \ldots$ in $\mathbb{N}$ such that $n_1 < n_2 < n_3 < \cdots$ and $f^{n_j}(x) \in V$ for all $j = 1, 2, 3, \ldots$. Moreover, this sequence may be chosen so that $\lim_{n \to \infty} f^{n_j}(x) = x$.*

4.4. Recurrent points

We can think of Proposition 4.8 in the following way. If x is a recurrent point, and if the system starts in the initial state x, then it evolves so as to be close to that state at some future time—in fact, there is even an infinity of future times at which the system is in a state as close to the initial state as we care to specify. Alternatively, a point being recurrent means that an infinite number of points in the orbit of the point "return" or "recur" to be as close to the point as we wish. In this section, the main aim is to discuss a class of transformations for which almost all points are recurrent. The class of transformations consists of those that preserve the lengths of basic sets, as discussed in Section 4.2. The lemmas that have been proved enable us to say quite a lot about the preponderance of recurrent points for certain types of transformations.

Theorem 4.9. *(Poincaré's Recurrence Theorem, second version).*[3] *Let S be a bounded interval and let f be a length-preserving transformation on S. Then, almost all points of S are recurrent.*

Proof. The open subintervals of S that have rational end points may be arranged in a sequence $U_1, U_2, U_3 \ldots$. Then, if V is an open subinterval of S and $x \in V$, there is $\ell \in \mathbb{N}$ such that $x \in U_\ell \subseteq V$. For each $\ell = 1, 2, 3, \ldots$ and $n = 2, 3, \ldots$, let

$$\mathcal{Z}_{\ell,n} = U_\ell \cap \left(\bigcap_{j=1}^{n-1} f^{-j}(U_\ell^c) \right).$$

As we shall see, the significance of the sets $\mathcal{Z}_{\ell,n}$ lies in the fact that the non-recurrent points can be expressed in terms of them. Note that $\mathcal{Z}_{\ell,n}$ is a basic set and, by Lemma 4.6 (applied to U_ℓ in place of A),

$$\mu(\mathcal{Z}_{\ell,n}) \leq \frac{\mu(S)}{n}. \tag{4.7}$$

Now let $\varepsilon > 0$. Let $m \in \mathbb{N}$ be such that $m > 1$ and $\mu(S)/m < \varepsilon$. Using (4.7) gives

$$\bigcap_{n=2}^{\infty} \mathcal{Z}_{\ell,n} \subseteq \mathcal{Z}_{\ell,m} \text{ and } \mu(\mathcal{Z}_{\ell,m}) < \varepsilon.$$

However, $\mathcal{Z}_{\ell,m}$ is a basic set and so it follows from the definition of a set of measure zero that

$$\bigcap_{n=2}^{\infty} \mathcal{Z}_{\ell,n} \text{ is a set of measure zero for all } \ell \in \mathbb{N}. \quad (4.8)$$

Let $\mathcal{Z}$ denote the set of all non-recurrent points, and let $x \in \mathcal{Z}$. By the definition of a recurrent point, there is some open subinterval V such that $x \in V$ and $f^n(x) \notin V$ for all $n \in \mathbb{N}$. Then, because of the way the intervals U_n have been chosen, one of them, say U_ℓ, will have the property that $x \in U_\ell \subseteq V$. Thus, $x \in U_\ell$ but $f^n(x) \notin U_\ell$ for all $n \in \mathbb{N}$. That is, $x \in U_\ell$ and $f^n(x) \in U_\ell^c$ for all n. So, we have

$$x \in \bigcap_{n=2}^{\infty} \left(U^\ell \cap \left(\bigcap_{j=1}^{n-1} f^{-n}(U_\ell^c) \right) \right) = \bigcap_{n=2}^{\infty} \mathcal{Z}_{\ell,n}.$$

We deduce that

$$\mathcal{Z} \subseteq \bigcup_{\ell=1}^{\infty} \left(\bigcap_{n=2}^{\infty} \mathcal{Z}_{\ell,n} \right). \quad (4.9)$$

Now from (4.8) we have that $\bigcap_{n=2}^{\infty} \mathcal{Z}_{\ell,n}$ is a set of measure zero, for all $\ell \in \mathbb{N}$. It now follows from (4.9) that $\mathcal{Z}$ has measure zero, because it is contained in the union of a sequence of sets of measure zero (see (2) of Proposition 3.12). Thus, $\mathcal{Z}$ is a set of measure zero and, as $\mathcal{Z}$ consists of all the non-recurrent points, every point not in $\mathcal{Z}$ must be recurrent, and the conclusion follows. □

4.5 Kac's result on average recurrence times

Suppose we have a length-preserving transformation on a bounded interval S. According to Poincaré, if U is a subinterval of positive length, almost all points of U will eventually return to U under iterations of the transformation applied to the point. For some points in U, the time taken to return to U (that is the smallest number of iterations) may be short, but for other points in U, the time taken to return to U may be long. A

4.5. Kac's result on average recurrence times 257

question answered by Kac[4] in 1947 is: on the average, how long does it take for a point of U to return to U under iterations of the transformation? This average is called the *average recurrence time* for the interval U and the transformation.

The question of average recurrence times may be considered heuristically, as follows. As the transformation is length-preserving, the probability $\mu(f^{-n}(U))/\mu(S)$ that $f^n(x)$ is in U is equal to the probability $\mu(U)/\mu(S)$ of U. Thus, if the probability of U is small, it will be "harder" for $f^n(x)$ to be in U. So, the time taken for x to return to U should be related to the probability of U occurring in the first place—the lower the probability of U, the longer it should take, in general, for x to return to U. That is, it seems that the average recurrence time for U is inversely related to the probability of U, and the latter is $\mu(U)/\mu(S)$. The simplest operation which could express this inverse relationship is that of taking reciprocals. So, perhaps the simplest we can hope for under this heuristic argument is that

$$\text{the average recurrence time for } U = \frac{\mu(S)}{\mu(U)}. \tag{4.10}$$

Kac proved that this is what actually occurs. Note that his result shows that the average recurrence time depends only upon $\mu(U)$ – not upon f, and not upon the position of U within S. We shall consider this problem in relation to length-preserving transformations on an interval, whereas Kac's analysis was more general. Before proceeding, note that (4.10) is readily confirmed in the case $U = S$. For, if $x \in S$, $f(x) \in S$. Thus, $\Theta_S = 1$, the average recurrence time for S is 1, and this also equals $\mu(S)/\mu(S) = 1$. Also, in the case where $U = \emptyset$, since $f^n(x) \notin \emptyset$ for all n, it is natural to interpret the recurrence time on $\emptyset$ to be ∞, so the "average recurrence time" for $\emptyset$ is ∞. However, when $U = \emptyset$, the right-hand side of (4.10) is $\mu(S)/0$, which we interpret as ∞. So, we see that in both these cases we have agreement with (4.10). These cases are extreme in the sense that in them S is either empty or consists of everything. An intermediate example is found by taking f to be the transformation on $[0, 1)$ given by $f(x) = x + 1/2$ for $0 \leq x < 1/2$ and $f(x) = x - 1/2$ for $1/2 \leq x < 1$, and taking $U = [0, 1/2)$. Then it is easy to check that f is length-preserving and that for all $x \in U$, $f(x) \notin U$ but $f^2(x) = x \in U$, so

the average recurrence time for $U = 2 = \frac{1}{1/2} = \frac{1}{\mu(U)}$.

It is instructive to consider the above in relation to coin tossing. If we have an unbiased coin that is tossed repeatedly, how long should it take, on average, before we obtain "heads"? The probability of getting heads on the first toss is $1/2$, the probability of getting heads on the second toss and not before is $1/4$, the probability of getting heads on the third toss and not before is $1/8$, and so on. The average number of tosses before getting heads therefore seems to be

$$1 \cdot \frac{1}{2} + 2 \cdot \frac{1}{4} + 3 \cdot \frac{1}{8} + 4 \cdot \frac{1}{16} + \cdots = 2 = \frac{1}{1/2}.$$

This gives a further confirmation of (4.10).

Now let us consider the above ideas more formally. Let S be a bounded interval and let $f : S \longrightarrow S$ be a length-preserving transformation on S. Let U be a given basic subset of S that has strictly positive length. As we saw in Poincaré's Recurrence Theorem, Theorem 4.7, almost all points x of U have the property that $f^n(x) \in U$ for some $n \in \mathbb{N}$. Hence, for almost all points x of U, there is a *minimum* or *least* value of $n \in \mathbb{N}$ such that $f^n(x) \in U$. Note that this minimum value of n will depend on the point x.

Definition Let S be a bounded interval and let $f : S \longrightarrow S$ be a transformation on S. Let U be a subset of S. Then θ_U is the function $\theta_U : S \longrightarrow \mathbb{N} \cup \{\infty\}$ given as follows:

$$\theta_U(x) = \begin{cases} \infty, \text{ if } f^n(x) \notin U \text{ for all } n \in \mathbb{N}; \\ \min\{n : n \in \mathbb{N} \text{ and } f^n(x) \in U\}, \text{otherwise.} \end{cases} \quad (4.11)$$

Then, $\theta_U(x)$ is called the *recurrence time of x relative to U and f*, or simply the *recurrence time of x*.

Note that in the case when the transformation is length-preserving, Poincaré's Recurrence Theorem 4.7 is equivalent to saying: $\theta_U(x) < \infty$ for almost all $x \in U$.

4.5. Kac's result on average recurrence times

Lemma 4.10. *Let S be a bounded interval. Then the following hold.*

(1) Let (A_n) be a sequence of basic subsets of S such that $A_{n+1} \subseteq A_n$ for all n and $\bigcap_{n=1}^{\infty} A_n$ is a set of measure zero. Then, $\lim_{n \to \infty} \mu(A_n) = 0$.

(2) Let B be a basic set, let (B_n) be a sequence of pairwise disjoint basic sets, and let W be a set of measure zero such that $B = W \cup (\bigcup_{n=1}^{\infty} B_n)$. Then, $\mu(B) = \sum_{n=1}^{\infty} \mu(B_n)$.

Proof. (1) First, assume that each basic set A_n is a finite union of *closed* intervals. Then the intersection of closed sets is closed (see [1, p. 314] or [14, p. 63] for example). Thus, $\bigcap_{n=1}^{\infty} A_n$ is closed and bounded. Let $\varepsilon > 0$. Then, as $\bigcap_{n=1}^{\infty} A_n$ is a set of measure zero, there is a sequence (W_n) of open intervals such that

$$\bigcap_{n=1}^{\infty} A_n \subseteq \bigcup_{n=1}^{\infty} W_n \text{ and } \sum_{n=1}^{\infty} \mu(W_n) < \varepsilon. \quad (4.12)$$

Now in (4.12), (W_n) is an open covering of the closed bounded set $\bigcap_{n=1}^{\infty} A_n$. So, by the Heine-Borel Theorem ([1, p. 321] or [14, p. 65]), there is $\ell \in \mathbb{N}$ such that

$$\bigcap_{n=1}^{\infty} A_n \subseteq \bigcup_{n=1}^{\ell} W_n.$$

Taking complements gives

$$\bigcup_{n=1}^{\infty} A_n^c \supseteq \bigcap_{n=1}^{\ell} W_n^c.$$

But the complement of a set is open if and only if the set is closed ([1, p. 313] or [14, p. 63]). So, the sets A_n^c are open and the sets W_n^c are closed. Hence, (A_n^c) is an open covering of the closed and bounded set $\bigcap_{n=1}^{\infty} W_n^c$. Noting that the sets A_n^c increase as n increases, and again using the Heine-Borel Theorem, there is $k \in \mathbb{N}$ such that

$$A_k^c = \bigcup_{n=1}^{k} A_n^c \supseteq \bigcap_{n=1}^{\ell} W_n^c.$$

Taking complements again gives

$$A_k \subseteq \bigcup_{n=1}^{\ell} W_n. \qquad (4.13)$$

Using (4) of Proposition 3.10, and using (4.12) and (4.13) we have

$$\mu(A_k) \leq \sum_{n=1}^{\ell} \mu(W_n) \leq \sum_{n=1}^{\infty} \mu(W_n) < \varepsilon.$$

It now follows that as $\mu(A_k) < \varepsilon$,

$$\ell \geq k \implies A_\ell \subseteq A_k \implies \mu(A_\ell) \leq \mu(A_k) \implies \mu(A_\ell) < \varepsilon.$$

Thus, by the definition of a convergent sequence, $\lim_{n\to\infty} \mu(A_n) = 0$. This proves the result in the case where each basic set A_n is a finite union of closed intervals.

In general, there may be at least one set A_n that is not a finite union of closed intervals. However, given a bounded interval, there is a closed subinterval of it whose length is as close as we wish to the length of the given interval. So, for each of the basic sets A_n, there is a basic set B_n such that

B_n is a finite union of disjoint closed intervals,

$$B_n \subseteq A_n \text{ and } \mu(B_n) > \left(1 - \frac{1}{n}\right)\mu(A_n). \qquad (4.14)$$

This implies that $\bigcap_{n=1}^{\infty} B_n \subseteq \bigcap_{n=1}^{\infty} A_n$, and as $\bigcap_{n=1}^{\infty} A_n$ is a set of measure zero, $\bigcap_{n=1}^{\infty} B_n$ is also a set of measure zero. Because each B_n is a finite union of closed intervals, the partial result proved above shows that $\lim_{n\to\infty} \mu(B_n) = 0$. Returning again to (4.14), we see that

$$0 \leq \lim_{n\to\infty} \mu(A_n) \leq \lim_{n\to\infty} \frac{\mu(B_n)}{1 - 1/n} = \frac{\lim_{n\to\infty} \mu(B_n)}{\lim_{n\to\infty}(1 - 1/n)} = \frac{0}{1} = 0,$$

which implies that $\lim_{n\to\infty} \mu(A_n) = 0$.

4.5. Kac's result on average recurrence times 261

(2) Take the sets B, B_n as given and put $A_n = B \cap (\bigcup_{k=1}^{n} B_k)^c$ for $n = 1, 2, \ldots$. Then each set A_n is basic, $A_1 \supseteq A_2 \supseteq \cdots$ and, as $B = W \cup (\bigcup_{n=1}^{\infty} B_n)$, we have $\bigcap_{n=1}^{\infty} A_n \subseteq W$. Consequently, (A_n) is non-increasing and $\bigcap_{n=1}^{\infty} A_n$ is a set of measure zero. By (1), $\lim_{n \to \infty} \mu(A_n) = 0$. Now,

$$\mu(B) = \mu\left(A_n \cup \left(\bigcup_{k=1}^{n} B_k\right)\right)$$
$$= \mu(A_n) + \mu\left(\bigcup_{k=1}^{n} B_k\right)$$
$$= \mu(A_n) + \sum_{k=1}^{n} \mu(B_k).$$

Letting $n \to \infty$ gives $\mu(B) = \sum_{k=1}^{\infty} \mu(B_k)$. □

As shall be seen, in a "random" dynamical system with a transformation f that is length-preserving, (4.16) in the following result can be interpreted as saying that the average time it takes for a point in an interval U to return to U under iteration by f is $\mu(S)/\mu(U)$.

Theorem 4.11. *(Kac's Theorem (1947)) Let S be a bounded interval and let $f : S \longrightarrow S$ be a length-preserving transformation. Let U be a basic subset of S of positive length. Assume that there is a subset $\mathcal{Z}$ of S that is a set of measure zero and is such that*

$$x \in S \text{ and } x \notin \mathcal{Z} \implies \text{there is } n \in \mathbb{N} \text{ such that } f^n(x) \in U. \quad (4.15)$$

Let $\theta_U : S \longrightarrow \mathbb{N} \cup \{\infty\}$ be the function as given in (4.11) by

$$\theta_U(x) = \begin{cases} \infty, \text{if } f^n(x) \notin U \text{ for all } n \in \mathbb{N}; \\ \min\{n : n \in \mathbb{N} \text{ and } f^n(x) \in U\}, \text{otherwise.} \end{cases}$$

Then,

$$\frac{1}{\mu(U)} \left(\sum_{n=1}^{\infty} n\, \mu\big(\{x : x \in U \text{ and } \theta_U(x) = n\}\big)\right) = \frac{\mu(S)}{\mu(U)}. \quad (4.16)$$

Comments on Kac's Theorem and Its Proof The proof depends on calculating relationships between the probabilities of various recurrence times. It reveals a quantitative connection between the probability that a point is in U and will return to U after exactly n iterations for $n = 1, 2, \ldots$, and the probability that a point is not in U but will return to U after exactly k iterations. The precise relationship between the probabilities is expressed in (4.21) below. Then, what we can term to be the "randomness" condition (4.15) is used to help calculate $\mu(U^c)$, and the equation $\mu(S) = \mu(U) + \mu(U^c)$ is applied to get a formula for $\mu(S)$ in terms of the probabilities of recurrence times of points in U only. The calculation as presented here is influenced by Kac's original approach in 1947 [8], similarly followed in other discussions of it in [2, 4, 11], but incorporating some other ideas. Note that although Kac assumed that the transformation is invertible, this is not an essential assumption. Whereas Kac and subsequent writers on this topic seem to have assumed a knowledge of measure theory, the previous work in this section means that we can discuss Kac's result, albeit in more restricted circumstances, without prior knowledge of measure theory. Also, the usual assumption for proving Kac's Theorem is that the system is ergodic, a concept we do not define at this stage. However, assumption (4.15) is weaker than ergodicity, so that the result here applies to some non-ergodic systems. Note that in (4.16), the left-hand side can be interpreted as the average recurrence time for points in U, so that this average recurrence time is $\mu(S)/\mu(U)$.

Proof. Let $A_0 = B_0 = U$, and for $k = 1, 2, 3, \ldots$ put

$$A_k = \{x : x \in U \text{ and } \theta_U(x) = k\}, B_k = \{x : x \in U^c \text{ and } \theta_U(x) = k\}.$$

Then, it is easy to see from the definition of θ_U that $A_0 = U$, $A_1 = U \cap f^{-1}(U)$ and for $k = 2, 3, \ldots$,

$$A_k = U \cap f^{-1}(U^c) \cap \cdots \cap f^{-(k-1)}(U^c) \cap f^{-k}(U). \qquad (4.17)$$

Also, $B_0 = U$ and for $k = 1, 2, \ldots$

$$B_k = U^c \cap f^{-1}(U^c) \cap \cdots \cap f^{-(k-1)}(U^c) \cap f^{-k}(U). \qquad (4.18)$$

4.5. Kac's result on average recurrence times

It is clear from (4.17) and (4.18) that A_k and B_k are finite intersections of basic sets and so are basic. For $k = 1, 2, \ldots$ we have from (4.17) and (4.18) that

$$\begin{aligned} A_{k+1} &= U \cap f^{-1}(U^c) \cap \cdots \cap f^{-k}(U^c) \cap f^{-(k+1)}(U) \\ &= U \cap f^{-1}\left(U^c \cap f^{-1}(U^c) \cap \cdots \cap f^{-(k-1)}(U^c) \cap f^{-k}(U)\right) \\ &= U \cap f^{-1}(B_k), \end{aligned} \qquad (4.19)$$

and we observe that this also holds for $k = 0$. Again, for $k = 1, 2, \ldots$ we have from (4.19) and (4.20) that

$$\begin{aligned} B_{k+1} &= U^c \cap f^{-1}(U^c) \cap \cdots \cap f^{-k}(U^c) \cap f^{-(k+1)}(U) \\ &= U^c \cap f^{-1}\left(U^c \cap f^{-1}(U^c) \cap \cdots \cap f^{-(k-1)}(U^c) \cap f^{-k}(U)\right) \\ &= U^c \cap f^{-1}(B_k), \end{aligned} \qquad (4.20)$$

and we can see that this also holds for $k = 0$. It follows from (4.19) and (4.20) that for $k = 0, 1, 2, \ldots$,

$$\begin{aligned} \mu(B_k) &= \mu(f^{-1}(B_k)) \\ &= \mu\left(f^{-1}(B_k) \cap U\right) + \mu\left(f^{-1}(B_k) \cap U^c\right) \\ &= \mu(A_{k+1}) + \mu(B_{k+1}). \end{aligned}$$

Thus, for all $m \geq k + 1$,

$$\begin{aligned} \mu(B_k) &= \mu(A_{k+1}) + \mu(B_{k+1}) \\ &= \mu(A_{k+1}) + \mu(A_{k+2}) + \mu(B_{k+2}) = \cdots \\ &= \sum_{n=k+1}^{m} \mu(A_n) + \mu(B_m). \end{aligned} \qquad (4.21)$$

Now (B_n) is a sequence of disjoint basic subsets of S and $\mu(S) < \infty$. Consequently, $\sum_{n=1}^{\infty} \mu(B_n) < \infty$, and we deduce that $\lim_{n \to \infty} \mu(B_n) = 0$. Thus, using (4.21) and letting $m \to \infty$, we have that for $k = 0, 1, 2, \ldots$,

$$\mu(B_k) = \sum_{n=k+1}^{\infty} \mu(A_n). \qquad (4.22)$$

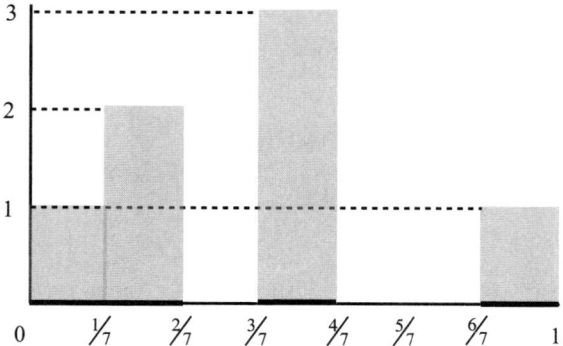

Figure 4.7. The figure illustrates Kac's formula for the transformation τ on $[0, 1)$ given by $\tau(x) = \text{frac}(x + 1/7)$ and the set $U = [0, 2/7) \cup [3/7, 4/7) \cup [6/7, 1)$ indicated by the bold line segments on the horizontal axis. The values $\theta_U(x)$ are plotted against the values of $x \in U$. Note that $\mu(U) = 4/7$. The transformation satisfies the conditions of Theorem 4.11, and τ moves points in $[0, 1)$ to the right through a distance of $1/7$, with points that become greater than 1 going back to the left-hand side of the interval. We have the recurrence time θ_U in this case given by

$$\theta_U(x) = \begin{cases} 1, & \text{for } x \in [0, 1/7) \cup [6/7, 1); \\ 2, & \text{for } x \in [1/7, 2/7); \\ 3, & \text{for } x \in [3/7, 4/7). \end{cases}$$

The equation

$$\frac{1}{\mu(U)} \left(\sum_{n=1}^{\infty} n\mu(B_n) \right) = \frac{7}{4} \left(\frac{2}{7} + \frac{2}{7} + \frac{3}{7} \right) = \frac{7}{4},$$

verifies Kac's formula in this case, with $7/4$ being the average recurrence time.

Observe that by (4.15) and the definition of the sets A_k and B_k, there are sets Y, W of measure zero such that $U = Y \cup (\bigcup_{k=1}^{\infty} A_k)$ and $U^c = W \cup (\bigcup_{k=1}^{\infty} B_k)$. Then by (2) of Lemma 4.10,

$$\sum_{k=1}^{\infty} \mu(A_k) = \mu(U) \text{ and } \sum_{k=1}^{\infty} \mu(B_k) = \mu(U^c). \tag{4.23}$$

4.5. Kac's result on average recurrence times

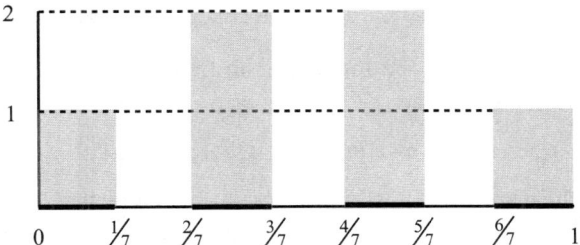

Figure 4.8. As in Figure 4.7, take $\tau(x) = \text{frac}(x + 1/7)$. Also, put $V = [0, 1/7) \cup [2/7, 3/7) \cup [4/7, 5/7) \cup [6/7, 1)$. The values $\theta_V(x)$ are plotted against the values of $x \in V$. Note that $\mu(U) = \mu(V) = 4/7$. Whereas θ_U in Figure 4.7 took on the values 1, 2 and 3, θ_V takes on the narrower range of values 1 and 2. This is because V is "spread out" more evenly throughout $[0, 1)$. Kac's Formula continues to hold, of course, but whereas the standard deviation in the case of V is $\sqrt{5}/4$ it has the larger value $\sqrt{11}/4$ in the case of U (see Section 4.7).

Using (4.22) and (4.23), and using the fact mentioned in Section 1.3, Chapter 1 that, for a double sequence of positive terms, the order of summation may be interchanged, we have

$$\mu(S) = \mu(U) + \mu(U^c)$$
$$= \sum_{n=1}^{\infty} \mu(A_n) + \sum_{k=1}^{\infty} \mu(B_k)$$
$$= \sum_{n=1}^{\infty} \mu(A_n) + \sum_{k=1}^{\infty} \sum_{n=k+1}^{\infty} \mu(A_n)$$
$$= \sum_{k=0}^{\infty} \sum_{n=k+1}^{\infty} \mu(A_n)$$
$$= \sum_{n=1}^{\infty} \sum_{k=0}^{n-1} \mu(A_n)$$
$$= \sum_{n=1}^{\infty} n\mu(A_n).$$

If this equation is divided by $\mu(U)$, the conclusion (4.16) follows. □

Kac's Theorem can be interpreted in terms of average recurrence times, as follows. The sets

$$\{x : x \in U \text{ and } \theta_U(x) = n\}, \text{ for } n = 1, 2, \ldots$$

partition the whole of U, except for a set of measure zero. Also, on the set $\{x : x \in U \text{ and } \theta_U(x) = n\}$, θ_U is the constant n by definition, so that $n\mu\big(\{x : x \in U \text{ and } \theta_U(x) = n\}\big)$ is the "area under the graph" of θ_U on the set $\{x : x \in U \text{ and } \theta_U(x) = n\}$. Thus

$$\sum_{n=1}^{\infty} n\, \mu\big(\{x : x \in U \text{ and } \theta_U(x) = n\}\big)$$

is the *total* "area under the graph" of the function θ_U on S. Now, the average of a real-valued function g on an interval $[a, b]$ is taken to be $\int_a^b g(x)dx/(b-a)$. So, by analogy with this, we make the following definition.

Definition The *average value of θ_U over U* is

$$\frac{1}{\mu(U)} \sum_{n=1}^{\infty} n\, \mu\big(\{x : x \in U \text{ and } \theta_U(x) = n\}\big).$$

Note that in a probabilistic context this sum would often be called the *expectation* of θ_U.

Using this definition, we can interpret Kac's Theorem as the statement that the average value of the recurrence time θ_U over U equals $\mu(S)/\mu(U)$.

4.6 Applications to the Kronecker and Borel systems

In order to apply Kac's Theorem to a set U, this set must be "attainable" from almost all points of the system. This requirement is expressed by (4.15), so we need to check it in any application of Kac's Theorem.

4.6. Applications to the Kronecker and Borel systems

Let α be an irrational number and define the function $\tau_\alpha : [0, 1) \longrightarrow [0, 1)$ by
$$\tau_\alpha(x) = \text{frac}(\alpha + x).$$
The transformation τ_α is length-preserving by Theorem 4.4. It was noted in Theorem 2.11 that if U is a non-empty open interval, then
$$x \in [0, 1) \Longrightarrow \text{frac}(x + n\alpha) \in U \text{ for some } n \in \mathbb{N}. \tag{4.24}$$
Now, by (2) of Proposition 2.1, $\text{frac}(\alpha + \text{frac}(n\alpha + x)) = \text{frac}(x + (n+1)\alpha)$. Taking $n = 1$ it follows that $\tau_\alpha^2(x) = \text{frac}(\alpha + \text{frac}(\alpha + x)) = \text{frac}(\alpha + 2x)$, and that, in general,
$$\tau_\alpha^n(x) = \text{frac}(n\alpha + x),$$
a result commented upon in (2.42) of Chapter 2. Thus, from (4.24), for the Kronecker dynamical system $([0, 1), \tau_\alpha)$ and for every basic subset U of $[0, 1)$ of positive length, we see that condition (4.15) is satisfied, so that Kac's Theorem applies to this system.

We now show that Kac's Theorem also applies to the Borel system $([0, 1), f)$, where f is the transformation given by
$$f(x) = \begin{cases} 2x, & \text{if } 0 \leq x < 1/2; \\ 2x - 1, & \text{if } 1/2 \leq x < 1. \end{cases}$$
We use the Recurrence Theorem 3.18. This says that there is a subset $\mathcal{Z}$ of measure zero of $[0, 1)$ with the following property: if $c_1, c_2, \ldots, c_r \in \{0, 1\}$ is a given sequence of binary symbols, and if $x \notin \mathcal{Z}$ and has a binary expansion given by
$$x = 0.d_1 d_2 d_3 \ldots, \tag{4.25}$$
then there is $n \in \mathbb{N}$ such that $d_n = c_1, d_{n+1} = c_2, \ldots, d_{n+r-1} = c_r$. What does this mean in terms of the system $([0, 1), f)$?

Well, there are two possibilities for the expansion in (4.25)—either $d_1 = 0$ or $d_1 = 1$. If $d_1 = 0$ we have
$$x = \sum_{j=2}^\infty \frac{d_j}{2^j} \in [0, 1/2),$$

268 Chapter 4. Recurrence

so that
$$f(x) = 2x = \sum_{j=2}^{\infty} \frac{d_j}{2^{j-1}} = \sum_{j=1}^{\infty} \frac{d_{j+1}}{2^j}.$$

In other words, if x has the binary expansion (4.25) with $d_1 = 0$, then the binary expansion of $f(x)$ is

$$f(x) = 0.d_2 d_3 d_4 \ldots. \tag{4.26}$$

Similarly, if $d_1 = 1$ in (4.25), we have $x \in [1/2, 1)$, $f(x) = 2x - 1$, and we find again that $f(x) = 0.d_2 d_3 d_4 \ldots$. Thus, for every $x \in [0, 1)$, if its binary expansion is given by (4.25), then the binary expansion of $f(x)$ is given by (4.26). Repeated application of this gives that for $n \in \mathbb{N}$, if the binary expansion of x is given by (4.25), then the binary expansion of $f^n(x)$ is given by

$$f^n(x) = 0.d_{n+1} d_{n+2} d_{n+3} \ldots. \tag{4.27}$$

Now, let U be an open subinterval of $[0, 1)$ of positive length. For some $r \in \mathbb{N}$ and some $k \in \{1, 2, \ldots, 2^r\}$ we will have

$$\left[\frac{k-1}{2^r}, \frac{k}{2^r} \right) \subseteq U.$$

But then, by Exercise 3 of Chapter 3 (and from the proof of Theorem 3.4), there are binary digits $c_1, c_2, \ldots, c_r \in \{0, 1\}$ such that

$$\left[\sum_{j=1}^{r} \frac{c_j}{2^j}, \sum_{j=1}^{r} \frac{c_j}{2^j} + \frac{1}{2^r} \right) = \left[\frac{k-1}{2^r}, \frac{k}{2^r} \right) \subseteq U. \tag{4.28}$$

Let $x \in [0, 1)$ with $x \notin \mathcal{Z}$, and let the binary expansion of x be given by (4.25). By (4.27) and the Recurrence Theorem 3.18, there is $n \in \mathbb{N}$ such that

$$f^n(x) = 0.c_1 c_2 \ldots c_r d_{n+r+1} d_{n+r+2} \ldots.$$

But by Theorem 3.4, and using (4.28), this means that

$$f^n(x) \in \left[\sum_{j=1}^{r} \frac{c_j}{2^j}, \sum_{j=1}^{r} \frac{c_j}{2^j} + \frac{1}{2^r} \right) \subseteq U. \tag{4.29}$$

4.6. Applications to the Kronecker and Borel systems 269

Note that whereas Poincaré's Recurrence Theorems 4.7 and 4.9 do apply to the system $([0, 1), f)$, they are not enough to deduce statement (4.29). This is because each of these results enables us to deduce that, for almost all points $x \in U$, there is $n \in \mathbb{N}$ such that $f^n(x) \in U$, but they do not give this conclusion under the weaker assumption that $x \in [0, 1)$. In the conclusion (4.29) no assumption about the interval U was made other than that it had a positive length. We can formally express the conclusion (4.29) to the preceding argument as follows.

Theorem 4.12. *Consider the Borel dynamical system* $([0, 1), f)$, *where*

$$f(x) = \begin{cases} 2x, & \text{if } x \in [0, 1/2); \\ 2x - 1, & \text{if } 1/2 \le x < 1. \end{cases}$$

Then for almost all points $x \in [0, 1)$, *the orbit* $x, f(x), f^2(x), f^3(x), \ldots$ *of x under f is dense in* $[0, 1)$. *That is, every open interval of* $[0, 1)$ *contains a point of the form* $f^n(x)$ *for some* $n \in \mathbb{N}$.

This shows that for a basic subset U of positive length, condition (4.15), necessary for Kac's Theorem, applies. Also the binary transformation f is length-preserving by Theorem 4.3. Thus, all conditions are satisfied and Kac's Theorem applies to the binary system $([0, 1), f)$ for every basic subset U of positive length.

A coin tossing experiment corresponds to choosing a number x in $[0, 1)$ and looking at its successive binary digits, thinking of a 0 as "heads" (say) and of a 1 as "tails". So, for example, if we ask "on the average, having obtained heads, tails, tails, heads consecutively, how long do we need to keep tossing before we again obtain heads, tails, tails, heads consecutively?" This is equivalent to asking the following question for the Borel system $([0, 1), f)$: "if $U = [6/16, 7/16]$ and for $x \in U$, on the average, how large does n have to be to first get $f^n(x) \in U$?" Kac's Theorem gives the answer to this as

$$\text{"average value for } n\text{"} = \frac{1}{\mu(U)} = \frac{1}{1/16} = 16.$$

4.7 The standard deviation of recurrence times

Now, let us consider the question of the *deviation from the average value* in Kac's Theorem. So, we have a bounded interval S, a length-preserving transformation $f : S \longrightarrow S$, a basic subset U of S of positive length and we assume that there is a subset $\mathcal{Z}$ of S that is a set of measure zero and is such that

$$x \notin \mathcal{Z} \implies f^n(x) \in U, \text{ for some } n \in \mathbb{N}. \tag{4.30}$$

Also, $\theta_U : S \longrightarrow \mathbb{N} \cup \{\infty\}$ is the function with $\theta_U(x)$ the minimum value of $n \in \mathbb{N}$ such that $f^n(x) \in U$, if such an n exists, and $\theta_U(x) = \infty$, otherwise. In these circumstances, Kac's Theorem tells us that the average value of θ_U over U is $\mu(S)/\mu(U)$. However, this result does not tell us how much the values taken by θ_U may deviate from the average value.

Note that $f(x) \in S$ if $x \in S$, so that in this case $\theta_S(x) = 1$ for all $x \in S$, and the average value of θ_S over S is $\mu(S)/\mu(S) = 1$. Thus, θ_S does not deviate from its average value at any point, for θ_S is the constant function 1 on S. However, if U is a proper basic subset of S, we would not expect this to occur—rather we would expect that some points in U will leave U under iteration by f and may take a long time to return to U (especially if the length of U is small), while other points may return to U after a smaller number of iterations by f. So, especially if the length of U is small, perhaps we might expect that substantial deviation of $\theta_U(x)$ from the average value of θ_U occurs for many values of x.

In general, if we wish to estimate how much $\theta_U(x)$ deviates from the average value $\mu(S)/\mu(U)$ of θ_U, we might simply look at the number $|\theta_U(x)-\mu(S)/\mu(U)|$. However, for a given x, this number is most likely difficult to determine, so rather than try to do this we ask: *on the average* how much does $\theta_U(x)$ deviate from $\mu(S)/\mu(U)$? Now A_n was defined in the proof of Kac's Theorem as

$$A_n = \{x : x \in U \text{ and } \theta_U(x) = n\}.$$

4.7. The standard deviation of recurrence times

It was seen that $A_1 = U \cap f^{-1}(U)$, and that
$$A_n = U \cap f^{-1}(U^c) \cap \cdots \cap f^{-(n-1)}(U^c) \cap f^{-n}(U), \text{ when } n > 1.$$
Thus, A_n is a basic set and θ_U is the constant n on each set A_n.

Definition The *standard deviation* of θ_U over U is

$$\sqrt{\frac{1}{\mu(U)} \sum_{n=1}^{\infty} \left| n - \frac{\mu(S)}{\mu(U)} \right|^2 \mu(A_n)}. \tag{4.31}$$

The standard deviation is an estimate of the average deviation of θ_U from its average value. Looking at the standard deviation is a common way of measuring how much the individual elements of a set of data vary from their average value (see [6] for example). If the function θ_U is constant, say $\theta_U(x) = n$ for almost all $x \in S$, then $\mu(A_m) = 0$ for all $m \neq n$, $n = \mu(S)/\mu(U)$, and the standard deviation is then the square root of $|n - \mu(S)/\mu(U)|\mu(A_n)/\mu(U) = 0$. Thus, if θ_U has a constant finite value except on some set of measure zero, the standard deviation is zero. It is easy to see that the converse is also true.

The aim now is to calculate the standard deviation of θ_U. Using the notation in the proof of Kac's Theorem, as mentioned above, we have:

$$\sum_{n=1}^{r} \left| n - \frac{\mu(S)}{\mu(U)} \right|^2 \mu(A_n) = \sum_{n=1}^{r} \left(n^2 - 2n \frac{\mu(S)}{\mu(U)} + \frac{\mu(S)^2}{\mu(U)^2} \right) \mu(A_n)$$

$$= \sum_{n=1}^{r} n^2 \mu(A_n) - 2 \frac{\mu(S)}{\mu(U)} \sum_{n=1}^{r} n \mu(A_n)$$

$$+ \frac{\mu(S)^2}{\mu(U)^2} \sum_{n=1}^{r} \mu(A_n). \tag{4.32}$$

Now, (4.30) gives $U = (U \cap \mathcal{Z}) \cup \left(\bigcup_{n=1}^{\infty} A_n \right)$ and, as $\mathcal{Z}$ is a set of measure zero, so too is $U \cap \mathcal{Z}$. So, by (2) of Lemma 4.10,

$$\mu(U) = \sum_{n=1}^{\infty} \mu(A_n). \tag{4.33}$$

Kac's Theorem 4.11 says $\sum_{n=1}^{\infty} n\mu(A_n) = \mu(S)$. If we use this together with (4.32) and (4.33) we see that

$$\sum_{n=1}^{\infty} \left| n - \frac{\mu(S)}{\mu(U)} \right|^2 \mu(A_n) = \sum_{n=1}^{\infty} n^2 \mu(A_n) - \frac{\mu(S)^2}{\mu(U)}, \qquad (4.34)$$

where this allows for the possibility that the sums on each side of (4.34) may be infinite. Thus, to calculate the standard deviation of θ_U as given by (4.31), we need to calculate the sum $\sum_{n=1}^{\infty} n^2 \mu(A_n)$. First we need the following two results.

Lemma 4.13. *Let (β_n) be a non-increasing sequence of non-negative real numbers such that $\sum_{n=1}^{\infty} \beta_n < \infty$. Then,*

$$\sum_{n=1}^{\infty} n(\beta_{n-1} - \beta_n) < \infty \text{ and } \lim_{n \to \infty} n\beta_n = 0.$$

Proof. Let $\sum_{n=1}^{\infty} \beta_n < \infty$ and observe that for all $r \geq 2$,

$$\sum_{n=2}^{r} n(\beta_{n-1} - \beta_n) = \sum_{n=2}^{r} \left((n-1)\beta_{n-1} - n\beta_n \right)$$

$$+ \sum_{n=2}^{r} \beta_{n-1}$$

$$= \beta_1 - r\beta_r + \sum_{n=1}^{r-1} \beta_n$$

$$\leq \beta_1 + \sum_{n=1}^{\infty} \beta_n < \infty. \qquad (4.35)$$

As (β_n) is non-increasing, the sum on the left is one of non-negative terms, and it follows that

$$\sum_{n=1}^{\infty} n(\beta_{n-1} - \beta_n) < \infty.$$

4.7. The standard deviation of recurrence times

Also, from (4.35) we see that

$$\sum_{n=2}^{r} n(\beta_{n-1} - \beta_n) = \beta_1 - r\beta_r + \sum_{n=1}^{r-1} \beta_n,$$

and by letting $r \to \infty$, the facts that

$$\sum_{n=1}^{\infty} \beta_n < \infty$$

and

$$\sum_{n=1}^{\infty} n(\beta_{n-1} - \beta_n) < \infty$$

imply that $\lim_{r \to \infty} r\beta_r$ exists. Put $\ell = \lim_{r \to \infty} r\beta_r$. Then $\ell \geq 0$ and if $\ell > 0$, there is r_0 such that for all $r > r_0$, $r\beta_r > \ell/2$. But then we have

$$\sum_{n=1}^{\infty} \beta_n \geq \sum_{r=r_0+1}^{\infty} \beta_r > \frac{\ell}{2} \sum_{r=r_0+1}^{\infty} \frac{1}{r} = \infty,$$

where we have used the fact that the sequence $(1/n)$ is not summable (see [1, p. 71] or [14, p. 69]). This contradiction establishes that

$$\lim_{n \to \infty} n\beta_n = 0. \qquad \square$$

Lemma 4.14. *Let $(\beta_n)_{n=-1}^{\infty}$ be a non-increasing sequence of non-negative real numbers such that $(\beta_{n-1} - \beta_n)_{n=0}^{\infty}$ is a non-increasing sequence and $\sum_{n=1}^{\infty} \beta_n < \infty$. Then, $\left(n^2(\beta_{n-2} - 2\beta_{n-1} + \beta_n)\right)$ is a summable sequence of non-negative terms and*

$$\sum_{n=1}^{\infty} n^2(\beta_{n-2} - 2\beta_{n-1} + \beta_n) = \beta_{-1} + 2\sum_{n=0}^{\infty} \beta_n.$$

Proof. As $(\beta_{n-1} - \beta_n)$ is non-increasing we have, for $n \geq 1$,

$$\beta_{n-2} - 2\beta_{n-1} + \beta_n = (\beta_{n-2} - \beta_{n-1}) - (\beta_{n-1} - \beta_n) \geq 0,$$

so $(n^2(\beta_{n-2} - 2\beta_{n-1} + \beta_n))$ is a sequence of non-negative terms. Now, for $r \geq 2$,

$$\sum_{n=1}^{r} n^2(\beta_{n-2} - 2\beta_{n-1} + \beta_n)$$

$$= \sum_{n=-1}^{r-2} (n+2)^2 \beta_n - 2 \sum_{n=0}^{r-1} (n+1)^2 \beta_n + \sum_{n=1}^{r} n^2 \beta_n$$

$$= \beta_{-1} + 2\beta_0 + \left[\sum_{n=1}^{r-2} \left((n+2)^2 - 2(n+1)^2 + n^2 \right) \beta_n \right]$$
$$- 2r^2 \beta_{r-1} + r^2 \beta_r + (r-1)^2 \beta_{r-1}$$

$$= \beta_{-1} + 2 \left(\sum_{n=0}^{r-2} \beta_n \right) - r^2 \beta_{r-1} + r^2 \beta_r - (2r-1)\beta_{r-1}$$

$$= \beta_{-1} + 2 \left(\sum_{n=0}^{r-2} \beta_n \right) - r^2(\beta_{r-1} - \beta_r) - (2r-1)\beta_{r-1} \quad (4.36)$$

$$\leq \beta_{-1} + 2 \sum_{n=0}^{r-2} \beta_n. \quad (4.37)$$

As (4.37) holds for all r and as $\sum_{n=1}^{\infty} \beta_n < \infty$, we see that

$$\sum_{n=1}^{\infty} n^2 (\beta_{n-2} - 2\beta_{n-1} + \beta_n) < \infty.$$

Now it follows from Lemma 4.13 that $\lim_{r \to \infty} r\beta_r = 0$ and consequently, $\lim_{r \to \infty} (2r-1)\beta_{r-1} = 0$. Consequently, if we let $r \to \infty$ and take limits in (4.36), we have that $\lim_{r \to \infty} r^2(\beta_{r-1} - \beta_r)$ exists, and equals ℓ, say. Clearly, $\ell \geq 0$ and, in fact, $\ell = 0$. For, if $\ell > 0$, there is r_0 such that $r^2(\beta_{r-1} - \beta_r) > \ell/2$ for all $r > r_0$. But then,

$$\sum_{r=1}^{\infty} r(\beta_{r-1} - \beta_r) \geq \sum_{r=r_0+1}^{\infty} r(\beta_{r-1} - \beta_r) > \frac{\ell}{2} \sum_{r=r_0+1}^{\infty} \frac{1}{r} = \infty.$$

Here, the divergence of $\sum_{r=1}^{\infty} r(\beta_{r-1} - \beta_r)$ contradicts a conclusion of Lemma 4.13, so that we must have $\ell = 0$. Thus, letting $r \to \infty$ in (4.36)

4.7. The standard deviation of recurrence times   275

and taking limits gives, this time,

$$\sum_{n=1}^{\infty} n^2(\beta_{n-2} - 2\beta_{n-1} + \beta_n) = \beta_{-1} + 2\sum_{n=0}^{\infty} \beta_n.$$

□

Theorem 4.15. *Let S be a bounded interval, let $f : S \longrightarrow S$ be a length-preserving transformation, and let U be a basic subset of S of positive length. Also, let $\theta_U : S \longrightarrow \mathbb{N} \cup \{\infty\}$ be the function with $\theta_U(x)$ the minimum value of $n \in \mathbb{N}$ such that $f^n(x) \in U$, if such an n exists, and $\theta_U(x) = \infty$, otherwise. Assume that*

$$\sum_{n=1}^{\infty} \mu\bigl(U^c \cap f^{-1}(U^c) \cap \cdots \cap f^{-n}(U^c)\bigr) < \infty. \qquad (4.38)$$

Then (4.30) holds, Kac's Theorem 4.11 applies, and the standard deviation of θ_U over U is finite and equals

$$\sqrt{-\frac{\mu(S)^2}{\mu(U)^2} + 3\frac{\mu(S)}{\mu(U)} - 2 + \frac{2}{\mu(U)} \sum_{n=1}^{\infty} \mu\bigl(U^c \cap f^{-1}(U^c) \cap \cdots \cap f^{-n}(U^c)\bigr)}. \qquad (4.39)$$

If in addition $\mu(U^c \cap f^{-1}(U^c)) = 0$, the standard deviation of θ_U over U equals

$$\sqrt{-\frac{\mu(S)^2}{\mu(U)^2} + 3\frac{\mu(S)}{\mu(U)} - 2}. \qquad (4.40)$$

Proof. Put $C_{-1} = S$, and for $k = 0, 1, 2, \ldots$ put

$$C_k = U^c \cap f^{-1}(U^c) \cap \cdots \cap f^{-k}(U^c). \qquad (4.41)$$

Then, each set C_k is basic, $C_{k+1} \subseteq C_k$ for all k, and (4.38) shows that $\sum_{k=1}^{\infty} \mu(C_k) < \infty$, so that $\lim_{k \to \infty} \mu(C_k) = 0$. If follows that $\bigcap_{k=1}^{\infty} C_k$ is a set of measure zero. Observe that if $x \in U^c$ but $x \notin \bigcap_{k=1}^{\infty} C_k$, then $f^n(x) \in U$ for some n. On the other hand, by Poincaré's Recurrence Theorem 4.7, there is a set $\mathcal{Z}$ of measure zero such that if

$x \in U \cap \mathcal{Z}^c$, then $f^n(x) \in U$ for some n. So, we see that $\mathcal{Z} \cup (\bigcap_{k=1}^{\infty} C_k)$ is a set of measure zero, and every x outside this set has the property that $f^n(x) \in U$ for some $n \in \mathbb{N}$. Thus, (4.30) holds.

We calculate the standard deviation of θ_U by calculating $\sum_{n=1}^{\infty} n^2 \mu(A_n)$ and then using (4.34). As before, put $B_0 = U$ and for $k \geq 1$ put

$$B_k = U^c \cap f^{-1}(U^c) \cap \cdots \cap f^{-(k-1)}(U^c) \cap f^{-k}(U). \quad (4.42)$$

Note that each set B_k is basic. Observe from (4.41) and (4.42) that

$$B_{k+1} = U^c \cap f^{-1}(U^c) \cap \cdots \cap f^{-k}(U^c) \cap f^{-(k+1)}(U), \text{ and}$$

$$C_{k+1} = U^c \cap f^{-1}(U^c) \cap \cdots \cap f^{-k}(U^c) \cap f^{-(k+1)}(U^c).$$

It follows that for $k = -1, 0, 1, 2, \ldots$,

$$B_{k+1} = C_k \cap f^{-(k+1)}(U) \text{ and } C_{k+1} = C_k \cap f^{-(k+1)}(U^c),$$

which gives

$$\mu(C_k) = \mu(C_{k+1}) + \mu(B_{k+1}). \quad (4.43)$$

Now, using (4.21), we see that for all $k = 0, 1, 2, \ldots$,

$$\mu(B_k) = \mu(A_{k+1}) + \mu(B_{k+1}). \quad (4.44)$$

It follows from (4.43) and (4.44) that for all $k = 1, 2, 3, \ldots$,

$$\mu(A_k) = \mu(B_{k-1}) - \mu(B_k)$$
$$= (\mu(C_{k-2}) - \mu(C_{k-1})) - (\mu(C_{k-1}) - \mu(C_k))$$
$$= \mu(C_{k-2}) - 2\mu(C_{k-1}) + \mu(C_k). \quad (4.45)$$

Put $\beta_k = \mu(C_k)$, for $k = -1, 0, 1, 2, \ldots$. Then from (4.41), (β_k) is non-increasing and (4.45) shows that $(\beta_{k-1} - \beta_k)$ is also non-increasing. Also, (4.38) means that $\sum_{k=-1}^{\infty} \beta_k < \infty$. These observations together with Lemma 4.14 and (4.45) now give

$$\sum_{n=1}^{\infty} n^2 \mu(A_n) = \sum_{n=1}^{\infty} n^2 (\beta_{n-2} - 2\beta_{n-1} + \beta_n) = \beta_{-1} + 2 \sum_{n=0}^{\infty} \beta_n < \infty.$$
$$(4.46)$$

4.7. The standard deviation of recurrence times

Now, we see from (4.34) that the standard deviation is finite. In fact, (4.46) gives

$$\sum_{n=1}^{\infty} n^2 \mu(A_n) = \mu(S) + 2\mu(U^c)$$

$$+ 2 \sum_{n=1}^{\infty} \mu\left(U^c \cap f^{-1}(U^c) \cap \cdots \cap f^{-n}(U^c)\right)$$

$$= 3\mu(S) - 2\mu(U)$$

$$+ 2 \sum_{n=1}^{\infty} \mu\left(U^c \cap f^{-1}(U^c) \cap \cdots \cap f^{-n}(U^c)\right). \quad (4.47)$$

Putting this in (4.34) gives

$$\frac{1}{\mu(U)} \left(\sum_{n=1}^{\infty} \left| n - \frac{\mu(S)}{\mu(U)} \right|^2 \mu(A_n) \right) = -\frac{\mu(S)^2}{\mu(U)^2} + 3\frac{\mu(S)}{\mu(U)}$$

$$-2 + \frac{2}{\mu(U)} \sum_{n=1}^{\infty} \mu\left(U^c \cap f^{-1}(U^c) \cap \cdots \cap f^{-n}(U^c)\right),$$

and the expression (4.39) for the standard deviation follows. In the case when $\mu\left(U^c \cap f^{-1}(U^c)\right) = 0$, (4.40) is immediate from (4.39). □

The formula (4.47) was first obtained by Blum and Rosenblatt [3], and they also showed that if $\sum_{n=1}^{\infty} n^2 \mu(A_n) < \infty$, then

$$\sum_{n=1}^{\infty} \mu\left(U^c \cap f^{-1}(U^c) \cap \cdots \cap f^{-n}(U^c)\right) < \infty.$$

Thus, (4.38) is a necessary and sufficient condition in order for the standard deviation in Theorem 4.15 to be finite. The formula (4.39) for the standard deviation appears to have been stated explicitly first by Kasteleyn [9, p. 822].

How should we interpret the formula (4.39) for the standard deviation? In the case when $\mu(U^c \cap f^{-1}(U^c)) = 0$,

$$\mu(S) \geq \mu(U^c) + \mu(f^{-1}(U^c)) = 2\mu(U^c) = 2\mu(S) - 2\mu(U),$$

so that $\mu(U) \geq \mu(S)/2$. It follows that the simpler formula for the standard deviation in (4.40) can only occur when $\mu(U) \geq \mu(S)/2$. When $\mu(U^c \cap f^{-1}(U^c)) = 0$, (4.40) shows that the standard deviation varies in accordance with the values of the expression $-x^{-2} + 3x^{-1} - 2$ as $1/2 \leq x \leq 1$. We see that the standard deviation is 0 when $\mu(U) = \mu(S)$ or when $\mu(U) = \mu(S)/2$. Also, the standard deviation has a maximum of $1/2$ when $\mu(U) = 2\mu(S)/3$. In this special case the standard deviation depends *only* on $\mu(U)$, and this dependence is illustrated in Figure 4.9.

We still consider the case when $\mu(U^c \cap f^{-1}(U^c)) = 0$. If $F = U^c \cap f^{-1}(U^c)$, then F is basic and $\mu(F) = 0$, so F is finite and we have

$$U^c \subseteq F \cup f^{-1}(U). \tag{4.48}$$

We see from (4.48) that if $x \in U^c$ and $x \notin F$, then $f(x) \in U$; that is $\theta_U(x) = 1$. But if $x \in U \cap f^{-1}(U^c)$, we have $f(x) \in U^c$ and we see from (4.48) that either $f(x) \in F$ or $f(x) \in f^{-1}(U)$. Thus, if $x \in U \cap f^{-1}(U^c)$, either $x \in f^{-1}(F)$, or $f^2(x) \in U$ in which case $\theta_U(x) = 2$. If $x \in U \cap f^{-1}(U)$ then $\theta_U(x) = 1$. We conclude from these observations that for $x \notin F \cup f^{-1}(F)$, $\theta_U(x) = 1$ or $\theta_U(x) = 2$. Because f is length-preserving, in this case we deduce that $F \cup f^{-1}(F)$ is finite and that, for x outside this set, $\theta_U(x)$ is 1 or 2.

In the general case, formula (4.39) shows that the value of the standard deviation of θ_U depends upon how the expression $-x^{-2}+3x^{-1}-2$, where $x = \mu(U)/\mu(S)$, balances out against the sum in (4.38). As a function of $\mu(U)$, the quadratic expression is negative on $(0, \mu(S)/2)$, positive on $(\mu(S)/2, \mu(S))$, increasing on $(0, 2\mu(S)/3]$, decreasing on $[2\mu(S)/3, \mu(S)]$ and has zeros at $\mu(S)/2$ and $\mu(S)$. The summation in (4.38) is decreasing as a function of U. So, on $(0, 2\mu(S)/3]$, the standard deviation derives from the quadratic expression and the series expression, which are respectively increasing and decreasing in U. On $[2\mu(S)/3, \mu(S)]$ both the quadratic expression and the series expression are decreasing in U, reinforcing our intuition that as U gets bigger the standard deviation of θ_U should get smaller. Figures 4.7 and 4.8 also suggest this, at least in the case of translations on $[0, 1)$.

4.7. The standard deviation of recurrence times

If U is a basic set and (4.38) holds, and if V is a basic set with $U \subseteq V$ we have $V^c \subseteq U^c$ and $\sum_{n=1}^{\infty} \mu \left(V^c \cap f^{-1}(V^c) \cap \cdots \cap f^{-n}(V^c) \right) < \infty$. That is, V also satisfies (4.38), and the conclusions of Theorem 4.15 apply with V in place of U and mean that θ_V has finite standard deviation also. Furthermore, the conclusions of Theorem 4.15 remain the same if U is replaced by $f^{-1}(U)$, and θ_U and $\theta_{f^{-1}(U)}$ have the same standard deviation. This is clear because f is length-preserving.

We previously observed, when we made the definition of the standard deviation, that the standard deviation is 0 if and only if θ_U is constant, except perhaps for a set of measure zero. Theorem 4.15 shows that when $\mu(U^c \cap f^{-1}(U^c)) = 0$, and when $\mu(U) = \mu(S)/2$ or $U = S$, then the standard deviation of θ_U is zero. However, if $\mu(U^c \cap f^{-1}(U^c)) = 0$ we have just seen that $\theta_U(x) = 1$ or $\theta_U(x) = 2$ except for a finite number of points in U. So, if $\mu(U^c \cap f^{-1}(U^c)) = 0$ and $\mu(U) = \mu(S)/2$ or $U = S$, either $\theta_U = 1$ or $\theta_U = 2$, except maybe for a set of measure zero. The case $\theta_U = 1$ corresponds to when $U = S$, for we know that $f(x) \in S$ for all $x \in S$. The case when $\theta_U = 2$ corresponds to $\mu(U) = \mu(S)/2$. This can be checked independently (see Exercise 9).

In the case of the system $([0, 1), \tau_\alpha)$, where $\tau_\alpha(x) = \text{frac}(\alpha + x)$ and α is irrational, it is Exercise 7 to prove that Theorem 4.15 always applies, and that the standard deviation relative to a basic subset of positive length is finite. In fact, Theorem 4.15 may apply to the system $([0, 1), \tau_\alpha)$ even when α is rational, as indicated in Figures 4.7 and 4.8. This Section concludes with some examples relating to the standard deviation of the recurrence time.

Example 4.16. Let $q \in \mathbb{N}$ with $q > 1$ and put $\tau(x) = \text{frac}(1/q + x)$, for $0 \le x < 1$. Then $\tau^n(x) = \text{frac}(n/q + x)$, for all $0 \le x < 1$ and $n \in \mathbb{N}$. Let $U = [0, 1/q)$. If $x \in [0, 1)$, there is $k \in \{0, 1, \ldots, q-1\}$ such that $x \in [k/q, (k+1)/q)$. Let $j \in \{1, 2, \ldots, q-k\}$. Then,

$$\tau^j(x) = \text{frac}\left(\frac{j}{q} + x\right) = \text{frac}\left(\frac{j+k}{q} + x - \frac{k}{q}\right). \qquad (4.49)$$

If $j < q - k$, $j + k < q$, and we see that $(j+k)/q \in [1/q, 1 - 1/q)$

280   Chapter 4. Recurrence

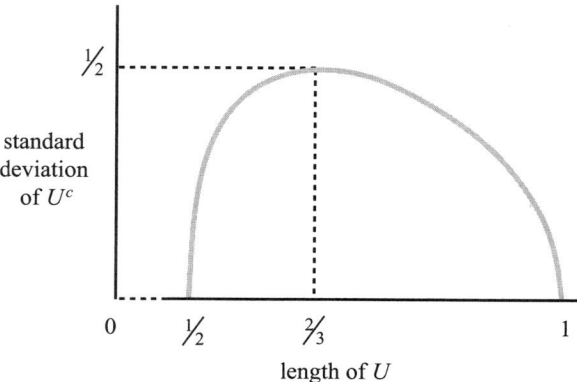

Figure 4.9. A consequence of Theorem 4.15 is that when $\mu(U^c \cap f^{-1}(U^c)) = 0$, $\mu(U) \geq \mu(S)/2$ and the standard deviation of the recurrence time over U depends on $\mu(U)$ only, and not upon the "position" of U within S. Also, in this case, θ_U can take only the values 1 or 2 (with maybe a finite number of exceptions). When $S = [0, 1)$ and $\mu(U^c \cap f^{-1}(U^c)) = 0$, the graph shows the standard deviation of θ_U on the vertical axis, plotted against the length of the basic set U on the horizontal axis. Thus, from (4.40), we see that the graph is that of the function on $[1/2, 1]$ given by

$$x \longmapsto \sqrt{-\frac{1}{x^2} + \frac{3}{x} - 2},$$

where x is in place of $\mu(U)$. The graph indicates that as $\mu(U)$ increases from $1/2$, the standard deviation of θ_U increases relatively quickly from a value 0, attaining a maximum of $1/2$ at $\mu(U) = 2/3$. The standard deviation then decreases more slowly until it vanishes when $\mu(U) = 1$.

and $x - k/q \in [0, 1/q)$. So, we see from (4.49) that

$$\tau^j(x) = \text{frac}\left(\frac{j}{q} + x\right) = \frac{j}{q} + x \in [1/q, 1),$$

so that $\tau^j(x) \notin U$. But if $j = q - k$,

$$\text{frac}\left(\frac{j}{q} + x\right) = \text{frac}\left(1 + x - \frac{k}{q}\right) = x - \frac{k}{q} \in [0, 1/q),$$

so that $\tau^j(x) \in U$. Thus, for $x \in [k/q, (k+1)/q)$, $\tau^j(x) \notin U$ for

4.7. The standard deviation of recurrence times 281

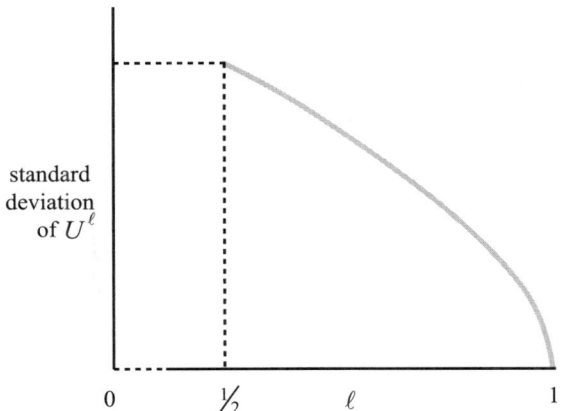

Figure 4.10. The graph is of the standard deviation of θ_{U_ℓ} over U_ℓ for $1/2 \leq \ell < 1$, where $U_\ell = [1 - \ell, 1)$ and the underlying transformation f is given by

$$f(x) = \begin{cases} 2x, & \text{if } 0 \leq x < 1/2; \\ 2x - 1, & \text{if } 1/2 \leq x < 1. \end{cases}$$

Note the superficial similarity with the right-hand side of the graph in Figure 4.9, even though here, in contrast to Figure 4.9, we have that $\mu(U_\ell^c \cap f^{-1}(U_\ell^c)) > 0$. The graph illustrates that for $\ell > 1/2$, as U_ℓ increases, the standard deviation of θ_U decreases.

$j \in \{1, 2, \ldots, q - k - 1\}$, but $\tau^{q-k}(x) \in U$. This shows that the system and U satisfy condition (4.30) and that $\theta_U(x) = q$ for all $x \in U$. These facts become obvious when we remember the geometric meaning of τ, for all τ does is move the point to the right by a distance of $1/q$ in $[0, 1)$. Since θ_U is constant, the standard deviation must be 0. It can be checked directly that (4.39) applies, and that it gives a standard deviation of 0, as expected (see Exercise 10).

Example 4.17. Let $0 \leq a < 1/2$, $V = [0, a)$ and let f be the transformation on $[0, 1)$ given by $f(x) = \text{frac}(2x)$. Then,

$$V \cap f^{-1}(V) = [0, a] \cap \left(\left[0, \frac{a}{2}\right] \cup \left[\frac{1}{2}, \frac{a+1}{2}\right) \right) = \left[0, \frac{a}{2}\right],$$

and

$$\begin{aligned}
V \cap f^{-1}(V) \cap f^{-2}(V) &= V \cap f^{-1}\left(V \cap f^{-1}(V)\right) \\
&= [0, a] \cap f^{-1}\left(\left[0, \frac{a}{2}\right]\right) \\
&= [0, a) \cap \left(\left[0, \frac{a}{4}\right) \cup \left[\frac{1}{2}, \frac{a+2}{4}\right)\right) \\
&= \left[0, \frac{a}{4}\right).
\end{aligned}$$

In general, this method will give

$$V \cap f^{-1}(V) \cap \cdots \cap f^{-n}(V) = \left[0, \frac{a}{2^n}\right),$$

so that

$$\begin{aligned}
\sum_{n=1}^{\infty} \mu\left(V \cap f^{-1}(V) \cap \cdots \cap f^{-n}(V)\right) &= \sum_{n=1}^{\infty} \mu\left(\left[0, \frac{a}{2^n}\right)\right) \\
&= \sum_{n=1}^{\infty} \frac{a}{2^n} = a < \infty.
\end{aligned}$$

If we put $U = V^c = [a, 1)$, then $U^c = V$, $\mu(U) = 1 - a$ and we see from Theorem 4.15 that the standard deviation of θ_U over U is finite and that it equals

$$\sqrt{-\frac{1}{(1-a)^2} + \frac{3}{1-a} - 2 + \frac{2a}{1-a}} = \frac{\sqrt{a(3-4a)}}{1-a}.$$

Thus, if $1/2 \leq \ell < 1$ and U_ℓ is the interval of length ℓ given by $U_\ell = [1 - \ell, 1)$, the standard deviation of θ_{U_ℓ} over U_ℓ is finite and equals

$$\frac{\sqrt{(1-\ell)(4\ell - 1)}}{\ell}.$$

Figure 4.10 illustrates how the standard deviation of θ_{U_ℓ} over U_ℓ varies as a function of ℓ.

Exercises

1. Let S be an interval of $\mathbb{R}$ and let A be a subset of S that is of measure zero. Prove that the complement of A in S is dense in S.

2. Let f be the transformation on $[0, 1)$ given by $f(x) = 2x$ for $0 \le x < 1/2$ and $f(x) = 2x - 1$, for $1/2 \le x < 1$. Let $J = [0, 1/2)$ and let $K = [1/2, 1)$.
   (i) If $x \in K$, prove that there is $n \in \mathbb{N}$ such that $f^n(x) \in J$.
   (ii) If $x \in J$ and $x > 0$, prove that there is $n \in \mathbb{N}$ such that $f^n(x) \in K$.
   (iii) If $x \in [0, 1)$ prove that there is $n \in \mathbb{N}$ such that $f^n(x) = 0$ if and only if there is $k \in \{0, 1, \ldots, 2^n - 1\}$ such that $x = k/2^n$.
   (iv) If $x \in J$ with $x > 0$ and $y \in K$ let
   $$\rho(x) = \min\{n : n \in \mathbb{N} \text{ and } f^n(x) \in K\} \text{ and}$$
   $$\phi(y) = \min\{n : n \in \mathbb{N} \text{ and } f^n(y) \in J\}.$$
   Calculate the average value of ρ over J and the average value of ϕ over K.

3. Let g be the function on $[0, 1/3)$ given by $g(x) = \sqrt{3x}$. Find an increasing function $h : [1/3, 1] \longrightarrow [0, 1]$ such that if f is the transformation on $[0, 1]$ given by $f(x) = g(x)$ for $0 \le x < 1/3$ and $f(x) = h(x)$ for $1/3 \le x \le 1$, then f is length-preserving. Illustrate the ideas involved with graphs and pictures, as appropriate.

4. Let S be an interval, let U be a subset of S, and let $f : S \longrightarrow S$. If $x \in S$ and $n \in \mathbb{N}$, prove that
   $$x \in f^{-1}(U^c) \cap f^{-2}(U^c) \cap \cdots \cap f^{-n}(U^c)$$
   $$\Longleftrightarrow \Theta_U(x) \in \{n + 1, n + 2, \ldots\} \cup \{\infty\}.$$

5. Using the example of Figure 4.6, or one of your own devising, show that there may exist a dynamical system (S, f) with a subset U with the following property: if $n \in \mathbb{N}$, there is $x \in U$ such that $x, f(x), \ldots, f^{n-1}(x) \in U$ but $f^n(x), f^{n+1}(x), \ldots \notin U$. Then, give an example of such a situation where the transformation f is length-preserving.

6. Apply Kac's Theorem to the Borel system $([0, 1), f)$, where $f(x) = \operatorname{frac}(2x)$ and with $U = [3/8, 1/2)$. Show that U consists precisely of those numbers in $[0, 1)$ that have $0, 1, 1$ respectively as the first three digits in their binary expansions. Apply Kac's Theorem and explain the conclusion in terms of a coin tossing experiment.

7. Let $\alpha \in (0, 1)$ be irrational, and let U be a basic subset of $[0, 1)$ with $0 < \mu(U) < 1$. Let τ_α be the transformation $x \mapsto \text{frac}(x + \alpha)$. Prove that there is $n \in \mathbb{N}$ such that

$$\mu(U^c \cap f^{-1}(U^c) \cap \cdots \cap f^{-n}(U^c)) = 0.$$

Deduce that Theorem 4.15 applies to the system $([0, 1), \tau_\alpha)$ and that the standard deviation of θ_U over U is finite.

8. If $\beta_0, \beta_1, \ldots, \beta_r$ are given numbers, where $r \geq 1$, prove that

$$\sum_{j=0}^{r-1} \beta_j = r\beta_r + \sum_{j=1}^{r} j(\beta_{j-1} - \beta_j).$$

Hence formulate alternative conditions to those in Theorem 4.15 that ensure that the standard deviation is finite. Give an example to show that there is a real-valued sequence (β_n) such that $\lim_{n\to\infty} n\beta_n = 0$ but $\sum_{n=1}^{\infty} n(\beta_{n-1} - \beta_n) = \infty$.

9. In the situation of Theorem 4.15, when $\mu(U^c \cap f^{-1}(U^c)) = 0$ and $\mu(U) = \mu(S)/2$, we have that the standard deviation is 0. Prove that for $x \in U$, $\theta_U(x) = 2$ for all but a finite number of points in U. Compare this conclusion with the discussion in the text.

10. Let $q \in \mathbb{N}$ with $q \geq 2$, and consider the dynamical system $([0, 1), \tau_{1/q})$ where $\tau_{1/q}$ is $x \mapsto \text{frac}(x + 1/q)$. Let $U = [0, 1/q)$. It was shown in Example 4.16 that $\theta_U(x) = q$ for all $x \in U$, so that the standard deviation of θ_U is 0. Verify this fact alternatively by a calculation using the formula (4.39).

Investigations

1. Theorems 4.3 and 4.4 describe two different types of piecewise linear length-preserving transformations on $[0, 1]$. This suggests the question: *what is the general form of a piecewise linear length-preserving transformation on $[0, 1]$?* It seems likely that the general form of such a transformation will be a "combination" of transformations of the two types.

2. Let $J_1, J_2, \ldots, J_n$ be consecutive disjoint subintervals of $[0, 1]$ and assume that $\cup_{j=1}^{n} J_j = [0, 1]$. For each subinterval $J_1, J_2, \ldots, J_{n-1}$ let f_j be a strictly increasing continuous function on J_j whose range on J_j is $[0, 1]$ except maybe for one or two of the endpoints. Investigate whether there is a strictly increasing continuous function f_n on J_n such that the transformation f on $[0, 1]$ as given below is defined and is length-preserving.

$$f(x) = f_j(x), \text{ for } x \in J_j.$$

Note that the case $n = 2$ has been discussed in Figure 4.5.

3. If $\alpha \in [0, 1)$ let τ_α be the transformation on $[0, 1)$ given by $\tau_\alpha(x) = \text{frac}(\alpha + x)$. Now, let $q \in \mathbb{N}$, put $\tau = \tau_{1/q}$, and let U be a subinterval of $[0, 1)$ having length $1/q$. Show that Kac's Theorem, and Theorem 4.15 on the standard deviation, apply to U and the system $([0, 1), \tau)$. Calculate the function θ_U explicitly for the system and from this calculate the standard deviation of θ_U over U. Now, investigate what happens if the interval U has length greater than $1/q$, and investigate what happens when a single interval is replaced by a finite number of disjoint intervals, carrying out some explicit calculations.

4. If $\alpha \in [0, 1)$ let τ_α be the translation on $[0, 1)$ given by $\tau_\alpha(x) = \text{frac}(\alpha+x)$, and let U be a subinterval of $[0, 1)$ having length α. Find conditions on U and α that ensure that Kac's Theorem applies to τ_α. Are these conditions necessary? If not, can you find necessary and sufficient conditions on α and U in order that Kac's Theorem applies?

5. Is it possible to give an example of a system that satisfies the requirements of Kac's Theorem but does not satisfy the requirement in Theorem 4.15 that

$$\sum_{n=1}^{\infty} \mu(U^c \cap f^{-1}(U^c) \cap \cdots \cap f^{-n}(U^c)) < \infty ?$$

6. Figures 4.7 and 4.8 indicate that the standard deviation of θ_U, at least in some cases, depends not only upon the length of U, but also upon how U is "distributed" in the whole space. Verify the values of the standard deviation of θ_U and θ_V given in Figures 4.7 and 4.8, and investigate in the case of translations on $[0, 1)$ the general question of how the standard deviation of θ_U depends upon the "distribution" of U within the whole space $[0, 1)$. Then, consider cases other than translations.

Chapter 4. Recurrence

Notes

1. [Page 239] The concept of a measure-preserving transformation is more common in the literature than is the concept of a length-preserving transformation as presented in Chapter 4. The discussion here is restricted to basic sets on an interval, and this family of sets is not closed under countable unions. So, the ideas are developed without reference to measure theory and the countable additivity of the length function. This means that the discussion is more general than measure theory in one sense, but more restricted in that in this chapter, discussion is essentially restricted to dynamical systems on intervals.

2. [Page 248] The French mathematician Henri Poincaré (1854–1912) is noted for his deep and wide ranging work in many different areas of mathematics. In particular, his Recurrence Theorem has important implications in physics and has received extensive discussion and generalization within mathematics (see [7], for example).

3. [Page 255] This type of recurrence theorem, when originally found by Poincaré, raised important and paradoxical issues concerning the foundations of statistical mechanics in the 1890s. W. Boltzmann's work on systems of molecules in motion proposed that in such a system there is associated a function that decreases with time and reflects the "state of disorder" of the system. Boltzmann called this function the H-function which, when replaced by $-H$, is now known as the entropy. As H decreases with time, this can only mean, apparently, that disorder in the system increases with time, perhaps attaining or approaching an equilibrium state of disorder. A further implication seemed to be that such processes could only go in one direction and were therefore called *irreversible*. Now, Joseph Liouville (1809–1882) had shown that the transformation describing the changing state over time of the system in $2n$-dimensional space (where n is the number of molecules) is one that preserves the *volume* of the space—and such a transformation in 1 dimension is what we have called *length-preserving*. Because of Liouville's result, a generalized version of Poincaré's Theorem applies in this higher dimensional space. The point is that in this context Poincaré's Theorem seems to lead to an opposite conclusion from Boltzmann's—namely, that far from attaining an "irreversible" state of equilibrium, the system of molecules will return over time so as to be *arbitrarily close to its original state*. See [5] and [10] for a discussion and clarification of the issues.

4. [Page 257] Mark Kac (1914–1984), Polish-American mathematician noted for his many-sided work, but especially for his work on probability theory and its connections with statistical mechanics and physics.

Bibliography

[1] R. G. Bartle and D. R. Sherbert, *Introduction to Real Analysis*, third edition, Wiley, New York, 2000.

[2] A Berger, *Chaos and Chance*, de Gruyter, New York 2001.

[3] J. R. Blum and J. I. Rosenblatt, *On the moments of recurrence time*, Journ. Math. Sci. (Delhi), **2** (1967), 1–6.

[4] J. R. Brown, *Ergodic Theory and Topological Dynamics*, Academic Press, New York 1976.

[5] P. and T. Ehrenfest, *Enclykopädie de mathematicschen Wissenschaften*, volume IV 2 II, No.6, Teubner, Leipzig 1912; translated as *The Conceptual Foundations of the Statistical Approach in Mechanics*, by M. J. Moravcsik, Cornell University Press, New York 1959.

[6] W. Feller, *An Introduction to Probability Theory and its Applications*, Third Edition, volume I, Wiley, New York 1968.

[7] H. Furstenburg, *Recurrence in Ergodic Theory and Combinatorial Number Theory*, Princeton University Press, Princeton 1981.

[8] M. Kac, *On the notion of recurrence in discrete stochastic processes*, Bull. Amer. Math. Soc., **53** (1947), 1002–1010.

[9] P. W. Kasteleyn, *Variations on a theme by Marc Kac*, Journ. Stat. Phys., **46** (1987), 811–827.

[10] M. J. Klein, *Paul Ehrenfest: volume 1, The Making of a Theoretical Physicist*, North-Holland, Amsterdam 1970, reprinted 1985, especially Chapter 6.

[11] K. Petersen, *Ergodic Theory*, Cambridge Studies in advanced mathematics 2, Cambridge University Press, 1983.

[12] H. Poincaré, *Sur le problème des trois corps et les équations de la dynamique, §8.Usage des invariantes intégraux*, Acta Math. **13**(1890).

[13] H. Poincaré, *Les Méthodes Nouvelles de la Mécanique Céleste*, vol.I, Gauthier-Villars, Paris 1892; translated as *New Methods of Celestial Mechanics* and edited with an introduction by Daniel L. Goroff, volume I, American Institute of Physics, Woodbury NY, 1993.

[14] K. A. Ross, *Elementary Analysis: the theory of calculus*, Springer, New York, 1980.

Chapter 5 Flow Chart

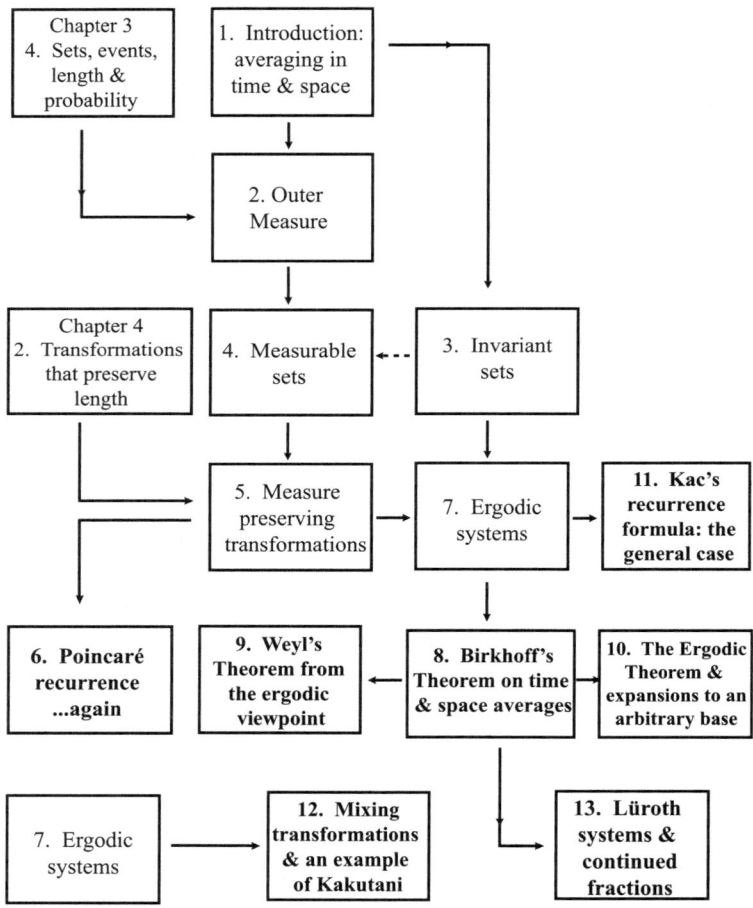

In the diagram, the sections in bold are those with the main results. The unbroken arrows indicate the sequences of sections necessary to get to the results of subsequent sections. The dotted arrows indicate that a preceding section is complementary to the following section, or that it may be useful to the following section without it being essential.

Chapter 5
Averaging in Time and Space

> All things began in order, so shall they end, and so shall they begin again; according to the ordainer of order and mystical mathematics of the city of heaven.   *Sir Thomas Browne* (1605-1682)

5.1 Introduction: averaging in time and space

This chapter of the book is different from the preceding Chapters 2, 3 and 4. Its role is primarily expository, and introduces more advanced ideas, concepts, results and topics that proceed naturally from the preceding chapters. In particular, there is an exposition of parts of measure theory, and this is used to revisit some earlier topics and ideas, and to view them through the powerful lens measure theory provides. Some proofs are included, but most are omitted. An aim is also, in part, to present some motivations and underlying ideas in measure theory and ergodic theory that may not be found in the standard sources. The main theme is that of averaging in dynamical systems, with Birkhoff's Individual Ergodic Theorem as the main general result.

Results concerning averaging have been seen in earlier chapters. For example, Borel's Theorem 3.22 asserts that for almost all $x \in [0, 1)$, and as $n \longrightarrow \infty$, the average number of 1s appearing in the first n binary digits of x has a limit equal to $1/2$. Similarly, the Normal Numbers Theorem 3.32 is a corresponding statement for the average occurrence of a given finite sequence of 0s and 1s in the binary expansion of a number. Like comments can be made for Weyl's Theorem 2.20 and the results

290   Chapter 5. Averaging in Time and Space

dealing with Benford-type phenomena in Sections 3.15–3.17 of Chapter 3 (see also [4, 7, 12, 14, 21, 30, 34]). All of these results concern averaging in various particular dynamical systems. Birkhoff's Theorem is a result concerning averaging in general dynamical systems.

Recall that if S is a set, a function $f : S \longrightarrow S$ is called a *transformation* on S, and the pair (S, f) is called an (abstract) *dynamical system*. Then, if $x \in S$, the sequence of points $x, f(x), f^2(x), \ldots$ is called the *orbit* of x. If A is a subset of S, the *characteristic function of* A is the function $\chi_A : S \longrightarrow \{0, 1\}$ given by

$$\chi_A(x) = \begin{cases} 1, & \text{if } x \in A; \\ 0, & \text{if } x \notin A. \end{cases}$$

Now, let (S, f) be a dynamical system and let A be a subset of S. Then if $x \in S$ and $j \in \mathbb{N}$,

$$f^{j-1}(x) \in A \iff \chi_A(f^{j-1}(x)) = 1,$$

and $f^{j-1}(x) \notin A \iff \chi_A(f^{j-1}(x)) = 0.$

Thus, if the system starts in a state x, the sum

$$\sum_{j=1}^{n} \chi_A\left(f^{j-1}(x)\right)$$

is the *number of occasions*, over the passage of $n - 1$ units of time, that the state of the system is in A. Hence, the average

$$\frac{1}{n}\left(\sum_{j=1}^{n} \chi_A\left(f^{j-1}(x)\right)\right)$$

is the *proportion* of time, over the passage of the first $n - 1$ units of time, that the system is in a state in A. If the system (S, f) is "random", we would expect that the sequence of points $x, f(x), f^2(x), \ldots$ would be spread evenly throughout S, the set of all possible states, not favoring any part of S over any other. That is, we would expect the asymptotic

5.1. Introduction: averaging in time and space

proportion of time that the system spends in states in A to be the same as the proportion that A occupies within the total set S of states—in the case where S is an interval and A is a basic subset of S as discussed in 3.4, this proportion is $\mu(A)/\mu(S)$, where $\mu(A)$ is the length of A. This proportion may be thought of as the "space average" of A within S. The ratio $\mu(A)/\mu(S)$ may also be regarded as the probability that a randomly chosen state of the system is in A. So, in a random system on an interval, we would expect that

$$\lim_{n\to\infty} \frac{1}{n} \left(\sum_{j=1}^{n} \chi_A\left(f^{j-1}(x)\right) \right) = \frac{\mu(A)}{\mu(S)}, \tag{5.1}$$

where $\mu(A)$ denotes the length of the basic set A. In fact, it is in general expecting too much for this to hold for *all* $x \in S$, but under certain conditions it does hold for all x outside a set of measure zero.

Finding conditions on a dynamical system (S, f) that ensure that equation (5.1), or its equivalent for more general settings than an interval, holds for "most" values of x is a crucial problem in dynamical systems. A satisfactory answer was first given by the American mathematician George Birkhoff in 1931, when he proved a result now known as the Individual Ergodic Theorem. There seems to be no easy route to Birkhoff's Theorem, which is why earlier chapters of this work used various special techniques to obtain results such as Borel's Theorem, that may be regarded as special applications of Birkhoff's Theorem. Also, even when one wants to use Birkhoff's Theorem, verifying that the conditions for the application of the Theorem are satisfied often poses a serious problem in itself.

A full discussion of Birkhoff's Theorem requires a knowledge of measure theory. In this chapter, the essential measure theory is described —this involves extending the notion of the family of basic sets on an interval to that of the family of *measurable sets* on a set. The concept of the *measure* of a measurable set corresponds to the length of a basic set. The notion of a measurable set is motivated by a problem in measure theory relating to the invariant subsets of an irrational translation on the interval $[0, 1)$. In this general setting, the concept of a *measure-preserving*

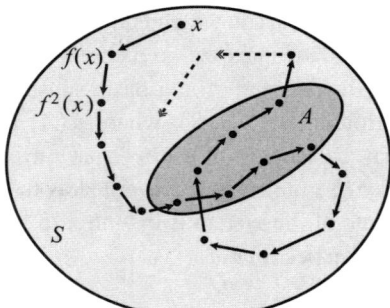

Figure 5.1. The figure depicts the orbit of a point x in a dynamical system (S, f). A given subset of S is depicted as A. In the figure, $x, f(x), f^2(x), f^3(x), f^4(x), f^5(x) \notin A$, $f^6(x), f^7(x), f^8(x), f^9(x) \in A$, $f^{10}(x), f^{11}(x), f^{12}(x), f^{13}(x) \notin A$, and so on. The question is: in the long run, what proportion of time do the points $x, f(x), f^2(x), \ldots$ in the orbit of x spend in the set A? See also Figure 5.2

transformation on a set extends the concept of a length-preserving transformation on an interval, introduced in Chapter 4. Birkhoff's Theorem is described and then applied to shed further light on Weyl's Theorem, Borel's Theorem, the Normal Numbers Theorem, and Poincaré recurrence. It should be noted that for certain systems, Birkhoff's Theorem contains a quantitative refinement of Poincaré's Recurrence Theorem of Chapter 4. Since Poincaré's Theorem is the statement that if (S, f) is a system where S is a bounded interval and f is a length-preserving transformation, then for a subinterval J of positive length, for almost all values of x in J we have $f^n(x) \in J$ for some $n \in \mathbb{N}$. Equivalently, for almost all $x \in J$,

$$\sum_{j=2}^{n} \chi_J \left(f^{j-1}(x) \right) > 0, \quad \text{for some} \quad n \in \mathbb{N}. \tag{5.2}$$

However, if (5.1) holds,

$$\lim_{n \to \infty} \frac{1}{n} \left(\sum_{j=1}^{n} \chi_J \left(f^{j-1}(x) \right) \right) = \frac{\mu(J)}{\mu(S)} > 0, \tag{5.3}$$

and this implies (5.2). In fact, (5.3) *quantifies* the statement (5.2), by telling us the proportion of time that the points in the orbit x, $f(x)$, $f^2(x)$, ... spend in A, whereas (5.2) in itself simply says that $f^n(x)$ will be in J for some $n \in \mathbb{N}$ with $n \geq 2$.

Some applications of Birkhoff's Theorem to other dynamical systems, not necessarily accessible using the approaches in Chapters 2, 3 and 4 that avoided measure theory, are presented in Section 5.12. Suitable background references and ones that provide complete proofs for this chapter are: [1, 2, 8, 17, 20, 28] for Sections 5.2 to 5.5 on measure theory, [3, 9, 11, 18, 26, 33, 35, 36, 39] for Sections 5.6 to 5.11 on dynamical systems and ergodic theory, [9, 18, 23, 24, 26, 27, 33] for mixing and Kakutani's example in Section 5.12, and [11, 29] for background on Lüroth systems and continued fractions, discussed in Section 5.13. More specific references will be given as required.

5.2 Outer measure

Recall that a basic set is a subset of $\mathbb{R}$ that is a finite union of intervals. Each basic set B has a length $\mu(B)$. The properties of basic sets and the length function were discussed in Section 3.4 of Chapter 3, where we also introduced the more general concept of an additive set function on the family of basic subsets of an interval. In this section we describe how the notion of an additive set function can be introduced in a more general context, together with associated ideas. Suitable references for the material of this section are [1, 17, 20].

Definition Let S be a set and let $\mathcal{B}$ be a family of subsets of S. Then $\mathcal{B}$ is an *algebra of subsets of* S, or simply an *algebra of sets*, if the following conditions hold.

(1) $\emptyset, S \in \mathcal{B}$;
(2) if $B \in \mathcal{B}$, then $B^c \in \mathcal{B}$; and
(3) if $B, C \in \mathcal{B}$, then $B \cup C \in \mathcal{B}$.

It is an immediate consequence of this definition and Proposition 3.7 that in the case where $\mathcal{B}$ is the family of basic subsets of an interval, $\mathcal{B}$ is an algebra of sets.

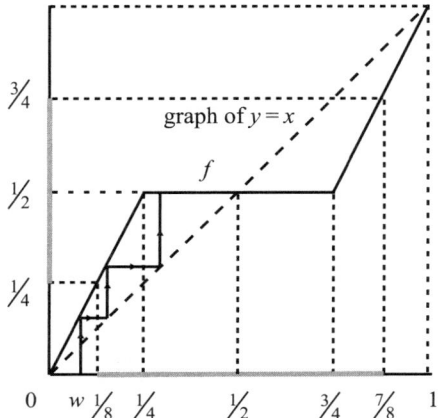

Figure 5.2. The figure depicts the system $([0, 1], f)$ where

$$f(x) = \begin{cases} 2x, & \text{if } 0 \le x < 1/4; \\ 1/2, & \text{if } 1/4 \le x < 3/4; \\ 2x - 1, & \text{if } 3/4 \le x \le 1. \end{cases}$$

As well, the graph of the identity function given by the line $y = x$ is depicted. The fixed points of f are 0, $1/2$ and 1, where the graphs intersect. Let us put $U = [1/4, 3/4]$. Then, if $f^k(x) \in U$ for some $k \in \mathbb{N}$, $f^n(x) = 1/2$ for all $n > k$. Also, as depicted in the figure, $f^{-1}(U) = [1/8, 7/8]$. In fact we have

$$f^{-k}(U) = \left[\frac{1}{2^{k+2}}, 1 - \frac{1}{2^{k+2}}\right].$$

Thus, $\cup_{k=0}^{\infty} f^{-k}(U) = (0, 1)$, and for each $x \in (0, 1)$ there is $k_0 \in \mathbb{N}$ such that $f^k(x) = 1/2$ for all $k > k_0$. In the case of the indicated point $w \in (0, 1)$, the figure illustrates successive points in the orbit of w and how there is $k_0 \in \mathbb{N}$ such that $f^{k_0}(x) \in U$. In the figure, $w \notin U$, $f(w) \notin U$, $f^2(w) \in U$, $f^3(w) = 1/2$ and $f^k(w) = 1/2$ for all $k \ge 3$. In fact, for every $w \in (0, 1)$, if $A \subseteq (0, 1)$ we have

$$\lim_{n \to \infty} \frac{1}{n} \left(\sum_{k=1}^{n} \chi_A(f^{k-1}(w)) \right) = \begin{cases} 1, & \text{if } 1/2 \in A; \\ 0, & \text{if } 1/2 \notin A. \end{cases}$$

We can also calculate the average number of iterations required for a point in $(0, 1)$ to become equal to the fixed point $1/2$. This average number is

$$\sum_{k=1}^{\infty} k\mu(f^{-k+1}(U) \cap f^{-k+2}(U)^c) = \sum_{k=1}^{\infty} \frac{k}{2^k} = 2.$$

5.2. Outer measure

Definition Let S be a set and let $\mathcal{B}$ be an algebra of subsets of S. A function $\nu : \mathcal{B} \longrightarrow [0, \infty]$ is called an *additive set function* on $\mathcal{B}$ if it satisfies the following condition: if $B, C \in \mathcal{B}$ and $B \cap C = \emptyset$, then

$$\nu(B \cup C) = \nu(B) + \nu(C).$$

If ν is an additive set function, we can think of $\nu(A)$ as assigning a notion of length, or area, or volume, to the set A. Note that in Proposition 3.10 we saw that the function μ, which assigns to a basic set its length, is an additive set function, which we continue to denote by μ. More generally, we saw in Proposition 3.10 that if v is a non-decreasing and real-valued function on an interval J, the associated function $\mu_v : \mathcal{B} \longrightarrow [0, \infty]$ is an additive set function on $\mathcal{B}$, the algebra of basic subsets of J. We now show how the notion of outer measure may be introduced for an arbitrary subset of the given set S.

Definition Let S be a set, let $\mathcal{B}$ be an algebra of subsets of S, and let ν be an additive set function on $\mathcal{B}$. Then, if A is a subset of the set S, the *outer measure* of A is defined to be the infimum, taken over sequences (V_n) of sets in $\mathcal{B}$, of the set

$$\left\{ \sum_{n=1}^{\infty} \nu(V_n) : A \subseteq \bigcup_{n=1}^{\infty} V_n \right\}. \tag{5.4}$$

The outer measure of A is denoted by $\nu_{out}(A)$.

When we have the situation that $A \subseteq \bigcup_{n=1}^{\infty} V_n$, A is being roughly "approximated" from the outside by $\bigcup_{n=1}^{\infty} V_n$, and in terms of ν this approximation is made precise by taking the infimum in (5.4). Hence the term *outer* measure. Note that it may happen that the outer measure of a set is ∞. However, if A is a subset of S such that $\mu_{out}(A) < \infty$, it follows from (5.4) and the definition of the infimum that for each $\varepsilon > 0$ there is a sequence (V_n) in $\mathcal{B}$ such that

$$A \subseteq \bigcup_{n=1}^{\infty} V_n \text{ and } \sum_{n=1}^{\infty} \nu(V_n) < \nu_{out}(A) + \varepsilon. \tag{5.5}$$

Figure 5.3. In 2-dimensional space $\mathbb{R}^2$, the family $\mathcal{B}$ of sets that are finite unions of half-open rectangles of the form $[a, b) \times [c, d)$ is an algebra of subsets of $\mathbb{R}^2$. Also, there is an additive set function ν on $\mathcal{B}$ such that $\nu([a, b) \times [c, d)) = (b - a)(d - c)$ (see Exercise 3). Thus, if C is a rectangle in $\mathcal{B}$, $\nu(C)$ is the usual area of C, and the same comment applies for every set in $\mathcal{B}$. Figure 5.3 illustrates the definition of the outer measure for subsets of $\mathbb{R}^2$, using the algebra $\mathcal{B}$ and the additive set function ν on $\mathcal{B}$. In the figure, let A denote the subset of $\mathbb{R}^2$ enclosed by the elliptical curve. In each case, A is covered by the squares and rectangles indicated. On the right, a finer set of rectangles has been used to cover A. In each case, the sum of the areas of the rectangles is at least as great as the outer measure of A. As the rectangles on the right form a "finer" mesh, we see that the sum of areas of rectangles on the left is at least the sum of the areas of the rectangles on the right, and each of these sums is at least the outer measure of A. As we take finer and finer meshes of rectangles to cover A, the sums of the areas of the rectangles in the meshes get closer and closer to the actual value of the outer measure of A—in fact, for the set A depicted in the figure, we would expect the outer measure of A to be the usual area of A.

Note that if $A \subseteq S$, $\nu_{out}(A) \leq \nu(S)$. In the special case when S is an interval and ν is the additive set function μ on the algebra of basic sets, the outer measure of some subset of S can be ∞ only if the interval is unbounded. As every basic set is a finite union of intervals, in calculating the outer measure of a subset A of $\mathbb{R}$ in (5.4), it suffices to take the infimum over all sequences (V_n) of *intervals* such that $A \subseteq \cup_{n=1}^{\infty} V_n$. Also, in this case the intervals (V_n) in (5.4) may be taken to be disjoint (see Exercise 4).

Although it is by no means easy or even possible to calculate the outer measure of an arbitrary subset of S, some properties of the outer measure are relatively easy to prove. For example, it follows from the definition of the outer measure that if $A \subseteq S$, then $0 \leq \mu_{out}(A) \leq \infty$, and that if $A, B \subseteq S$ and $A \subseteq B$ then $\mu_{out}(A) \leq \mu_{out}(B)$. The

5.2. Outer measure

following proposition mentions these facts, and describes a property of the outer measure under countable unions of sets.

Proposition 5.1. *Let S be a set, let $\mathcal{B}$ be an algebra of subsets of S, and let ν be an additive set function on $\mathcal{B}$. Then the following statements hold.*

(1) If A is a subset of S, $0 \leq \nu_{out}(A) \leq \infty$.
(2) If A, B are subsets of S and $A \subseteq B$, then $\nu_{out}(A) \leq \nu_{out}(B)$.
(3) If (A_n) is a sequence of subsets of S, then

$$\nu_{out}\left(\bigcup_{n=1}^{\infty} A_n\right) \leq \sum_{n=1}^{\infty} \nu_{out}(A_n).$$

Given a set A in $\mathcal{B}$, there are now two concepts which apparently assign a measure to A: the length or measure $\nu(A)$ of A, and the outer measure $\nu_{out}(A)$ of A. Are these two always equal on $\mathcal{B}$? The following example shows the answer is "no".

Example Let $\mathbb{N}$ denote the set of positive integers, and let $\mathcal{B}$ be the family of all subsets B of $\mathbb{N}$ such that either B or B^c is finite. Then $\{n\} \in \mathcal{B}$ for all $n \in \mathbb{N}$, and it is readily checked that $\mathcal{B}$ is an algebra of subsets of $\mathbb{N}$. Note that as $B \cup B^c = \mathbb{N}$, it is not possible that both B and B^c are finite. Thus, we may define $\nu : \mathcal{B} \longrightarrow \{0, 1\}$ by

$$\nu(B) = \begin{cases} 0, & \text{if } B \text{ is finite;} \\ 1, & \text{if } B^c \text{ is finite.} \end{cases}$$

Then ν is an additive set function on $\mathcal{B}$. Now, $\mathbb{N} = \bigcup_{n=1}^{\infty} \{n\}$, so we see that

$$\nu_{out}(\mathbb{N}) \leq \sum_{n=1}^{\infty} \nu(\{n\}) = \sum_{n=1}^{\infty} 0 = 0 < 1 = \nu(\mathbb{N}).$$

Thus, we see that $\nu_{out}(\mathbb{N}) < \nu(\mathbb{N})$ so that $\nu \neq \nu_{out}$ on $\mathcal{B}$.

Now, it may be rather confusing if the additive set function ν on $\mathcal{B}$ is used to construct an outer measure ν_{out} on all the subsets of S, but the

298   Chapter 5. Averaging in Time and Space

values of ν_{out} on $\mathcal{B}$ turn out to be possibly different from the values of ν on $\mathcal{B}$, as occurs in the preceding example. Rather, we would like ν_{out} to *extend* the values of ν on $\mathcal{B}$. For this, we need an extra condition on ν.

Definition Let S be a set, let $\mathcal{B}$ be an algebra of subsets of S, and let ν be an additive set function on $\mathcal{B}$. We call ν a *measure* if, when (A_n) is a sequence of sets in $\mathcal{B}$ such that $\bigcup_{n=1}^{\infty} A_n \in \mathcal{B}$ and $A_m \cap A_n = \emptyset$ for all $m \neq n$, then

$$\nu\left(\bigcup_{n=1}^{\infty} A_n\right) = \sum_{n=1}^{\infty} \nu(A_n).$$

This is called the *countable additivity property* of ν. A measure on an algebra of sets may be called a *countably additive set function* on the algebra. If ν is a measure on an algebra $\mathcal{B}$ of subsets of a set S and $B \in \mathcal{B}$, $\nu(B)$ will be called the *measure* of B.

The following result says that a measure ν on an algebra of subsets of a set S extends to an outer measure on the family of all subsets of S.

Proposition 5.2. *Let S be a set, let $\mathcal{B}$ be an algebra of subsets of S, and let ν be a measure on $\mathcal{B}$. Then, if $A \in \mathcal{B}$, $\nu_{out}(A) = \nu(A)$.*

When ν is a measure on the algebra $\mathcal{B}$, $\nu_{out}(A) = \nu(A)$ for all $A \in \mathcal{B}$, and it is not unreasonable to regard the concept of outer measure as an attempt to extend the properties of ν on $\mathcal{B}$ to *all* subsets of S. However, note that this is an attempt only, and that the outer measure ν_{out} on all subsets is generally not as well behaved as is the measure ν on the sets of $\mathcal{B}$. For example, for a subset A of S, by Proposition 5.2 we have

$$\nu_{out}(S) = \nu_{out}(A) + \nu_{out}(A^c),$$

for all sets A in $\mathcal{B}$, but this is generally not so for sets not in $\mathcal{B}$ [37, pp. 298–299].

Definition Suppose that we have a set, an algebra of subsets of the set, and a measure on the algebra. Then, a subset of S whose outer measure is zero is called a *set of measure zero*. If a certain statement involving elements of the set is true for all elements except for those in some set of

5.2. Outer measure

measure zero, we may say that the statement is true *almost everywhere*, or that it is true for *almost all* elements of the set.

Note that in the case when S is an interval, this definition of a set of measure zero coincides with the definition of a set of measure zero in (3.27) of Chapter 3. For, as noted earlier, in this case the infimum in (5.4) may be taken over sequences (V_n) of intervals.

In the general case, sets of measure zero have corresponding properties to sets of measure zero on the real line. For, if B is a set of measure zero and $A \subseteq B$, then by (2) of Proposition 5.1, $\mu_{out}(A) \leq \mu_{out}(B) = 0$, so that $\mu_{out}(A) = 0$ and it follows from the definition that A has measure zero. Also, if (A_n) is a sequence of sets, each of which has measure zero, then by (3) of Proposition 5.1,

$$\mu_{out}\left(\bigcup_{n=1}^{\infty} A_n\right) \leq \sum_{n=1}^{\infty} \mu_{out}(A_n) = 0 + 0 + 0 + \cdots = 0,$$

and we deduce that $\bigcup_{n=1}^{\infty} A_n$ has measure 0. So, we have the following result, corresponding to the real line case in (2) and (3) of Proposition 3.12.

Proposition 5.3. *A subset of a set of measure zero is also a set of measure zero. A set that is a union of a sequence of sets of measure zero is also a set of measure zero.*

Sets of measure zero are important in that "exceptional" sets, where points are not recurrent, or where time averages do not exist, or where "chaos" occurs, or where certain series diverge, appear under the right circumstances as sets of measure zero. A finite subset of the real line is a set of measure zero, a subset of the real line that is the union of a sequence of points has measure zero, the set of all rational numbers has measure zero, and the Cantor set (which is a "fractal" subset of [0, 1) and is not the union of any sequence of points) has measure zero, as discussed in Figure 3.4. The set of irrational numbers is *not* a set of measure zero, nor is an interval of positive length (Exercise 6 of Chapter 3).

5.3 Invariant sets

Let S be a set and let f be a transformation on S. Thus, we can think of S as the set of states of the dynamical system (S, f). Recall that

$$f^{-1}(A) = \{x : x \in S \text{ and } f(x) \in A\}.$$

Now, it may happen that $A \subseteq f^{-1}(A)$. We can think of this as meaning that every state in A will "cause" a state in A at the next stage. On the other hand it may happen that $f^{-1}(A) \subseteq A$. We can think of this as saying that every cause of a state in A is in A. That is, the set of causes of states in A is contained within A itself. The case when $A = f^{-1}(A)$ is important, and is given a special definition.

Definition Let (S, f) be a dynamical system. Then, a subset A of S is said to be f-*invariant*, or simply *invariant*, if $A = f^{-1}(A)$.

The idea in the definition is that when f^{-1} is applied to the set A, A does not change; that is, A is *invariant*. In view of the above comments we can think of A being an invariant set as meaning: the set of all states that cause a state in A is A itself or, that A is "independent" of the rest of the system. Note that $\emptyset = f^{-1}(\emptyset)$ and that $S = f^{-1}(S)$, which shows that $\emptyset$ and S are always invariant sets. So, interest lies in whether there are any invariant sets A that are (essentially) neither $\emptyset$ nor S.

Proposition 5.4. *Let S be a set and let $f : S \longrightarrow S$ be a transformation on S. Let A be a subset of S. Then the following hold.*

(1) If $A \subseteq f^{-1}(A)$, then $f : A \longrightarrow A$. That is, (A, f) is a dynamical system.

(2) If $f^{-1}(A) \subseteq A$ and if $f^n(x) \in A$ for some $x \in S$ and some $n \in \mathbb{N}$, then $x \in A$.

(3) A is invariant if and only if A^c is invariant.

(4) If $A = f^{-1}(A)$, then both (A, f) and (A^c, f) are dynamical systems.

(5) If (A, f) and (A^c, f) are dynamical systems, then $A = f^{-1}(A)$.

5.3. Invariant sets

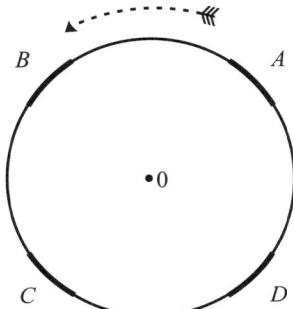

Figure 5.4. The figure depicts the circle $\mathbb{T}$ of center 0 and radius 1, whose set description is $\{(x, y) : x, y \in \mathbb{R} \text{ and } x^2 + y^2 = 1\}$. The arrow indicates the transformation ρ of $\mathbb{T}$ that rotates points of $\mathbb{T}$ counterclockwise through an angle of $\pi/2$. We see that if A, B, C, D are the indicated arcs, then $\rho(A) = B$, $\rho(B) = C$, $\rho(C) = D$ and $\rho(D) = A$. Also, if $Z = A \cup B \cup C \cup D$, then $\rho(Z) = \rho^{-1}(Z) = Z$, and so Z is ρ-invariant. In fact, as ρ^4 is the identity transformation on $\mathbb{T}$, for every subset X of $\mathbb{T}$, it can be seen that $X \cup \rho(X) \cup \rho^2(X) \cup \rho^3(X)$ is ρ-invariant.

Proof. (1) Let $A \subseteq f^{-1}(A)$ and let $x \in A$. Then, $x \in f^{-1}(A)$, which gives $f(x) \in A$. That is, $f : A \longrightarrow A$.

(2) Let $f^{-1}(A) \subseteq A$ and let $f^n(x) \in A$ for some $x \in S$ and some $n \in \mathbb{N}$. Then

$$f^n(x) \in A \implies f^{n-1}(x) \in f^{-1}(A) \implies f^{n-1}(x) \in A.$$

Repeating this step as needed, and using an induction argument, we deduce that $x \in A$.

(3) We have

$$f^{-1}(A) = A \iff f^{-1}(A)^c = A^c \iff f^{-1}(A^c) = A^c.$$

(4) If $A = f^{-1}(A)$, $A^c = f^{-1}(A^c)$, by part (3). Then, by part (1) applied first to A and then to A^c, we deduce that $f : A \longrightarrow A$ and $f : A^c \longrightarrow A^c$. That is, (A, f) and (A^c, f) are dynamical systems.

(5) Let $f : A \longrightarrow A$ and $f : A^c \longrightarrow A^c$. Because $f : A \longrightarrow A$, we have $A \subseteq f^{-1}(A)$. Also, because $f : A^c \longrightarrow A^c$, $A^c \subseteq f^{-1}(A^c)$. But

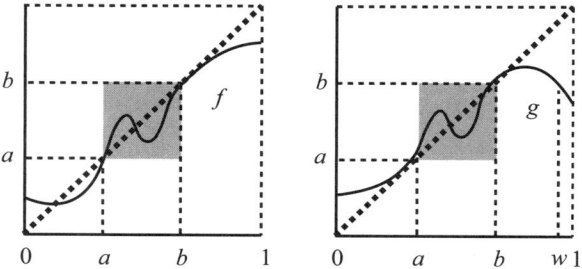

Figure 5.5. The figure depicts the graphs of functions $f, g : [0, 1] \longrightarrow [0, 1]$. On the left, the set $[a, b]$ is f-invariant—for, graphically, one can see that $x \in [a, b] \iff f(x) \in [a, b]$, expressing the invariance. On the right, function g is similar to function f, except that g is not increasing on $[b, 1]$, but "turns back on itself", as shown. Now we see that $x \in [a, b] \implies g(x) \in [a, b]$, but we also see that there is $w \notin [a, b]$ with $g(w) \in [a, b]$. This means that $g^{-1}([a, b]) \neq [a, b]$ so that $[a, b]$ is *not* g-invariant.

then we have

$$A^c \subseteq f^{-1}(A^c) \implies (A^c)^c \supseteq \left(f^{-1}(A^c)\right)^c \implies A \supseteq f^{-1}(A).$$

So $A \subseteq f^{-1}(A)$ and $A \supseteq f^{-1}(A)$, and it follows that $A = f^{-1}(A)$.
□

The simple Proposition 5.4 makes the significance of invariant sets clearer. For, if (S, f) is a dynamical system and A is an invariant subset of S such that $A \neq \emptyset$ and $A \neq S$, then the dynamical system "splits up" into two distinct and proper dynamical subsystems, (A, f) and (A^c, f). So, in this case, the problem of understanding and analyzing the behavior of the system (S, f) reduces to the problem of understanding and analyzing the behaviors of the smaller subsystems (A, f) and (A^c, f). On the other hand, if the system (S, f) does not split into proper disjoint systems (A, f) and (A^c, f)—that is, if there are no invariant sets other than $\emptyset$ and S, then it is what we could term an "irreducible" dynamical system, or an "ergodic" system (a formal definition will be given later).

There is a way of constructing invariant sets in dynamical systems, as described in the following result.

5.4. Measurable sets

Proposition 5.5. *Let S be a set, let f be a transformation on S, and let A be a subset of S. Then $\bigcap_{n=1}^{\infty} \left(\bigcup_{j=n}^{\infty} f^{-j}(A) \right)$ is an f-invariant set.*

Proof. Put $D = \bigcap_{n=1}^{\infty} \left(\bigcup_{j=n}^{\infty} f^{-j}(A) \right)$. First, observe that

$$D = \{x : x \in S \text{ and } f^j(x) \in A \text{ for infinitely many } j \in \mathbb{N}\}.$$

Now, if $f^j(x) \in A$ for infinitely many j, we see that $f^{j-1}(f(x)) \in A$ for infinitely many $j \geq 2$. Thus, if $x \in D$, $f(x) \in D$. On the other hand, if $f^j(f(x)) \in A$ for infinitely many j, we see that $f^{j+1}(x) \in A$ for infinitely many j. So, if $f(x) \in D$, $x \in D$. Thus, $x \in D \iff f(x) \in D$, so D is f-invariant. □

Although Proposition 5.5 shows how we can "construct" invariant sets, that does not mean it is straightforward to give specific examples of invariant sets apart from ∅ and S. This is because, in general, we need to understand the system very well before being able to describe explicitly a set of the form $\bigcap_{n=1}^{\infty} \left(\bigcup_{j=n}^{\infty} f^{-j}(A) \right)$.

5.4 Measurable sets

Let A be a given subset of some interval S. The subset A causes a given subset of C of S to be split into two disjoint pieces, $A \cap C$ and $A^c \cap C$, whose union is C. As these pieces are disjoint, it might be expected that the sum of the outer measures of these two pieces equals the outer measure of C. This is indeed the case when A and C are basic sets, as discussed in section 3.4. However, this is not always the case. Therefore, we single out the sets A such that for *every* subset C of S, the outer measure of C is the sum of the outer measures of $A \cap C$ and $A^c \cap C$. These sets are called *measurable*. Suitable references for the material of this section are [8] for the real line case, and [1, 2, 17, 20, 28] for the general case.

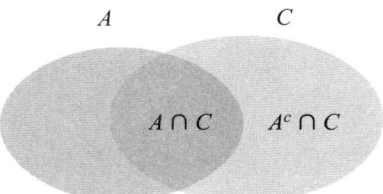

Figure 5.6. The figure indicates what it means for the set A to be measurable. It means that *whatever the subset C is*, the outer measure of C is the outer measure of the darker-shaded set $A \cap C$, plus the outer measure of the lighter-shaded set $A^c \cap C$.

Definition Let S be a set, let $\mathcal{B}$ be an algebra of subsets of S, and let ν be a measure on $\mathcal{B}$. Then a subset A of S is called *measurable* if, for all subsets C of S,

$$\nu_{out}(C) = \nu_{out}(A \cap C) + \nu_{out}(A^c \cap C). \qquad (5.6)$$

Constantin Carathéodory made the above definition of a measurable set in 1914 [10]. Since then, it has frequently been the subject of comment or defense in a way unusual for the definition of a mathematical concept. For example, in their *Real and Abstract Analysis* [20, p.127] Hewitt and Stromberg write: "How Carathéodory came to think of this definition seems mysterious, since it is not in the least intuitive. Carathéodory's definition has many useful implications". Also, Paul Halmos comments in his *Measure Theory* [17, p.44]: "It is rather difficult to get an understanding of the meaning of ... measurability except through familiarity with its implications ...The greatest justification of this apparently complicated concept is, however, its possibly surprising but absolutely complete success as a tool in proving the important and useful extension theorem".

These comments on Carathéodory's definition of a measurable set perhaps arise, in part anyway, from a lack of motivation in the definition. However, both Bressoud [8, p. 140] and Halmos [17, p. 44], comment upon the definition and interpret its meaning in terms of "cutting" or "splitting" the subsets of the whole space. Recently, it has been shown that if one considers the problem of calculating the outer measure of a

5.4. Measurable sets

subset of the interval $[0, 1)$ that is invariant under an irrational translation, then Carathéodory's definition arises from the problem [31]. Specifically, recall that $\text{frac}(x)$ denotes the fractional part of a number x, let α be an irrational number, and let τ denote the function mapping $[0, 1)$ into $[0, 1)$ that is $x \longmapsto \text{frac}(x + \alpha)$. Then it is proved in [31, Theorem 5.1] that if A is a subset of $[0, 1)$ that is invariant under τ, and if

$$\mu_{out}(C) = \mu_{out}(A \cap C) + \mu_{out}(A^c \cap C) \text{ for all } \textit{intervals } C, \quad (5.7)$$

then either A or A^c has to be a set of measure zero. Here, note the resemblance between (5.6) and (5.7). Also, note that the results in [31] are formulated in terms of the unit circle $\mathbb{T} = \{z : z \in \mathbb{C} \text{ and } |z| = 1\}$ and a rotation ρ on $\mathbb{T}$—that is, a transformation on $\mathbb{T}$ for which there is $a \in \mathbb{T}$ with $\rho(z) = az$ for all $a \in \mathbb{C}$. However, as we saw in 2.2, especially in Figures 2.1 and 2.2, these two settings are equivalent, and the results in [31] relating to $\mathbb{T}$ and ρ have entirely equivalent statements in terms of $[0, 1)$ and τ.

Which sets are measurable? It is not possible to give a simple characterization of all the measurable sets, other than the definition! However, note that for subsets A, C of S,

$$\nu_{out}(C) = \nu_{out}\big((A \cap C) \cup (A^c \cap C)\big) \leq \nu_{out}(A \cap C) + \nu_{out}(A^c \cap C).$$

Thus, to prove that A is measurable, it suffices to prove that for all subsets C of S,

$$\nu_{out}(A \cap C) + \nu_{out}(A^c \cap C) \leq \nu_{out}(C).$$

The following result identifies sets of measure zero and sets in $\mathcal{B}$ as being measurable.

Proposition 5.6. *Let S be a set, let $\mathcal{B}$ be an algebra of subsets of S, and let there be given a measure on $\mathcal{B}$. Then the following statements hold.*
(1) Every set of measure zero is measurable.
(2) Every set in $\mathcal{B}$ is measurable.

The results of this section so far indicate that the outer measure, when restricted to the measurable sets, has properties we would expect, given our notion of the outer measure as a generalized concept of length. Thus, when the set A is measurable, we generally write $\nu(A)$ instead of $\nu_{out}(A)$. Note that we *never* write $\nu(A)$ unless it is either assumed or known that A is measurable.

Definition If A is a measurable set, $\nu(A)$ is called the *measure* of A. In the case when S is an interval, the measure $\mu(A)$ of a measurable subset A of S may be called the *Lebesgue measure* of A.

The following result shows that the measurable sets form an algebra, but have properties extending beyond that of merely being an algebra, and that the function ν on the measurable sets is still a measure on the algebra of measurable sets.

Theorem 5.7 (the Carathéodory Extension Theorem). *Let S be a given set, let $\mathcal{B}$ be an algebra of subsets of S, and let $\nu : \mathcal{B} \longrightarrow [0, \infty]$ be a measure on $\mathcal{B}$. A particular case of this is when S is an interval, $\mathcal{B}$ consists of the basic subsets of S, and ν is the measure μ that assigns to an interval its usual length. In either case, let $\mathcal{M}$ denote all the measurable subsets of S. Then the following statements hold.*

(1) $S, \emptyset \in \mathcal{M}$.

(2) If $A \in \mathcal{M}$, $A^c \in \mathcal{M}$.

(3) If (A_n) is a sequence of sets in $\mathcal{M}$, then $\bigcup_{n=1}^{\infty} A_n \in \mathcal{M}$ and $\bigcap_{n=1}^{\infty} A_n \in \mathcal{M}$.

(4) If (A_n) is a sequence of sets in $\mathcal{M}$ such that $A_j \cap A_k = \emptyset$ if $j \neq k$, then

$$\nu\left(\bigcup_{n=1}^{\infty} A_j\right) = \sum_{n=1}^{\infty} \nu(A_j).$$

(5) If (A_n) is a non-decreasing sequence of sets in $\mathcal{M}$,

$$\nu\left(\bigcup_{n=1}^{\infty} A_j\right) = \lim_{n \to \infty} \nu(A_n).$$

(6) If (A_n) is a non-increasing sequence of sets in $\mathcal{M}$, and if $\nu(A_1) < \infty$,

$$\nu\left(\bigcap_{n=1}^{\infty} A_n\right) = \lim_{n \to \infty} \nu(A_n).$$

Let us conclude with some comments which are designed to put the work of this section in perspective.

Definition Let S be a given set, and let $\mathcal{M}$ be a family of subsets of S. Then if $\mathcal{M}$ satisfies properties (1), (2), (3) of Theorem 5.7, we say that $\mathcal{M}$ is a *σ-algebra* of subsets of S. Here, the letter σ indicates that the algebra $\mathcal{M}$ is closed under *countable* unions and intersections, not just under finite unions and intersections as required for the case of an algebra.

The results in this section may be summarized in the following way: if S is a set, if $\mathcal{B}$ is an algebra of subsets of S, and if ν is a measure on S, then the associated family $\mathcal{M}$ of measurable sets is a σ-algebra, $\mathcal{M}$ contains $\mathcal{B}$, and the restriction of the outer measure to $\mathcal{M}$ is a measure on $\mathcal{M}$ which extends the measure ν on $\mathcal{B}$.

5.5 Measure-preserving transformations

In the introduction to Chapter 4, it was pointed out that the transformations that preserve length can be regarded as coming from systems that are seemingly "stationary". That is, however long we observe the behavior of the system, its future from that point on looks much like the past. In Chapter 4 we called a transformation f mapping an interval S into itself *length-preserving* when, for every basic subset B of S, $f^{-1}(B)$ is also basic and $\mu(f^{-1}(B)) = \mu(B)$. We now need to put these ideas in a more general context, where the role of the basic sets is now assumed by the measurable sets, and instead of considering transformations that only preserve length, we consider those transformations that preserve measure. The approach is to assume we have a transformation that preserves the measure of sets in an algebra, and show that if we extend this measure to the σ-algebra generated by the algebra, then the transformation preserves the measure of the sets in this σ-algebra. As standard sources do not always discuss this in detail, proofs are included. First, we introduce the notion of the σ-algebra generated by a family of sets.

Lemma 5.8. *Let $\mathcal{A}$ be a family of subsets of a set S and let Υ denote the family of all σ-algebras of subsets of S that contain the given family $\mathcal{A}$. Then the family Υ is non-empty, so we may put*

$$\mathcal{M} = \bigcap \{\mathcal{F} : \mathcal{F} \in \Upsilon\}.$$

Then, $\mathcal{M}$ is a σ-algebra of subsets of S, $\mathcal{M}$ contains $\mathcal{A}$, and in fact if $\mathcal{C}$ is any σ-algebra of subsets of S such that $\mathcal{C}$ contains $\mathcal{A}$, then $\mathcal{C}$ contains $\mathcal{M}$. That is, $\mathcal{M}$ is the smallest σ-algebra of subsets of S that contains the given family $\mathcal{A}$.

Proof. The family of *all* subsets of S is clearly a σ-algebra that contains $\mathcal{A}$. So, Υ is non-empty. It is easy to check that $\mathcal{M}$ is a σ-algebra. For example, let (A_n) be a sequence of sets in $\mathcal{M}$, and let $\mathcal{F}$ be a σ-algebra of sets that contains $\mathcal{A}$. Then, (A_n) is a sequence of sets in the σ-algebra $\mathcal{F}$, so that $\bigcup_{n=1}^{\infty} A_n \in \mathcal{F}$. Hence, $\bigcup_{n=1}^{\infty} A_n \in \cap \{\mathcal{F} : \mathcal{F} \in \Upsilon\} = \mathcal{M}$. The other σ-algebra properties may be proved in like manner. □

Lemma 5.8 makes the following definitions possible.

Definition Let $\mathcal{A}$ be a family of subsets of a set S. Then the smallest σ-algebra that contains $\mathcal{A}$ is called the σ-algebra *generated by* $\mathcal{A}$. In the case when S is an interval of real numbers, the basic sets generate a σ-algebra of subsets of S, called the *Borel sets*.

If $\mathcal{B}$ is an algebra of subsets of a set S and ν is a measure on $\mathcal{B}$, then the σ-algebra generated by $\mathcal{B}$, as described in Lemma 5.8, is a subfamily of the σ-algebra of measurable sets. However, in general, the σ-algebra generated by $\mathcal{B}$ is a proper subfamily of the measurable sets.

Definition Let S be a given set, let $\mathcal{B}$ be an algebra of subsets of S, and let $\nu : \mathcal{B} \longrightarrow [0, \infty]$ be a measure on $\mathcal{B}$. Let $f : S \longrightarrow S$ be a given transformation. Then f is called ν-*invariant on* $\mathcal{B}$ if, for every set B in $\mathcal{B}$, $f^{-1}(B)$ is also in $\mathcal{B}$ and $\nu(f^{-1}(B)) = \nu(B)$. In this case, f might also be said to be ν-*measure-preserving*, or simply *measure-preserving*, if the measure is understood. The following result tells us when a transformation is measure-preserving on a generated σ-algebra.

5.5. Measure-preserving transformations

Theorem 5.9. *Let S be a set, let $\mathcal{B}$ be an algebra of subsets of S, let $v : \mathcal{B} \longrightarrow [0, \infty]$ be a measure on $\mathcal{B}$ with $v(S) < \infty$, and let v also denote the extension of v to the σ-algebra of measurable subsets of S, as described by the Carathéodory Extension Theorem 5.7. Let $\mathcal{M}$ be the σ-algebra generated by $\mathcal{B}$ and let $f : S \longrightarrow S$ be a transformation such that for all $A \in \mathcal{B}$, $f^{-1}(A) \in \mathcal{M}$ and $v(f^{-1}(A)) = v(A)$. Then if $A \in \mathcal{M}$, $f^{-1}(A) \in \mathcal{M}$ and $v(f^{-1}(A)) = v(A)$. That is, f is a v-measure-preserving transformation on $\mathcal{M}$.*

Proof. Make the definition that

$$\mathcal{G} = \left\{ A : A \in \mathcal{M} \text{ and } f^{-1}(A) \in \mathcal{M} \right\}.$$

The family $\mathcal{G}$ contains $\mathcal{B}$, by assumption. Now, if $A \in \mathcal{G}$, $A \in \mathcal{M}$ so that $A^c \in \mathcal{M}$ and $f^{-1}(A^c) = f^{-1}(A)^c \in \mathcal{M}$. It follows that if $A \in \mathcal{G}$ then $A^c \in \mathcal{G}$. Also, if (A_n) is a sequence of sets in $\mathcal{G}$,

$$\bigcup_{n=1}^{\infty} A_n \in \mathcal{M} \text{ and } f^{-1}\left(\bigcup_{n=1}^{\infty} A_n\right) = \bigcup_{n=1}^{\infty} f^{-1}(A_n) \in \mathcal{M}.$$

So, the union of a sequence of sets in $\mathcal{G}$ is also in $\mathcal{G}$. Thus, $\mathcal{G}$ is a σ-algebra and $\mathcal{B} \subseteq \mathcal{G} \subseteq \mathcal{M}$. Since $\mathcal{M}$ is the smallest σ-algebra containing $\mathcal{B}$, we must have $\mathcal{G} = \mathcal{M}$. Thus, if $A \in \mathcal{M}$, $A \in \mathcal{G}$ and so, by the definition of $\mathcal{G}$, we have $f^{-1}(A) \in \mathcal{M}$.

Now, let $A \in \mathcal{M}$ and let (A_n) be a sequence of sets in $\mathcal{B}$ such that $A \subseteq \bigcup_{n=1}^{\infty} A_n$. Then,

$$f^{-1}(A) \subseteq f^{-1}\left(\bigcup_{n=1}^{\infty} A_n\right) = \bigcup_{n=1}^{\infty} f^{-1}(A_n),$$

and it follows from (3) of Proposition 5.1 that

$$v(f^{-1}(A)) = v_{out}(f^{-1}(A)) \leq \sum_{n=1}^{\infty} v(f^{-1}(A_n)) = \sum_{n=1}^{\infty} v(A_n).$$

Taking the infimum over all possible sequences (A_n), we have

$$v(f^{-1}(A)) \leq v_{out}(A) = v(A), \text{ for all } A \in \mathcal{M}. \tag{5.8}$$

Since this holds for all $A \in \mathcal{M}$, and since $\mathcal{M}$ is a σ-algebra, (5.8) also holds with A^c in place of A. So, for all $A \in \mathcal{M}$,

$$\nu(S) - \nu(f^{-1}(A)) = \nu(f^{-1}(A^c)) \le \nu(A^c) = \nu(S) - \nu(A). \quad (5.9)$$

Then (5.8) and (5.9) now give $\nu(A) \le \nu(f^{-1}(A)) \le \nu(A)$, so that $\nu(A) = \nu(f^{-1}(A))$ for all $A \in \mathcal{M}$.

□

5.6 Poincaré recurrence ... again

In 4.3 and 4.4, Poincaré's Recurrence Theorem and results on recurrent points on intervals were presented. The discussion there used the fact that, on the basic sets, the length function μ is an additive set function that is also length-preserving. The abstract measure theory setting, described in Sections 5.2 and 5.4, enables us to prove more general versions of these results, and with simpler arguments. A comparison of this discussion with the original approach used in obtaining Theorem 4.7 provides an initial indication of the power of measure theory.

Definition Let S be a set, let $\mathcal{M}$ be a σ-algebra of subsets of S, let $A \in \mathcal{M}$ and let there be given a measure on $\mathcal{M}$. Also, let $f : S \longrightarrow S$ be a transformation on S. Then a point x of A is said to be *recurrent with respect to A* if there is $n \in \mathbb{N}$ such that $f^n(x) \in A$. Also, a point x of A is said to be *infinitely recurrent with respect to A* if $f^n(x) \in A$ for infinitely many $n \in \mathbb{N}$.

Theorem 5.10. *(Poincaré, 1899) Let S be a set, let $\mathcal{M}$ be a σ-algebra of subsets of S and let $\nu : \mathcal{M} \longrightarrow [0, \infty]$ be a measure on $\mathcal{M}$ with $\nu(S) < \infty$. Let $f : S \longrightarrow S$ be a measure-preserving transformation on S. Then for every $A \in \mathcal{M}$, almost every point of A is recurrent with respect to A.*

Proof. Let $A \in \mathcal{M}$ and put

$$\mathcal{Z} = A \cap \left(\bigcap_{n=1}^{\infty} f^{-n}(A^c) \right).$$

5.6. Poincaré recurrence ... again

As $\mathcal{M}$ is an algebra and f is measure-preserving, $f^{-1}(A^c) \in \mathcal{M}$ for all $A \in \mathcal{M}$. So, as $\mathcal{M}$ is a σ-algebra, $\mathcal{Z} \in \mathcal{M}$. Now, for $j, k \in \{0, 1, 2, \ldots\}$ with $j < k$,

$$\begin{aligned} x \in f^{-j}(\mathcal{Z}) &\implies f^j(x) \in f^{-(k-j)}(A^c) \\ &\implies f^k(x) \in A^c \\ &\implies f^k(x) \notin A \\ &\implies f^k(x) \notin \mathcal{Z} \\ &\implies x \notin f^{-k}(\mathcal{Z}). \end{aligned}$$

Hence,

$$f^{-j}(\mathcal{Z}) \cap f^{-k}(\mathcal{Z}) = \emptyset, \quad \text{for all } j, k \in \{0, 1, 2, \ldots\} \text{ with } j \neq k.$$

We now have

$$\infty > \nu(S) \geq \nu\left(\bigcup_{n=1}^{\infty} f^{-n}(\mathcal{Z})\right) = \sum_{n=1}^{\infty} \nu\left(f^{-n}(\mathcal{Z})\right) = \sum_{n=1}^{\infty} \nu(\mathcal{Z}),$$
(5.10)

which establishes that $\nu(\mathcal{Z}) = 0$. So, $\mathcal{Z}$ is a subset of A, $\mathcal{Z} \in \mathcal{M}$, and $\mathcal{Z}$ has measure zero. But, if $x \in A$ and $x \notin \mathcal{Z}$, there is $n \in \mathbb{N}$ such that $x \notin f^{-n}(A^c)$. That is, if $x \in A$ and $x \notin \mathcal{Z}$, there is $n \in \mathbb{N}$ such that $f^n(x) \in A$. This shows that almost every point of A is recurrent with respect to A. □

Note that the countable additivity property of ν was used in (5.10). Theorem 5.10 includes Theorem 4.7 as a special case.

Theorem 5.11. *Let S be a set, let $\mathcal{M}$ be a σ-algebra of subsets of S, and let $\nu : \mathcal{M} \longrightarrow [0, \infty]$ be a measure on $\mathcal{M}$ with $\nu(S) < \infty$. Let $f : S \longrightarrow S$ be a measure-preserving transformation on S and let $A \in \mathcal{M}$. Then, almost every point x in A is infinitely recurrent with respect to A.*

Proof. For each $n \in \mathbb{N}$, f^n is a measure-preserving transformation on S. So, by Theorem 5.10, for each $n \in \mathbb{N}$, there is a subset set $\mathcal{Z}_n \in \mathcal{M}$

having measure zero, such that

$$x \notin \mathcal{Z}_n \implies \left(f^n\right)^r(x) \in A \text{ for some } r \in \mathbb{N}.$$

That is,
$$x \notin \mathcal{Z}_n \implies f^{rn}(x) \in A \text{ for some } r \in \mathbb{N}. \tag{5.11}$$

Now, put $\mathcal{Z} = \cup_{n=1}^\infty \mathcal{Z}_n$. Then $\mathcal{Z}$ has measure zero by Proposition 5.3, and it follows from (5.11) that for every $n \in \mathbb{N}$,

$$\begin{aligned}x \notin \mathcal{Z} &\implies x \notin \mathcal{Z}_n \\ &\implies f^{rn}(x) \in A \text{ for some } r \in \mathbb{N} \\ &\implies f^j(x) \in A \text{ for some } j \geq n.\end{aligned}$$

By successively taking $n = 1, 2, \ldots$, we see that if $x \notin \mathcal{Z}$, then $f^j(x) \in A$ for infinitely many $j \in \mathbb{N}$. □

5.7 Ergodic systems

If the initial state of a dynamical system (S, f) is x, the average time that the system will spend in a state in the subset A of S is

$$\lim_{n \to \infty} \frac{1}{n} \sum_{j=1}^n \chi_A\left(f^{j-1}(x)\right),$$

assuming that this limit exists. In this case, the limit is called a *time average* (relative to A and x). If we imagine a system that can be split into two parts, each part having no interaction whatsoever with the other part, for a given subset A, there is no reason to think that the time average for an initial state in one part of the system should have any relationship to the time average obtained from an initial state in the other part of the system. On the other hand, if it is *not* possible to split the system into two disjoint parts, there is reason to think that the time averages for different initial states will not be independent, since any state in the system perhaps may be reached, even if only in an approximate sense, from any given initial

5.7. Ergodic systems

state of the system. The precise formulation of these ideas depends upon what we mean by "splitting" the system into disjoint parts. Now if A is an invariant set for the system (S, f), we have by (4) of Proposition 5.4 that $f : A \longrightarrow A$ and $f : A^c \longrightarrow A^c$, so that (A, f) and (A^c, f) are dynamical systems in their own right. Thus, the invariant set A leads to a "splitting" of (S, f) into the disjoint systems (A, f) and (A^c, f). If the system does not allow such a splitting, then either A or A^c must be, at least approximately, the whole set S of possible states. Alternatively, either A or A^c must be, at least approximately, the empty set S. Systems for which this is the case are called *ergodic*. Here are formal definitions.

Definition Let S be a set, let $\mathcal{M}$ be a σ-algebra of subsets of S, and let ν be a measure on $\mathcal{M}$ with $\nu(S) < \infty$. Let $f : S \longrightarrow S$ be a measure-preserving transformation. Then the dynamical system (S, f) is said to be *ergodic* if the following condition holds.

$$A \in \mathcal{M} \text{ and } A \text{ is } f\text{-invariant} \implies \nu(A) = 0 \text{ or } \nu(A^c) = 0.$$

Note that $\nu(A^c) = 0$ is equivalent to $\nu(A) = \nu(S)$. If A, B are sets, we define $A \triangle B$ by putting

$$A \triangle B = (A \cap B^c) \cup (A^c \cap B.$$

A set $A \in \mathcal{M}$ is called *almost invariant* if $\nu(A \triangle f^{-1}(A)) = 0$.

The set $A \triangle B$ is sometimes called the *symmetric difference* of A and B. We have $A = B$ if and only if $A \triangle B = \emptyset$ and, in general, $A \triangle B$ measures how close A and B are to being equal—in the sense that the smaller $A \triangle B$ is, the "closer" are A and B to being equal. If A is invariant, then $A \triangle f^{-1}(A) = A \triangle A = \emptyset$. As $\nu(\emptyset) = 0$, we see that an invariant set is almost invariant.

Proposition 5.12. *Let S be a set, let $\mathcal{M}$ be a σ-algebra of subsets of S, let ν be a measure on $\mathcal{M}$ with $\nu(S) < \infty$, and let $f : S \longrightarrow S$ be a measure-preserving transformation on S. Then the following conditions are equivalent.*

(1) (S, f) is ergodic.

(2) If A is an invariant set in $\mathcal{M}$, then $\nu(A) = 0$ or $\nu(A^c) = 0$.
(3) If A is an almost invariant set in $\mathcal{M}$, then $\nu(A) = 0$ or $\nu(A^c) = 0$.
(4) For each set $A \in \mathcal{M}$ with $\nu(A) > 0$, for almost all $x \in S$ we have that $f^n(x) \in A$ for infinitely many $n \in \mathbb{N}$.

Ideas in the Proof The equivalence of (1) and (2) is immediate from the definition of an ergodic system. Also, as an invariant set is almost invariant, we see that (3) implies (2). To get an idea of how (2) implies (3), consider an almost invariant set $A \in \mathcal{M}$. Then, put

$$B = \bigcap_{n=1}^{\infty} \left(\bigcup_{j=n}^{\infty} f^{-j}(A) \right). \tag{5.12}$$

Using Proposition 5.5, we see from (5.12) that $B \in \mathcal{M}$ and that B is invariant. As (2) holds, we must have $\nu(B) = 0$ or $\nu(B^c) = 0$. On the other hand, as $\nu(A \Delta f^{-1}(A)) = 0$, it is possible to show that $\nu(A \Delta f^{-n}(A)) = 0$ for all $n \in \mathbb{N}$. That is, A is "approximately" $f^{-n}(A)$ for all $n \in \mathbb{N}$. In fact we can prove from (5.12) that $\nu(B \Delta A) = 0$, so that $\nu(A) = \nu(B)$. As we have seen that $\nu(B) = 0$ or $\nu(B^c) = 0$, we must have $\nu(A) = 0$ or $\nu(A^c) = 0$. Thus, (2) implies (3). (The details of this proof are in [39, p.27].) We have now seen that (1), (2), (3) are equivalent.

We now indicate why (4) is equivalent to each of (1), (2), (3). First, note that the set B in (5.12) is also given by

$$B = \left\{ x : x \in S \text{ and } f^n(x) \in A \text{ for infinitely many } n \right\}. \tag{5.13}$$

Now, let (1) hold and let $A \in \mathcal{M}$ with $\nu(A) > 0$. Then, as noted above, the set B in (5.12) and (5.13) is invariant and $B \in \mathcal{M}$. Thus, by (1), $\nu(B) = 0$ or $\nu(B^c) = 0$. Now, as $\nu(A) > 0$, by Theorem 5.11, almost every point of A is in B. That is, $\nu(A \cap B^c) = 0$, and we deduce that

$$0 < \nu(A) = \nu(A \cap B) + \nu(A \cap B^c) = \nu(A \cap B) \leq \nu(B).$$

As B is invariant and $\nu(B) > 0$, and as (1) holds, $\nu(B^c) = 0$, so that almost every point of S is in B, and then (5.13) establishes that (4) holds.

5.7. Ergodic systems

Thus, (1) implies (4). Conversely, let (4) hold, and let $A \in \mathcal{M}$ be an invariant set with $\nu(A) > 0$. Put $B = \bigcap_{n=1}^{\infty}\left(\bigcup_{j=n}^{\infty} f^{-j}(A)\right)$, as in (5.12). Then, as A is invariant, $A = f^{-j}(A)$ for all $j \in \mathbb{N}$, so that $B = A$. However, as (4) holds, by (5.13) we must have $\nu(B^c) = 0$, and so $\nu(A^c) = \nu(B^c) = 0$. Thus, (4) implies (1), and all conditions are equivalent.

Now in Proposition 5.12, note that condition (4) has appeared in a weaker form as the assumption (4.15) in Kac's Theorem 4.11. The assumption there was weaker in that it was assumed that for a *particular* set U, for almost all $x \in S$ it is the case that $f^n(x) \in U$ for some $n \in \mathbb{N}$ (see also the example in Figures 4.7 and 4.8). This assumption does not mean that the system is necessarily ergodic, whereas most statements of Kac's Theorem assume that the system is ergodic.

Example 5.13. Let $S = [0, 1)$, let $\mathcal{M}$ be the σ-algebra of Borel subsets of $[0, 1)$, let μ be the measure on $\mathcal{M}$ deriving from the usual notion of length, let α be an irrational number, and let τ_α be translation through α as defined in 2.5. That is, τ_α is the transformation $x \longmapsto \text{frac}(x + \alpha)$ on $[0, 1)$. Now, τ_α is linear with slope 1 on each of the intervals $[0, 1 - \alpha)$ and $[1 - \alpha, 1)$ (the graph of τ_α is illustrated in Figure 2.8). It follows that τ_α is length-preserving by Theorem 4.4, and so it is measure-preserving by Theorem 5.9. Then, it follows from results on the invariant sets of an irrational rotation [31, Theorem 5.3], interpreted so as to apply on $[0, 1)$, that if A is a measurable and τ_α-invariant set, then $\mu(A) = 0$ or $\mu(A^c) = 0$. Thus, $([0, 1), \tau_\alpha)$ is ergodic. Note that the discussion in [31] does not presuppose any knowledge of measure theory.

For the reader familiar with measure and integration theory, Hilbert space and functional analysis, here is an elegant alternative approach to proving that τ_α is ergodic (see [26, pp. 146–147] for more details). Let A denote a Borel subset of $[0, 1)$ that is τ_α-invariant, where α is a given irrational number. Also, for $n \in \mathbb{Z}$, let $e_n : [0, 1) \longrightarrow \mathbb{C}$ be the function given by

$$t \longmapsto e^{2\pi i n t}.$$

Then, $\{e_n : n \in \mathbb{Z}\}$ is a complete orthonormal set in the Hilbert space

$L^2([0, 1))$ of square integrable functions on $[0, 1)$, so if we put

$$c_n = \frac{1}{2\pi} \int_A e^{-2\pi i n t} dt,$$

we have

$$\chi_A = \sum_{n=-\infty}^{\infty} c_n e_n, \qquad (5.14)$$

where the series converges in $L^2([0, 1))$. Observe that

$$e_n \circ \tau_\alpha(t) = e^{2\pi i n \operatorname{frac}(t+\alpha)} = e^{2\pi i n (t+\alpha)} = e^{2\pi i n \alpha} e_n(t).$$

Thus, $e_n \circ \tau_\alpha = e^{2\pi i n \alpha} e_n$. Using this fact, and the invariance of A, we now have

$$\chi_A = \chi_{\tau_\alpha^{-1}(A)} = \chi_A \circ \tau_\alpha = \sum_{n=-\infty}^{\infty} c_n e_n \circ \tau_\alpha = \sum_{n=-\infty}^{\infty} c_n e^{2\pi i n \alpha} e_n. \qquad (5.15)$$

By the uniqueness of orthonormal expansions in a Hilbert space, we may compare coefficients in (5.14) and (5.15), and we see that $c_n = c_n e^{2\pi i n \alpha}$ for all $n \in \mathbb{Z}$. But α is irrational, so $e^{2\pi i n \alpha} \neq 1$ for all $n \in \mathbb{Z}$ with $n \neq 0$. So, $c_n = 0$ for all $n \neq 0$, and (5.14) gives $\chi_A = c_0$ in $L^2([0, 1)]$. Thus, χ_A is a constant function in $L^2([0, 1))$. As χ_A takes only the values 0 and 1, either $\chi_A(t) = 0$ for almost all $t \in [0, 1)$ or $\chi_A(t) = 1$ for almost all $t \in [0, 1)$. Thus, $\mu(A) = 0$ or $\mu(A) = 1$, so we deduce that τ_α is ergodic.

Example 5.14. Let $b \in \mathbb{N}$, $b \geq 2$ and let

$$f(t) = \operatorname{frac}(bt), \text{ for } 0 \leq t < 1.$$

Then $f : [0, 1) \longrightarrow [0, 1)$ so that $([0, 1), f)$ is a dynamical system. We may describe f alternatively by observing that the intervals $[(j-1)/b, j/b)$ for $j = 1, 2, \ldots, b$ form a partition of $[0, 1)$ and we have

$$f(t) = bt - j + 1, \text{ for } t \in \left[\frac{j-1}{b}, \frac{j}{b}\right). \qquad (5.16)$$

5.7. Ergodic systems

By Theorem 4.3 (see also Figure 4.4), f is a length-preserving transformation. So, by Theorem 5.9, f is measure-preserving. The transformation f is ergodic, and it is possible to give a proof of this in the spirit of [31], where it was done for irrational rotations. This argument is in [32] and uses only the concept of the outer measure. The following result can be proved: *let μ_{out} denote the usual outer measure function defined on all subsets of $[0, 1)$. Then if A is an f-invariant subset of $[0, 1)$, either $\mu_{out}(A) = 0$ or $\mu_{out}(A) = 1$*. This immediately gives that the transformation f is ergodic. The functional analysis approach, described above for an irrational rotation, also works for the transformation f (see [26, pp. 150–151] for more details). Let A be an f-invariant Borel subset of $[0, 1)$. Then, as in (5.14) we can write

$$\chi_A = \sum_{n=-\infty}^{\infty} c_n e_n, \text{ where } c_n = \int_A e^{-2\pi i n t} dt \text{ for all } n \in \mathbb{Z}. \quad (5.17)$$

Observe that for $j \in \{1, 2, \ldots, b\}$ and $t \in [(j-1)/b, j/b)$, by (5.16) we have

$$e_n \circ f(t) = e^{2\pi i n(bt-j+1)} = e^{2\pi i n b t} = e_{bn}(t).$$

Thus, $e_n \circ f = e_{bn}$. Going back to (5.17) we now have

$$\chi_A = \chi_{f^{-1}(A)} = \chi_A \circ f = \sum_{n=-\infty}^{\infty} c_n e_n \circ f = \sum_{n=-\infty}^{\infty} c_n e_{bn}. \quad (5.18)$$

Comparing coefficients in (5.17) and (5.18) we have

$$c_n = c_{bn}, \text{ for all } n \in \mathbb{Z}. \quad (5.19)$$

Now it is known from the theory of orthonormal expansions in Hilbert space that $\sum_{n=-\infty}^{\infty} |c_n|^2 < \infty$. So, $\lim_{n \to \pm\infty} c_n = 0$. Also, from (5.19), we have $c_n = c_{b^j n}$, for all $j \in \mathbb{N}$ and $n \in \mathbb{Z}$. Consequently, if $n \neq 0$,

$$c_n = \lim_{j \to \infty} c_n = \lim_{j \to \infty} c_{b^j n} = 0.$$

Then, from (5.18) we have $\chi_A = c_0$ in $L^2([0, 1))$, so χ_A is a constant and it follows as in the previous example that f is ergodic.

5.8 Birkhoff's Theorem on time and space averages

Let S be a given set and let $f : S \longrightarrow S$ be a transformation on S. Recall that in considering the dynamical system (S, f), we can think of $f^n(x)$ as the state of the system after n time units, given that the system started in state x. Let A be a given subset of S. Then, $\chi_A(x)$ measures whether the state of the system is or is not in A, since $x \in A \iff \chi_A(x) = 1$ and $x \notin A \iff \chi_A(x) = 0$. Now for $x \in S$, consider the expression

$$\frac{1}{n}\left(\sum_{j=1}^{n} \chi_A\left(f^{j-1}(x)\right)\right). \tag{5.20}$$

The quantity in (5.20) lies in $[0, 1]$. We may think of it as the proportion of time that the system is in a state in A, over the elapse of $n - 1$ time units, given that its initial state is x. Alternatively, (5.20) may be thought of as the *proportion* of the n points $x, f(x), f^2(x), \ldots, f^{n-1}(x)$ that belong to the given set A.

We are interested in the existence of the (asymptotic)*time average*; that is, in the existence of the limit

$$\lim_{n \to \infty} \frac{1}{n}\left(\sum_{j=1}^{n} \chi_A\left(f^{j-1}(x)\right)\right).$$

This limit, if it exists, represents the average amount, or the proportion, of time that the system will ultimately spend in A, given that it started off in the state x. If the system is random in its behavior, we would expect that the system will prefer no part of the set S of states to any other part of S. Randomness should mean that no part of the system receives preferred or privileged treatment. So in a random system, and for a given set A of states, we would expect that the states of the system would spend a proportion of time in A that is equal to the proportion that A occupies within the set of states as a whole. That is, the "average over time" which the system spends in states in A should equal the "average over space", which is the proportion of S that A occupies. This statement, known as

5.8. Birkhoff's Theorem on time and space averages 319

the "ergodic hypothesis", was formulated by Ludwig Boltzmann in the nineteenth century, but it proved difficult to find a mathematical proof. It is not at all obvious if or when the time average will exist, nor is it obvious that when it exists it will equal the space average. Birkhoff's Theorem[1] answers these questions.

Theorem 5.15 (Birkhoff's Individual Ergodic Theorem). *Let S be a set, let $\mathcal{M}$ be a σ-algebra of subsets of S and let $v : \mathcal{M} \longrightarrow [0, \infty)$ be a measure on $\mathcal{M}$, so that $v(S) < \infty$. Let $f : S \longrightarrow S$ be a measure-preserving transformation, and let A be a set in $\mathcal{M}$. Then, the limit*

$$\lim_{n \to \infty} \frac{1}{n} \left(\sum_{j=1}^{n} \chi_A \left(f^{j-1}(x) \right) \right) \tag{5.21}$$

exists for almost all points x in S. Moreover, if the system is ergodic, for all sets $A \in \mathcal{M}$ we have

$$\lim_{n \to \infty} \frac{1}{n} \left(\sum_{j=1}^{n} \chi_A \left(f^{j-1}(x) \right) \right) = \frac{v(A)}{v(S)},$$

for almost all $x \in S$.

There is no easy proof of Birkhoff's Theorem, and no proof is presented here. Birkhoff's original proof dates from 1931 [5]. Since then, various proofs have been given and may be found in [9, 16, 26, 33, 36, 39]. The main problem is to show that, for almost all x, the limit in (5.21) exists. When the system is ergodic, the value of the limit is found by interchanging the limit with the integration operation, a procedure justified by Lebesgue's Dominated Convergence Theorem. This requires a knowledge of integration theory as well as measure theory, such as is found in [1, 2, 8, 17, 20, 28]. The idea is that if the limit in (5.21) exists, put it equal to $g(x)$. If the limit in (5.21) does not exist, put $g(x) = 0$. Then, g is invariant in the sense that $g \circ f = g$. Because the system is assumed ergodic, we can prove that g is almost everywhere a constant, c say. The question is: what is the value of this constant c? The Dominated

Convergence Theorem justifies the following argument.

$$cv(S) = \int_S g\, dv$$

$$= \int_S \left(\lim_{n \to \infty} \frac{1}{n} \left(\sum_{j=1}^{n} \chi_A \circ f^{j-1}(x) \right) \right) dv(x)$$

$$= \lim_{n \to \infty} \frac{1}{n} \left(\int_S \left(\sum_{j=1}^{n} \chi_{f^{-(j-1)}(A)} \right) dv \right)$$

$$= \lim_{n \to \infty} \frac{1}{n} \sum_{j=1}^{n} v\left(f^{-(j-1)}(A) \right)$$

$$= \lim_{n \to \infty} \frac{1}{n} \sum_{j=1}^{n} v(A), \text{ as } v \text{ is } f\text{-invariant,}$$

$$= v(A).$$

Thus, $c = v(A)/v(S)$, as in Birkhoff's Theorem.

5.9 Weyl's Theorem from the ergodic viewpoint

A sequence (x_n) in the interval $[0, 1)$ is *uniformly distributed* in $[0, 1)$ if, for every subinterval J of $[0, 1)$, we have

$$\lim_{n \to \infty} \frac{1}{n} \left| \left\{ j : 1 \leq j \leq n \text{ and } x_j \in J \right\} \right| = \mu(J).$$

Weyl's Theorem 2.20 says that for every irrational number, the sequence of fractional parts of the multiples of the number is uniformly distributed in $[0, 1)$. Birkhoff's Theorem 5.15 leads to a new approach to Weyl's Theorem. For, if we let α be a given irrational number, and let $\tau_\alpha(x) = $ frac$(x + \alpha)$ as in Chapter 2, then τ_α is both measure-preserving and ergodic, by the discussion of Example 5.13. Observing that $\tau_\alpha^j(x) = $

5.9. Weyl's Theorem from the ergodic viewpoint

frac$(x + j\alpha)$ for all $x \in [0, 1)$ and all j by Proposition 2.9, we obtain from Birkhoff's Theorem that if J is a Borel measurable subset of $[0, 1)$ then

$$\lim_{n\to\infty} \frac{1}{n}\left|\left\{j : 1 \leq j \leq n \text{ and frac}(x + j\alpha) \in J\right\}\right| = \mu(J), \quad (5.22)$$

for almost all $x \in [0, 1)$.

The deduction of Weyl's Theorem from (5.22) is not immediate, for we do not know if we can put $x = 0$ in (5.22). We overcome this by observing that the set $\mathbb{Q}$ of rational numbers is countable (see the remarks in Section 1.3 of Chapter 1), and from this it follows that the set of all subintervals of $[0, 1)$ with rational endpoints may be arranged as a sequence (J_m). Then, by (5.22), for each m there is a subset $\mathcal{Z}_m$ of $[0, 1)$ of measure zero such that (5.22) holds with J_m in place of J and all $x \notin \mathcal{Z}_m$. The set $\mathcal{Z} = \cup_{m=1}^{\infty} \mathcal{Z}_m$ has measure zero and we see that if K is an interval with rational endpoints, and $x \notin \mathcal{Z}$, then

$$\lim_{n\to\infty} \frac{1}{n}\left|\left\{j : 1 \leq j \leq n \text{ and frac}(x + j\alpha) \in K\right\}\right| = \mu(K). \quad (5.23)$$

But, having observed this, let J be a subinterval of $[0, 1)$, and let $\varepsilon > 0$. Then there are subintervals K_1, K_2 with rational endpoints such that

$$K_1 \subseteq J \subseteq K_2, \ \mu(J) - \mu(K_1) < \varepsilon \text{ and } \mu(K_2) - \mu(J) < \varepsilon. \quad (5.24)$$

Thus, from (5.23) and (5.24) we have that for all $x \notin \mathcal{Z}$,

$$\mu(J) - \varepsilon < \mu(K_1)$$
$$= \lim_{n\to\infty} \frac{1}{n}\left|\left\{j : 1 \leq j \leq n \text{ and frac}(x + j\alpha) \in K_1\right\}\right|$$
$$\leq \liminf_{n\to\infty} \frac{1}{n}\left|\left\{j : 1 \leq j \leq n \text{ and frac}(x + j\alpha) \in J\right\}\right|$$
$$\leq \limsup_{n\to\infty} \frac{1}{n}\left|\left\{j : 1 \leq j \leq n \text{ and frac}(x + j\alpha) \in J\right\}\right|$$
$$\leq \lim_{n\to\infty} \frac{1}{n}\left|\left\{j : 1 \leq j \leq n \text{ and frac}(x + j\alpha) \in K_2\right\}\right|$$
$$= \mu(K_2)$$
$$< \mu(J) + \varepsilon. \quad (5.25)$$

As this holds for all $\varepsilon > 0$, we deduce that the lim sup and the lim inf appearing in (5.25) are both equal to $\mu(J)$. We deduce that for all $x \notin \mathcal{Z}$, and for all subintervals J of $[0, 1)$,

$$\lim_{n \to \infty} \frac{1}{n} \left| \left\{ j : 1 \leq j \leq n \text{ and } \operatorname{frac}(x + j\alpha) \in J \right\} \right| = \mu(J). \quad (5.26)$$

Now, let $[a, b)$ be a subinterval of $[0, 1)$ whose endpoints are such that $0 < a < b \leq 1$. Let $0 < \varepsilon < a$, and observe that for $u \in [0, 1)$,

$$\operatorname{frac}(\varepsilon + u) \in [a, b) \iff u \in [a - \varepsilon, b - \varepsilon). \quad (5.27)$$

As $\mathcal{Z}$ has measure zero, $\mathcal{Z}^c$ is dense in $[0, 1)$ (see Exercise 1, Chapter 4). Let $y \in (0, 1)$, and let $x \in \mathcal{Z}^c$ be such that $0 < y - x < a$, and put $\varepsilon = y - x$. For $n \in \mathbb{N}$ we have

$$\operatorname{frac}(y + n\alpha) = \operatorname{frac}(y - x + \operatorname{frac}(x + n\alpha)) = \operatorname{frac}(\varepsilon + \operatorname{frac}(x + n\alpha)).$$

Using (5.27) we have from this that

$$\operatorname{frac}(y + n\alpha) \in [a, b) \iff \operatorname{frac}(x + n\alpha) \in [a - \varepsilon, b - \varepsilon). \quad (5.28)$$

As $\mu([a, b)) = \mu([a - \varepsilon, b - \varepsilon)) = b - a$, and as $x \notin \mathcal{Z}$, we have from (5.26) and (5.28) that for all $0 < y < 1$,

$$\lim_{n \to \infty} \frac{1}{n} \left| \left\{ j : 1 \leq j \leq n \text{ and } \operatorname{frac}(y + j\alpha) \in [a, b) \right\} \right| = \mu([a, b)). \quad (5.29)$$

The possibility remains that $y = 0$. In this case, choose $x \in \mathcal{Z}^c$ such that $0 < 1 - x < a$. Then observe that

$$\operatorname{frac}(n\alpha) = \operatorname{frac}(1 - x + \operatorname{frac}(x + n\alpha)).$$

In this case we put $\varepsilon = 1 - x$ and proceed as before, deducing that (5.29) holds also for $y = 0$, and thus that (5.29) holds for all $y \in [0, 1)$. Now observe that (5.29) holds for every interval of the form $[a, 1)$, so it holds also for the complementary interval $[a, 1)^c = [0, a)$ in place of $[a, 1)$. It follows that (5.29) holds for *all* subintervals of $[0, 1)$ of the form $[a, b)$. But if an interval is of the form $[a, b]$, or $(a, b]$, or (a, b), it differs from

5.10. The Ergodic Theorem and expansions to an arbitrary base 323

$[a, b)$ by only one or two points. Also, as α is irrational, for a single point $c \in [0, 1)$ it is easily checked that there is at most one value of j such that $\text{frac}(x + j\alpha) = c$. Thus,

$$\lim_{n \to \infty} \frac{1}{n} \left| \{ j : 1 \leq j \leq n \text{ and } \text{frac}(x + j\alpha) = c \} \right| = \mu(\{c\}) = 0.$$

Weyl's Theorem, in the following form, now follows.

Theorem 5.16 (Weyl's Theorem). *Let α be an irrational number. Then for every $x \in [0, 1)$ and every subinterval J of $[0, 1)$,*

$$\lim_{n \to \infty} \frac{1}{n} \left| \{ \text{frac}(x + \alpha), \text{frac}(x + 2\alpha), \ldots, \text{frac}(x + n\alpha) \} \cap J \right| = \mu(J).$$

Further discussion of the systems associated with Weyl's Theorem may be found, for example, in [9, 26, 35, 36, 39].

5.10 The Ergodic Theorem and expansions to an arbitrary base

Borel's Theorem says that almost all numbers in $[0, 1)$ have an asymptotically equal number of 0s and 1s in their binary expansions. An approach to this result was discussed in 3.10, using the Rademacher functions. Also discussed was a refinement of Borel's Theorem in 3.12, which used the Walsh functions. It was pointed out in 3.10 that Borel's Theorem could be regarded as a result about the iterates of a function on $[0, 1)$, and that it was related to an averaging property of these iterates. Now these results from Chapter 3 on binary expansions are, in fact, consequences of Birkhoff's Theorem 5.15. However, the strength of Birkhoff's Theorem is such that it can give us these results, not just for the case of binary expansions, but for expansions to every base.

Let b be a natural number, $b \geq 2$. Then, as discussed in 3.2, for each $x \in [0, 1)$ there is a certain sequence $d_1(x), d_2(x), d_3(x), \ldots$ such that

$d_n(x) \in \{0, 1, 2, \ldots, b-1\}$ for all $n \in \mathbb{N}$

$$x = \sum_{n=1}^{\infty} \frac{d_n(x)}{b^n}. \tag{5.30}$$

The number $d_j(x)$ is called the jth digit of x to the base b, and we shall denote the sequence $(d_n(x))$ of these digits by $d(x)$. Thus, $d(x) = (d_n(x))$. Note that $d_1(x) = j$ if and only if $x \in [j/b, (j+1)/b)$. Also, note that $(d_n(x))$ does not contain an infinite sequence of consecutive terms equal to $b - 1$.

Let $f : [0, 1) \longrightarrow [0, 1)$ be the transformation given by $f(x) = \text{frac}(bx)$. Then, as in (5.16) we can say that if $j \in 1, 2, \ldots, b$ and $x \in [(j-1)/b, j/b)$,

$$f(x) = bx - j + 1.$$

Note that f is measure-preserving and ergodic, by Example 5.14.

Now, observe that if x is given as in (5.30), we have

$$bx = d_1(x) + \sum_{j=2}^{\infty} \frac{d_j(x)}{b^{j-1}} = d_1(x) + \sum_{j=1}^{\infty} \frac{d_{j+1}(x)}{b^j}.$$

Thus, as

$$\sum_{j=1}^{\infty} \frac{d_{j+1}(x)}{b^j} < \sum_{j=1}^{\infty} \frac{b-1}{b^j} = 1,$$

we have

$$f(x) = \text{frac}(bx) = \sum_{j=1}^{\infty} \frac{d_{j+1}(x)}{b^j}.$$

It follows that

$$d(f(x)) = (d_{n+1}(x)) = (d_2(x), d_3(x), \ldots).$$

More generally, if follows from this by an inductive argument that for all $r \in \{0\} \cup \mathbb{N}$ and $x \in [0, 1)$,

$$d(f^r(x)) = (d_{r+1}(x), d_{r+2}(x), \ldots). \tag{5.31}$$

5.10. The Ergodic Theorem and expansions to an arbitrary base

Let $k \in \{0, 1, \ldots, b-1\}$ be given. It follows from (5.31) that

$$d_{r+1}(x) = k \iff d_1(f^r(x)) = k \iff f^r(x) \in \left[\frac{k}{b}, \frac{k+1}{b}\right). \tag{5.32}$$

Put $J = [k/b, (k+1)/b)$. As f is an ergodic transformation on $[0, 1)$ we have that for almost all $x \in [0, 1)$,

$$\lim_{n \to \infty} \frac{1}{n} \left(\sum_{r=1}^{n} \chi_J \left(f^{r-1}(x) \right) \right) = \mu(J) = \frac{1}{b}. \tag{5.33}$$

If we now interpret (5.33) in the light of (5.32), we have the following result.

Theorem 5.17 (Borel's Theorem to base b). *Let* $b \in \mathbb{N}$ *with* $b \geq 2$. *For* $n \in \mathbb{N}$ *let* $d_n : [0, 1) \longrightarrow \{0, 1, 2, \ldots, b-1\}$ *be the function that assigns to x the nth digit of the expansion of x to the base b, and let* $k \in \{0, 1, 2, \ldots, b-1\}$ *be given. Then, for almost all* $x \in [0, 1)$,

$$\lim_{n \to \infty} \frac{1}{n} \left| \{r : 1 \leq r \leq n \text{ and } d_r(x) = k\} \right| = \frac{1}{b}.$$

That is, for almost all $x \in [0, 1)$, *the proportion of the first n digits in the expansion of x to the base b that are equal to k is asymptotically equal to* $1/b$.

This result is saying "in general, each of the digits $0, 1, 2, \ldots, b-1$ in the expansion of x to the base b occurs with a frequency of $1/b$". Since there are b possible digits, this shows that each digit occupies the "correct" proportion of the total, except for some exceptional numbers which are all in a particular set of measure zero. In the case $b = 2$, Theorem 5.17 gives us Borel's Theorem 3.22.

Theorem 5.17 concerns the frequency of occurrence of a *single digit* in the expansion of a number to a base. Birkhoff's Theorem also deals in a similar way with the question of the occurrence of a given finite number of digits. To see how, let $k_1, k_2, \ldots, k_s \in \{0, 1, 2, \ldots, b-1\}$ be given and put

$$J = \left[\sum_{j=1}^{s} \frac{k_j}{b^j}, \sum_{j=1}^{s} \frac{k_j}{b^j} + \frac{1}{b^s} \right).$$

Then, as seen in Theorem 3.4,

$$x \in J \iff d_j(x) = k_j \text{ for all } j = 1, 2, \ldots, s. \tag{5.34}$$

Also, from (5.31) we have

$$d_s(f^r(x)) = d_{r+s}(x). \tag{5.35}$$

Thus, if $s \in \mathbb{N}$ is given,

$$d_r(x) = k_1, d_{r+1}(x) = k_2, \ldots, d_{r+s-1}(x) = k_s$$
$$\iff d_1(f^{r-1}(x)) = k_1, \ldots, d_s(f^{r-1}(x)) = k_s, \text{ by (5.35)},$$
$$\iff f^{r-1}(x) \in J, \text{ by (5.34)},$$
$$\iff \chi_J\left(f^{r-1}(x)\right) = 1.$$

Hence,

$$\frac{1}{n}\left(\sum_{r=1}^{n} \chi_J\left(f^{r-1}(x)\right)\right) = \frac{1}{n}\bigg| \{r : 1 \leq r \leq n \\ \text{and } d_r(x) = k_1, d_{r+1}(x) = k_2, \ldots, d_{r+s-1}(x) = k_s\}\bigg|. \tag{5.36}$$

As the system is ergodic, and as $\mu(J) = 1/b^s$, Birkhoff's Theorem applied to (5.36) gives the following result.

Theorem 5.18 (The Normal Numbers Theorem to base b). *Let $b \in \mathbb{N}$ with $b \geq 2$. For $n \in \mathbb{N}$ let $d_n : [0, 1) \longrightarrow \{0, 1, 2, \ldots, b-1\}$ be the function that assigns to x the nth digit of the expansion of x to the base b, and let $k_1, k_2, \ldots, k_s \in \{0, 1, 2, \ldots, b-1\}$ be given. Then, for almost all $x \in [0, 1)$,*

$$\lim_{n \to \infty} \frac{1}{n}\bigg|\Big\{r : 1 \leq r \leq n \text{ and } d_r(x) = k_1, \ldots, d_{r+s-1}(x) = k_s\Big\}\bigg| = \frac{1}{b^s}.$$

That is, for almost all $x \in [0, 1)$, the proportion in which the sequence $k_1, k_2, \ldots, k_s$ appears in the expansion of x to the base b is asymptotically equal to $1/b^s$.

5.11. Kac's recurrence formula: the general case

There are b^s possible sequences of the form $k_1, k_2, \ldots, k_s$ where each $k_j \in \{0, 1, 2, \ldots, b-1\}$. Thus, the probability of the sequence $k_1, k_2, \ldots, k_s$ occurring at random is $1/b^s$. Thus, Theorem 5.18 shows that each sequence $k_1, k_2, \ldots, k_s$ appears the "correct" number of times in the expansion of a general number to the base b, except for some exceptional numbers which are all in a particular set of measure zero. In the case when $s = 1$, Theorem 5.18 reduces to Theorem 5.17.

Further discussion of the system associated with Borel's Theorem may be found, for example, in [9, 26, 35, 39].

5.11 Kac's recurrence formula: the general case

Kac's recurrence formula was stated for length-preserving transformations on an interval in Theorem 4.11. However, assuming the background in measure theory, it can now be stated in a more general setting as follows, with the accompanying formula for the standard deviation of recurrence times, corresponding to (4.39) in Theorem 4.15.

Theorem 5.19. *Let S be a set, let $\mathcal{M}$ be a σ-algebra of subsets of S, and let ν be a measure on $\mathcal{M}$ with $\nu(S) < \infty$. Let f be a measure-preserving transformation on S that is ergodic, and let $U \in \mathcal{M}$ with $\nu(U) > 0$. If $x \in S$ is such that $f^n(x) \in U$ for some n, put*

$$\Theta_U(x) = \min\{n : n \in \mathbb{N} \text{ and } f^n(x) \in U\}.$$

If $f^n(x) \notin U$ for all n, put $\Theta_U(x) = \infty$. Then, $\Theta_U(x) < \infty$ for almost all $x \in U$, and

$$\frac{1}{\nu(U)} \sum_{n=1}^{\infty} n \, \nu(\{x : x \in U \text{ and } \Theta_U(x) = n\}) = \frac{\nu(S)}{\nu(U)}.$$

That is, the average value of the recurrence time Θ_U over U is $\nu(S)/\nu(U)$. Also, if

$$\sum_{n=1}^{\infty} \nu\left(U^c \cap f^{-1}(U^c) \cap \cdots \cap f^{-n}(U^c)\right) < \infty,$$

the standard deviation of Θ_U over U, as defined in (4.31), is finite and equals

$$\sqrt{-\frac{\nu(S)^2}{\nu(U)^2} + 3\frac{\nu(S)}{\nu(U)} - 2 + \frac{2}{\nu(U)} \sum_{n=1}^{\infty} \nu\bigl(U^c \cap f^{-1}(U^c) \cap \cdots \cap f^{-n}(U^c)\bigr)}.$$

The proof of Theorem 5.19 proceeds exactly along the lines of the proofs of Theorems 4.11 and 4.15, but with the measure ν on $\mathcal{M}$ replacing the additive set function μ on the basic subsets of a bounded interval. Despite the greater generality, the assumption of a knowledge of measure theory means that the proof is easier—for, Lemma 4.10 now needs no prior discussion as it is immediate from the properties of the measure ν on $\mathcal{M}$. Further details may be found in [6, 9, 15, 22, 25].

5.12 Mixing transformations and an example of Kakutani

In this section, we discuss further dynamical systems and indicate some developments of the ideas in this work. First, we consider the notion of *mixing* in ergodic theory, a topic discussed in more detail in [9, 11, 17, 26, 33, 36, 39], for example. In order to express this idea more precisely, let S be a set, let $\mathcal{M}$ be a σ-algebra of subsets of S, let ν be a measure on $\mathcal{M}$ with $\nu(S) < \infty$, and let f be a measure-preserving transformation on S. Then if $A, B \in \mathcal{M}$ and $n \in \mathbb{N}$, $A \cap f^{-n}(B)$ is the set of points in A that appear in B after applying n iterations of f. Thus, the *proportion* of points in A that are in B after n iterations of f is $\nu(A \cap f^{-n}(B))/\nu(A)$. If we have mixing, for large n we would expect this to be about equal to the proportion of B within S. That is, for large n we might expect that $\nu(A \cap f^{-n}(B))/\nu(A)$ approximately equals $\nu(B)/\nu(S)$. So if we consider only the case where $\nu(S) = 1$, mixing seems to mean that in some sense $\nu(A \cap f^{-n}(B))$ should approximate $\nu(A)\nu(B)$ for large n. Alternatively, using the notion of independent events from 3.6, we could say that mixing means that for large n, the events A and $f^{-n}(B)$

5.12. Mixing transformations and an example of Kakutani

are "approximately" independent. These ideas motivate the following definitions.

Definition Let S be a set, let $\mathcal{M}$ be a σ-algebra of subsets of S, let ν be a measure on $\mathcal{M}$ with $\nu(S) = 1$, and let f be a ν-measure-preserving transformation on S. Then f is called *weakly mixing* if for all $A, B \in \mathcal{M}$,

$$\lim_{n\to\infty} \frac{1}{n} \sum_{k=0}^{n-1} |\nu(A \cap f^{-k}(B)) - \nu(A)\nu(B)| = 0. \tag{5.37}$$

Also, f is called *strongly mixing* if for all $A, B \in \mathcal{M}$,

$$\lim_{n\to\infty} \nu(A \cap f^{-n}(B)) = \nu(A)\nu(B).$$

A weakly mixing transformation is ergodic. For, if $f^{-1}(B) = B$, $f^{-k}(B) = B$ for all k, so we have that for all $A \in \mathcal{M}$, $\nu(A \cap f^{-k}(B)) = \nu(A \cap B)$ and from (5.37) we see that

$$\nu(A \cap B) = \nu(A)\nu(B), \text{ for all } A \in \mathcal{M}.$$

Putting $A = B$ gives $\nu(B) = \nu(B)^2$, so that $\nu(B) = 0$ or 1 and so f is ergodic. However, if f is ergodic it need not be weakly mixing—for an irrational translation on $[0, 1)$ is not weakly mixing, as discussed in Figure 5.7, but it is ergodic, as we saw in Example 5.13. But, if f is strongly mixing, then f is weakly mixing (see Exercise 13).

An example of a weakly mixing transformation that is not strongly mixing is due to Kakutani [24]. To describe this, let $\mathcal{B}$ be the σ-algebra of Borel subsets of $[0, 1)$ and, as usual, let μ be the usual (Lebesgue) measure on $\mathcal{B}$. Observe that the interval $[0, 1)$ may be written as

$$[0, 1) = \left[0, \frac{1}{2}\right) \cup \left[\frac{1}{2}, \frac{3}{4}\right) \cup \cdots \cup \left[1 - \frac{1}{2^n}, 1 - \frac{1}{2^{n+1}}\right) \cup \cdots,$$

where the intervals on the right-hand side of this equation are disjoint. Then, consider the transformation ψ on the unit interval $[0, 1)$ given by,

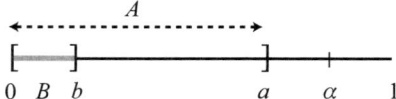

Figure 5.7. Let α be irrational with $0 < \alpha < 1$. We put $\beta = 1 - \alpha$, and note that β is irrational and that $0 < \beta < 1$. Letting τ_α denote the irrational translation $x \mapsto \text{frac}(x + \alpha)$, note that $\tau_\alpha^{-1} = \tau_\beta$. Let $0 < a, b < 1$ be chosen so that $0 < b < a < \alpha$ and $b < 1 - \alpha$. In the figure, A denotes the interval $[0, a]$ and B denotes the interval $[0, b]$. By Kronecker's Theorem 2.5, there is k such that $\text{frac}(k\beta) \in (a, \alpha)$, so that $b + \text{frac}(k\beta) < 1$. We now have from Proposition 2.9 that

$$\tau_\alpha^{-k}(B) = \tau_\beta^k([0, b]) = [\text{frac}(k\beta), b + \text{frac}(k\beta)],$$

and that

$$A \cap \tau_\alpha^{-k}(B) = [0, a] \cap [\text{frac}(k\beta), b + \text{frac}(k\beta)) = \emptyset.$$

This situation is illustrated in Figure 5.8.

for $n = 0, 1, 2, 3 \ldots$,

$$\psi(x) = x - 1 + \frac{1}{2^n} + \frac{1}{2^{n+1}}, \text{ for } x \in \left[1 - \frac{1}{2^n}, 1 - \frac{1}{2^{n+1}}\right).$$

The graph of ψ is illustrated in Figure 5.9. Using the approach for the class of piecewise linear transformations in Theorem 4.4, and using Theorem 5.9, it follows that ψ is measure-preserving. Also, ψ is one-to-one, and it is proved in [24] that ψ is ergodic. However, ψ is not weakly mixing, for if we take $A = (0, 1/2)$ and $B = (1/2, 1)$, we have $\psi(A) = B$ and $\psi(B) = A$. So, in this case,

$$\lim_{n \to \infty} \frac{1}{n} \sum_{j=1}^n |\mu(A \cap \psi^{-j}(B)) - \mu(A)\mu(B)| = \frac{1}{4}.$$

In order to construct Kakutani's example, we introduce the set

$$A = [0, 1) \cup \left(\bigcup_{n=0}^\infty \left[2 - \frac{1}{2^{2n}}, 2 - \frac{1}{2^{2n+1}}\right)\right).$$

5.12. Mixing transformations and an example of Kakutani 331

```
                A                 τ_α^{-k}(B)
  ←---------------------→         ←----→
  [       ]                       ]  [  ]
  0   B   b                       a   α   1
```

Figure 5.8. The figure illustrates that when $\text{frac}(k\beta) \in (a, \alpha)$, $A \cap \tau_\alpha^{-k}(B)$ is an interval that is disjoint from $A = [0, a]$. Now, this occurs for a positive proportion of values of k. For, by Weyl's Theorem 2.20,

$$\lim_{n\to\infty} \frac{1}{n}\left|\left\{k : 1 \le k \le n \text{ and } \text{frac}(k\beta) \in (a, \alpha)\right\}\right| = \alpha - a > 0.$$

Thus,

$$\frac{1}{n}\sum_{k=1}^{n}|\mu(A \cap \tau_\alpha^{-k}(B)) - \mu(A)\mu(B)| \ge \frac{1}{n}$$

$$\sum_{\substack{k=1 \\ \text{frac}(k\alpha) \in (a,\alpha)}}^{n} |\mu(A \cap \tau_\alpha^{-k}(B)) - \mu(A)\mu(B)|$$

$$= \frac{1}{n}\sum_{\substack{k=1 \\ \text{frac}(k\alpha) \in (a,\alpha)}}^{n} \mu(A)\mu(B)|$$

$$\to \mu(A)\mu(B)(\alpha - a), \text{ as } n \to \infty.$$

As $\mu(A)\mu(B)(\alpha - a) > 0$, this shows that τ_α is not weakly mixing. Thus, for irrational α, τ_α is ergodic but is not weakly mixing.

Kakutani's example is the transformation $\widetilde{\psi}$ on A given by

$$\widetilde{\psi}(x) = \begin{cases} \psi(x), & \text{if } x \in \bigcup_{n=1}^{\infty}\left[1 - \frac{1}{2^{2n-1}}, 1 - \frac{1}{2^{2n}}\right), \\ x + 1, & \text{if } x \in \bigcup_{n=0}^{\infty}\left[1 - \frac{1}{2^{2n}}, 1 - \frac{1}{2^{2n+1}}\right), \text{ and} \\ \psi(x - 1), & \text{if } x \in \bigcup_{n=0}^{\infty}\left[2 - \frac{1}{2^{2n}}, 2 - \frac{1}{2^{2n+1}}\right). \end{cases}$$

Then, the function $\widetilde{\psi}$ maps A into A and it is proved by Kakutani in [24] that the system $(A, \widetilde{\psi})$ is weakly mixing but not strongly mixing. The Borel systems provide examples of strongly mixing transformations— for each $b \in \mathbb{N}$ with $b \ge 2$, the transformation on $[0, 1)$ given by

332 Chapter 5. Averaging in Time and Space

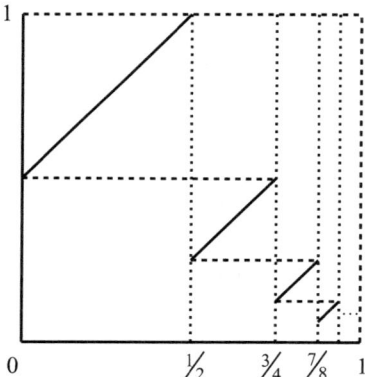

Figure 5.9. This illustrates the graph of the transformation ψ. Kakutani proved in [24] that ψ is a measure-preserving transformation on $[0, 1)$ that is one-to-one and ergodic, but it is not weakly mixing.

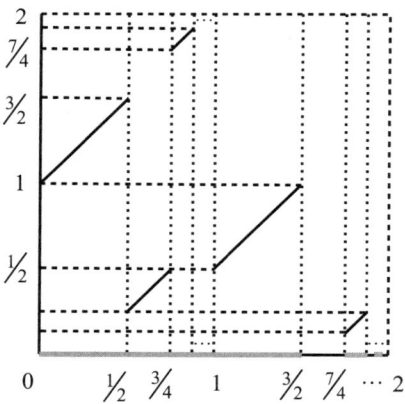

Figure 5.10. Illustrated is the graph of the transformation $\widetilde{\psi}$ whose domain is

$$[0, 1) \cup \left[1, \frac{3}{2}\right) \cup \left[\frac{7}{4}, \frac{15}{8}\right) \cup \cdots \cup \left[2 - \frac{1}{2^{2n}}, 2 - \frac{1}{2^{2n+1}}\right) \cup \cdots.$$

The transformation $\widetilde{\psi}$ is measure-preserving on its domain, and it is weakly mixing but not strongly mixing.

5.13 Lüroth transformations and continued fractions

$x \mapsto \text{frac}(bx)$ is strongly mixing [26, p.153]. For further discussion of mixing systems see [9, 18, 26, 33].

5.13 Lüroth transformations and continued fractions

Lüroth transformations were introduced by J. Lüroth [29] in 1883. A *Lüroth transformation* is a transformation of $[0, 1)$ or $[0, 1]$ that arises as follows: for $[0, 1)$ there is a partition $\{J_n : n \in A\}$ of $[0, 1)$ into intervals, where A is either $\mathbb{N}$ or a finite subset of $\mathbb{N}$ with at least two elements such that, on each interval J_n, f is an increasing linear function whose range is an interval with endpoints 0 and 1. There is a corresponding definition for $[0, 1]$. Thus, if the partition of $[0, 1)$ associated with a Lüroth transformation is finite, the transformation satisfies the conditions of Theorem 4.3. Lüroth transformations, and a wider class of generalized Lüroth transformations, are discussed by Dajani and Kraaikamp [11, pp. 36–50 and pp. 68–70]. We have already seen that a Lüroth transformation arises when we consider the binary expansion of numbers, for then the Lüroth transformation $x \mapsto \text{frac}(2x)$, which has two linear components, produces the digits in the binary expansion of a number x in $[0, 1)$, as described for the Borel system in 3.10.

We can prove, along the lines of Theorem 4.3 and using Theorem 5.9, or as in [11, p.43], that a Lüroth transformation is μ-measure-preserving. A Lüroth transformation is ergodic – this is proved in [11, p.68], where measure theory is used. For a Lüroth transformation with a finite number of linear components, a proof of ergodicity based only on the notion of outer measure is in [32], and it appears this proof may also work in the case of an infinite number of linear components.

A classic case of a Lüroth transformation with an infinite number of linear components is given by f, where

$$f(x) = \begin{cases} 0, & \text{if } x = 0; \\ k(k+1)x - k, & \text{for } x \in \left[\frac{1}{k+1}, \frac{1}{k}\right) \text{ and } k = 1, 2, \ldots. \end{cases}$$
(5.38)

The graph of this transformation is illustrated in Figure 5.11. For each $x \in [0, 1]$ such that $f^{n-1}(x) \neq 0$ for all $n = 1, 2, \ldots$, we associate a sequence $(d_n(x))$ in $\mathbb{N}$ as follows:

$$\text{if } f^{n-1}(x) \in \left[\frac{1}{k+1}, \frac{1}{k}\right) \text{ we put } d_n(x) = k. \tag{5.39}$$

Observe that

$$d_j(f^{n-1}(x)) = k \iff f^{j-1+n-1}(x) \in \left[\frac{1}{k+1}, \frac{1}{k}\right) \iff d_{j+n-1}(x) = k.$$

Thus,

$$d_j(f^{n-1}(x)) = d_{j+n-1}(x). \tag{5.40}$$

Now, take a particular x and let the associated sequence $(d_n(x))$ in $\mathbb{N}$, as given by (5.39), be denoted by (c_n). Thus, $d_n(x) = c_n$ for all n. Taking $n = 1$ and using (5.38) and (5.39) we have $f(x) = c_1(c_1 + 1)x - c_1$ so that

$$x = \frac{1}{c_1 + 1} + \frac{f(x)}{c_1(c_1 + 1)}. \tag{5.41}$$

Now, apply (5.41) to $f(x)$ in place of x and also use (5.40) to obtain

$$x = \frac{1}{c_1 + 1} + \frac{1}{c_1(c_1 + 1)}\left(\frac{1}{d_1(f(x)) + 1} + \frac{f^2(x)}{d_1(f(x))(d_1(f(x)) + 1)}\right)$$

$$= \frac{1}{c_1 + 1} + \frac{1}{c_1(c_1 + 1)}\left(\frac{1}{c_2 + 1} + \frac{f^2(x)}{c_2(c_2 + 1)}\right)$$

$$= \frac{1}{c_1 + 1} + \frac{1}{c_1(c_1 + 1)(c_2 + 1)} + \frac{f^2(x)}{c_1(c_1 + 1)c_2(c_2 + 1)}.$$

Continuing in this way we get

$$x = \frac{1}{c_1 + 1} + \cdots$$
$$+ \frac{1}{c_1(c_1 + 1) \cdots c_{n-1}(c_{n-1} + 1)(c_n + 1)}$$
$$+ \frac{f^n(x)}{c_1(c_1 + 1) \cdots c_n(c_n + 1)}.$$

5.13. Lüroth transformations and continued fractions

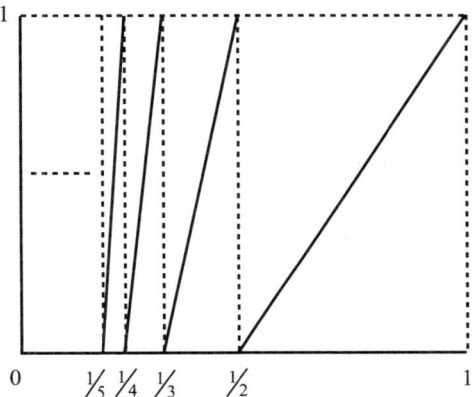

Figure 5.11. The graph of the particular Lüroth transformation f given by $f(0) = 0$ and $f(x) = k(k + 1)x - k$, for $1/(k + 1) \leq x < 1/k$ and $k = 1, 2, 3, \ldots$. The transformation f is ergodic and measure-preserving on $[0, 1)$.

As $f^n(x) \in [0, 1)$ and as each $c_j \geq 1$, it is clear that

$$\lim_{n \to \infty} \frac{f^n(x)}{c_1(c_1 + 1) \cdots c_n(c_n + 1)} = 0,$$

so we deduce that

$$x = \frac{1}{c_1 + 1} + \frac{1}{c_1(c_1 + 1)(c_2 + 1)} + \cdots$$
$$+ \frac{1}{c_1(c_1 + 1) \cdots c_{n-1}(c_{n-1} + 1)(c_n + 1)} + \cdots. \quad (5.42)$$

This is called the *Lüroth expansion of* x for the transformation given by (5.38).

If we now apply the Ergodic Theorem to this system, we deduce that for a given $k \in \mathbb{N}$, for almost all $x \in [0, 1)$, the relative frequency of the digit k in the expansion (5.42) is $1/k(k + 1)$. For, using (5.39) and the

Ergodic Theorem gives, for almost all x,

$$\lim_{n\to\infty} \frac{1}{n} |\{j : 1 \leq j \leq n \text{ and } c_j = k\}|$$
$$= \lim_{n\to\infty} \frac{1}{n} \left|\left\{j : 1 \leq j \leq n \text{ and } f^{j-1}(x) \in \left[\frac{1}{k+1}, \frac{1}{k}\right)\right\}\right|$$
$$= \mu\left(\left[\frac{1}{k+1}, \frac{1}{k}\right)\right)$$
$$= \frac{1}{k} - \frac{1}{k+1}$$
$$= \frac{1}{k(k+1)}.$$

The Ergodic Theorem also may be used to calculate the relative frequency of digits in the continued fractions expansion of a number. To introduce the appropriate dynamical system, define a transformation $f : [0, 1] \longrightarrow [0, 1]$ by $f(0) = 0$ and $f(x) = \text{frac}(1/x)$ for $0 < x \leq 1$. That is,

$$f(x) = \begin{cases} 0, & \text{if } x = 0; \\ \frac{1}{x} - \text{int}\left(\frac{1}{x}\right), & \text{if } 0 < x \leq 1. \end{cases}$$

The graph of f is depicted in Figure 5.12. We see that for $0 < x \leq 1$,

$$x = \frac{1}{\text{int}\left(\frac{1}{x}\right) + f(x)}.$$

If $f(x) \neq 0$, we may repeat this step with $f(x)$ in place of x to get

$$x = \frac{1}{\text{int}\left(\frac{1}{x}\right) + \dfrac{1}{\text{int}\left(\frac{1}{f(x)}\right) + f^2(x)}}.$$

Then, if $x, f(x), \ldots, f^{n-1}(x)$ are all non-zero, we may continue to ap-

5.13. Lüroth transformations and continued fractions

ply this step to $f^2(x), f^3(x), \ldots, f^{n-1}(x)$ to get

$$x = \cfrac{1}{\operatorname{int}\left(\frac{1}{x}\right) + \cfrac{1}{\operatorname{int}\left(\frac{1}{f(x)}\right) + \cfrac{\cdots}{\ddots \cfrac{}{+ \cfrac{1}{\operatorname{int}\left(\frac{1}{f^{n-1}(x)}\right) + f^n(x)}}}}}.$$

(5.43)

Now if $f^n(x) = 0$ in (5.43), then (5.43) takes a simplified form and we can see that x is rational. Thus, if x is irrational, $f^n(x) \neq 0$ for all n. In this case, we may let $n \to \infty$ in (5.43), and it is proved in [11, pp. 26–29], for example, that

$$x = \cfrac{1}{\operatorname{int}\left(\frac{1}{x}\right) + \cfrac{1}{\operatorname{int}\left(\frac{1}{f(x)}\right) + \cfrac{\cdots}{\ddots \cfrac{}{+ \cfrac{1}{\operatorname{int}\left(\frac{1}{f^{n-1}(x)}\right) + \cdots +}}}}}$$

(5.44)

In this case we call (5.44) the *continued fraction* expansion of x. Thus, almost all numbers in [0, 1], including all irrational numbers, have a continued fraction expansion. There is more to be said on continued fraction expansions—see [11, 19], for example.

Now, if x has a continued fraction expansion, we see from (5.44) that there is associated with x a sequence of positive integers

$$\operatorname{int}\left(\frac{1}{x}\right), \operatorname{int}\left(\frac{1}{f(x)}\right), \operatorname{int}\left(\frac{1}{f^2(x)}\right), \ldots, \operatorname{int}\left(\frac{1}{f^n(x)}\right), \ldots.$$

Given an positive integer k, the question then arises as to the proportion of integers that appear in this sequence and equal k.

338   Chapter 5. Averaging in Time and Space

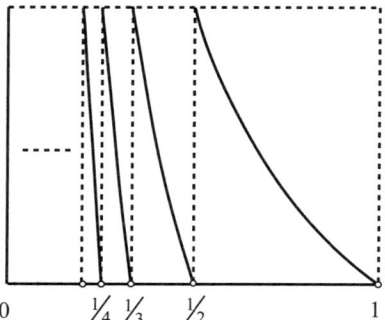

Figure 5.12. The graph is of the transformation f on $[0, 1]$ such that $f(0) = 0$ and

$$f(x) = \text{frac}(x) = \frac{1}{x} - \text{int}\left(\frac{1}{x}\right), \text{ for } 0 < x \leq 1.$$

As noted in (5.46), for x in the interval $(1/(n + 1), 1/n]$, $f(x) = 1/x - n$. The transformation f is decreasing on each interval $(1/(n + 1), 1/n]$, and maps this interval onto $[0, 1)$. The transformation somewhat resembles the piecewise linear transformations of Theorem 4.4 and Figure 4.5. However, if we let v be the function $x \longrightarrow \log(1 + x)$, and let μ_v be the corresponding additive set function as in Proposition 3.11, then by the Carathéodory Extension Theorem, μ_v extends to become a measure ν on the Borel subsets on $[0, 1]$. Then, the transformation f is ν-measure-preserving, as discussed in the main text. Note that if we assume a knowledge of integration theory as well as measure theory, we can show that

$$\nu(A) = \int_A \frac{dx}{1 + x}, \text{ for all Borel subsets } A \text{ of } [0, 1].$$

Observe that

$$(0, 1] = \bigcup_{k=1}^{\infty} \left(\frac{1}{k + 1}, \frac{1}{k}\right],$$

where this union is disjoint. Thus, if $x \in (0, 1]$, there is a unique $k \in \mathbb{N}$ such that $x \in (1/(k + 1), 1/k]$. We see that for $x \in [0, 1)$ and $n \in \mathbb{N}$,

$$x \in \left(\frac{1}{k + 1}, \frac{1}{k}\right] \iff k \leq \frac{1}{x} < k + 1 \iff \text{int}\left(\frac{1}{x}\right) = k. \quad (5.45)$$

5.13. Lüroth transformations and continued fractions 339

In particular, we see from (5.45) that

$$x \in \left(\frac{1}{k+1}, \frac{1}{k}\right]$$
$$\implies f(x)$$
$$= \frac{1}{x} - k. \quad (5.46)$$

We see from (5.46) that if $(a, b]$ is a subinterval of $[0, 1]$,

$$f^{-1}((a, b]) = \bigcup_{k=1}^{\infty} \left[\frac{1}{b+k}, \frac{1}{a+k}\right), \quad (5.47)$$

and this union is disjoint.

Now the function $x \longmapsto \log(1 + x)$ is increasing on $[0, 1)$. If J is the subinterval of $(0, 1]$ whose left endpoint is a and whose right endpoint is b and we put

$$\nu(J) = \frac{1}{\log 2} \left(\log(b+1) - \log(a+1)\right), \quad (5.48)$$

we see that $\nu([0, 1]) = 1$. Also, from Proposition 3.11 we see that ν is an additive set function on the family $\mathcal{B}$ of basic subsets of $[0, 1]$. Theorem 5.7 shows that ν extends to be a measure, also denoted by ν, on the σ-algebra of Borel subsets of $[0, 1]$. In fact, the measure ν is f-invariant on the Borel sets. To check this, by Theorem 5.9 it suffices to prove that for every subinterval J of $[0, 1]$, $\nu(f^{-1}(J)) = \nu(J)$. Now if the subinterval

J has endpoints a, b, we have from (5.47) and (5.48) that

$$\nu(f^{-1}(J)) = \sum_{k=1}^{\infty} \nu\left(\left[\frac{1}{b+k}, \frac{1}{a+k}\right)\right)$$

$$= \frac{1}{\log 2} \sum_{k=1}^{\infty} \left(\log\left(1 + \frac{1}{a+k}\right) - \log\left(1 + \frac{1}{b+k}\right)\right)$$

$$= \frac{1}{\log 2} \sum_{k=1}^{\infty} \left(\left[\log(a+k+1) - \log(b+k+1)\right]\right.$$

$$\left. - \left[\log(a+k) - \log(b+k)\right]\right)$$

$$= \frac{1}{\log 2} \lim_{s \to \infty} \sum_{k=1}^{s} (a_{k+1} - a_k),$$

$$= \frac{1}{\log 2} \lim_{s \to \infty} (a_{s+1} - a_1), \qquad (5.49)$$

where
$$a_k = \log(a+k) - \log(b+k). \qquad (5.50)$$

Now from (5.50) we have

$$\lim_{s \to \infty} a_s = \lim_{s \to \infty} \log\left(\frac{a+s}{b+s}\right) = \log \lim_{s \to \infty} \left(\frac{a+s}{b+s}\right) = \log 1 = 0.$$

Using this, we see from (5.48), (5.49) and (5.50) that

$$\nu(f^{-1}(J)) = \frac{1}{\log 2} \lim_{s \to \infty} (-a_1 + a_{s+1})$$

$$= \frac{-a_1}{\log 2}$$

$$= \frac{1}{\log 2} (\log(b+1) - \log(a+1))$$

$$= \nu(J).$$

Thus, f is a ν-invariant transformation. It is proved in [11, p.83], that the system $([0, 1], f)$ is ergodic, so that Birkhoff's Theorem may be applied,

5.13. Lüroth transformations and continued fractions

and we obtain a result of Paul Lévy[2] on the frequency of a given integer in the sequence $(\text{int}(1/f^n(x)))$ of integers appearing in the continued fraction expansion of x in (5.44).

Theorem 5.20. *Let* $f : [0, 1) \longrightarrow [0, 1)$ *be the transformation given by* $f(0) = 0$ *and*

$$f(x) = \text{frac}\left(\frac{1}{x}\right), \text{ for } 0 < x < 1.$$

The sequence

$$\text{int}\left(\frac{1}{x}\right), \text{int}\left(\frac{1}{f(x)}\right), \text{int}\left(\frac{1}{f^2(x)}\right), \ldots, \text{int}\left(\frac{1}{f^n(x)}\right), \ldots,$$

of positive integers is the sequence appearing in the continued fraction expansion of x *as described by (5.44). Then, for almost all numbers* $x \in [0, 1)$, *for every positive integer* k *we have*

$$\lim_{n \to \infty} \frac{1}{n} \left| \left\{ j : 1 \leq j \leq n \text{ and int}\left(\frac{1}{f^j(x)}\right) = k \right\} \right|$$
$$= \frac{1}{\log 2} \log\left(1 + \frac{1}{k(k+2)}\right).$$

Proof. Let $k \in \mathbb{N}$ and $x \in [0, 1]$ be given, and observe that, by (5.45),

$$f^n(x) \in \left(\frac{1}{k+1}, \frac{1}{k}\right] \iff \text{int}\left(\frac{1}{f^n(x)}\right) = k. \tag{5.51}$$

By Birkhoff's Theorem, we have from (5.51) that for almost all $x \in [0, 1]$, and all $k = 1, 2, \ldots$,

$$\lim_{n\to\infty} \frac{1}{n}\left|\left\{j : 1 \leq j \leq n \text{ and int}\left(\frac{1}{f^j(x)}\right) = k\right\}\right|$$
$$= \nu\left(\left(\frac{1}{k+1}, \frac{1}{k}\right]\right)$$
$$= \frac{1}{\log 2}\left(\log\left(1 + \frac{1}{k}\right) - \log\left(1 + \frac{1}{k+1}\right)\right)$$
$$= \frac{1}{\log 2}\log\left(\frac{(k+1)^2}{k(k+2)}\right)$$
$$= \frac{1}{\log 2}\log\left(1 + \frac{1}{k(k+2)}\right).$$

$\square$

The discussion here of the Lüroth transformations and continued fractions has been influenced by the work of Dajani and Kraaikamp [11], which is recommended. Note that [11] has a much more extensive discussion of these two topics, and other topics associated with Birkhoff's Ergodic Theorem, together with an extensive guide to further reading.

Exercises

1. Let S be an interval. Let A, B be subsets of S and let U, V be open subintervals of S such that $A \subseteq U$, $B \subseteq V$ and $U \cap V = \emptyset$. Let μ denote the usual length function on the algebra of basic subsets of S. Show from the definition of outer measure that
$$\mu_{out}(A \cup B) = \mu_{out}(A) + \mu_{out}(B).$$

2. Let S be a set and let $\mathcal{B}$ be an algebra of subsets of S. Let $\mathcal{F}$ be the σ-algebra generated by $\mathcal{B}$. Let ν be a measure on $\mathcal{B}$ with $\nu(S) < \infty$, and let ν_{out} be the associated outer measure on the subsets of S. The restriction of ν_{out} to $\mathcal{M}$, the σ-algebra of measurable sets, is denoted by ν. Recall that for sets A, B, $A \triangle B = (A^c \cap B) \cup (A \cap B^c)$. Let
$$\mathcal{G} = \left\{A \triangle E : A \in \mathcal{F}, E \subseteq S \text{ and } \nu_{out}(E) = 0\right\}.$$
Let $f : S \longrightarrow S$ be a given transformation such that for all $B \in \mathcal{B}$, $f^{-1}(B) \in \mathcal{F}$ and $\nu(f^{-1}(B)) = \nu(B)$.

(i) Prove that if $B \subseteq S$, $B \in \mathcal{G}$ if and only if there is a set $A \in \mathcal{F}$ such that $\nu_{out}(A \Delta B) = 0$.

(ii) Prove that $\mathcal{F} \subseteq \mathcal{G}$ and that $\mathcal{G}$ is a σ-algebra of measurable sets.

(iii) Prove that if $A \subseteq S$, then $\nu_{out}(f^{-1}(A)) \leq \nu_{out}(A)$.

(iv) If $A \in \mathcal{G}$ prove that $f^{-1}(A) \in \mathcal{G}$ and that $\nu(f^{-1}(A)) = \nu(A)$.

3. Let $\mathcal{B}$ denote the family of subsets of $\mathbb{R}^2$ consisting of all finite unions of half open rectangles of the type $J \times K$, where each of J, K is an interval of the form $[a, b)$ for $a, b \in \mathbb{R}$, or $(-\infty, b)$ for $b \in \mathbb{R}$, or $[a, \infty)$ for $a \in \mathbb{R}$, or $(-\infty, \infty)$. Prove that $\mathcal{B}$ is an algebra of subsets of $\mathbb{R}^2$, and prove that every set in $\mathcal{B}$ is a *disjoint* union of rectangles of the given type. If A is a rectangle of the form $J \times K$ put $\nu(A) = \mu(J)\mu(K)$, with the conventions that $0 \cdot \infty = 0$ and $\infty \cdot \infty = \infty$. That is, $\nu(A)$ is the usual area of the rectangle A. Then, if the set Z in $\mathcal{B}$ is such that $Z = \cup_{j=1}^{n} A_j$, where each A_j is a rectangle of the given type and the union is disjoint, put

$$\nu(Z) = \sum_{j=1}^{n} \nu(A_j).$$

Now, prove that this definition of $\nu(Z)$ is well defined in the sense that it is independent of the particular expression of Z as the union of the sets $A_1, \ldots, A_n$. Also, prove that ν is a measure on $\mathcal{B}$.

4. Let S be a set, let $\mathcal{B}$ be an algebra of subsets of S, and let ν be an additive set function on $\mathcal{B}$. Let ν_{out} denote the corresponding outer measure on the family of all subsets of S. Prove that, for every subset A of S, the outer measure $\nu_{out}(A)$ of A equals the infimum of

$$\left\{ \sum_{n=1}^{\infty} \nu(V_n) : V_n \in \mathcal{B} \text{ for all } n, \, V_m \cap V_n = \emptyset \text{ if } m \neq n \text{ and } A \subseteq \bigcup_{n=1}^{\infty} V_n \right\}.$$

That is, in the definition (5.4), prove that the sets in (V_n) may be taken to be disjoint.

5. Let S be a set, let $\mathcal{B}$ be an algebra of subsets of S, and let ν be an additive set function on $\mathcal{B}$. If $\nu_{out}(A) = \nu(A)$ for all $A \in \mathcal{B}$, prove that ν is a measure on $\mathcal{B}$.

6. Let S be a set and let ι denote the identity function on S. Let $f : S \longrightarrow S$ be a transformation on S such that for some $n \in \mathbb{N}$, $f^n = \iota$. Prove that f is one-to-one and onto, and prove that for every subset A of S, the set $A \cup f(A) \cup f^2(A) \cup \cdots \cup f^{n-1}(A)$ is f-invariant.

7. Let $f : [0, 1] \longrightarrow [0, 1]$ be the function given by
$$f(x) = \begin{cases} 4x^2/3, & \text{for } 0 \le x < 3/4; \\ x, & \text{for } 3/4 \le x \le 1. \end{cases}$$

Sketch the graph of f, and find a proper subinterval of $[0, 1]$ that is an invariant set for the dynamical system $([0, 1], f)$.

8. Let f be a transformation on a set S and let $A \subseteq S$.
   (i) If f maps onto S and A is f-invariant, prove that $f(A) = A$.
   (ii) If f is one-to-one and $f(A) = A$, prove that A is invariant.
   (iii) If f is one-to-one and onto, prove that A is invariant if and only if $f(A) = A$.

9. Let $f : [0, 1] \to [0, 1]$ be the function given by
$$f(x) = \begin{cases} 2x + 1/2, & \text{for } 0 \le x < 1/4; \\ 2x - 1/2, & \text{for } 1/4 \le x < 1/2; \\ 2x - 1, & \text{for } 1/2 \le x \le 1. \end{cases}$$

Prove that f is a μ-measure-preserving transformation on $[0, 1]$. Sketch the graph of the second iterate f^2 of f.

10. Let S be a set, let A be a subset of S, let $f : S \to S$ be a transformation of S and put
$$B = \left\{ x : x \in S \text{ and } \lim_{n \to} \frac{1}{n} \left(\sum_{j=1}^{n} \chi_A \left(f^{j-1}(x) \right) \right) \text{ exists} \right\}.$$

Prove that B is an invariant subset of S.

11. Let α be a rational number and let $\tau_\alpha : [0, 1) \longrightarrow [0, 1)$ be the transformation given by
$$\tau_\alpha(x) = \text{frac}(x + \alpha).$$

Describe an infinite invariant set for the dynamical system $([0, 1), \tau_\alpha)$ that is neither $[0, 1)$ nor the empty set. Deduce that, with the σ-algebra of Borel sets and the usual measure μ on the Borel sets, the system $([0, 1), \tau_\alpha)$ is not ergodic.

12. Let $S = \{1, 2, 3, 4, 5\}$, let $\mathcal{M}$ be the σ-algebra consisting of all subsets of S, and let $\mu : \mathcal{M} \to \{0, 1, 2, 3, 4, 5\}$ be the measure given by
$$\mu(A) = \text{ the number of points in } A,$$

for all subsets A of S. Let $f : S \to S$ be the transformation given by

$$f(1) = 2, f(2) = 3, f(3) = 4, f(4) = 5, \text{ and } f(5) = 1.$$

Show that the system (S, f) is ergodic, and write down the specific conclusions for the system (S, f) that derive from Birkhoff's theorem. Do we really need Birkhoff's Theorem to deduce these conclusions?

13. Let (a_n) be a convergent sequence of real numbers with limit ℓ. Prove that

$$\lim_{n \to \infty} \frac{a_1 + a_2 + \cdots + a_n}{n} = \ell.$$

Use this result to prove that, if f is a strongly mixing transformation as defined at the beginning of Section 5.12, then f is weakly mixing.

Investigations

1. This investigation is concerned with how the average length of time for a particular type of system to become "absolutely stable" depends upon the parameters of the system. Here, the term "absolutely stable" means that the system is in a state that does not change once this state is attained—that is, the state is a fixed point of the system.

    Let $a, c \in (0, 1/2)$ and let $b, d \in (1/2, 1)$. Then, define a transformation f on $[0, 1]$ by putting

    $$f(x) = \begin{cases} \frac{1}{c}\left(\frac{1}{2} - a\right)x + a, & \text{if } 0 \leq x < c; \\ \frac{1}{2}, & \text{if } c \leq x \leq d; \\ \frac{1}{2(1-d)}\Big[(2b - 1)x + 1 - 2bd\Big], & \text{if } d < x \leq 1. \end{cases}$$

    The function f is continuous and is linear on each interval $[0, c]$, $[c, d]$ and $[d, 1]$. Sketch the graph of f and verify that it is a transformation of $[0, 1]$. Identify the (unique) fixed point of f. Now, carry out a similar analysis for f as was carried out for the transformation of $[0, 1]$ described in Figure 5.2 Identify the set of points in $[0, 1]$ that become equal to the fixed point after applying a finite number of iterations by f. Then, investigate how the average number of iterations for such points in $[0, 1]$ to become equal to the fixed point varies with the parameters a, b, c, d.

2. In Section 4.2 of Chapter 4, two types of piecewise linear length-preserving transformations on an interval were described. These are all μ-measure-preserving, and some of these are known to be ergodic, as we have seen

in Section 5.7 in this chapter (see also [11, 24, 27, 31, 32]). The following questions arise.

(i) What is the most general type of piecewise linear transformation on a bounded interval that is also μ-measure-preserving? (For length-preserving transformations this is Investigation 1 in Chapter 4.)
(ii) Which piecewise linear and measure-preserving transformations on $[0, 1)$ (say) are ergodic?
(iii) If we can identify some of these as ergodic transformations, then Birkhoff's Theorem may be applied. Can the conclusions be interpreted in any way similar to Weyl's Theorem (in the case of translations) or Borel's Theorem (in the case of expansions to the base 2 or, more generally, to the base b)?

Notes

1. [Page 319] George Birkhoff (1884–1944) was an influential American mathematician, president of the American Mathematical Society from 1925–1926. His Ergodic Theorem mentioned here is also known as the *Individual Ergodic Theorem*, as it deals with the convergence of the sequence of averages

$$\left(n^{-1} \sum_{j=1}^{n} \chi_A(f^{j-1}(x)) \right)$$

for individual points x. A result related to Birkhoff's Theorem is the "Mean Ergodic Theorem" of John von Neumann [38], proved in 1932. If U is a unitary operator on a Hilbert space H, and if $x \in H$, there is a vector $z \in H$ such that $U(z) = z$ and

$$\lim_{n\to\infty} \frac{1}{n} \sum_{k=0}^{n-1} U^k x = z.$$

The connection with Birkhoff's Theorem comes from taking H to be the space $L^2(S, \nu)$ of square integrable functions on a set S, where there is a measure ν given, and being given a measure-preserving transformation f on S. Then, $h \mapsto h \circ f$ is unitary on $L^2(S, \nu)$, and von Neumann's result says that if $h \in L^2(S, \nu)$, there is a function $g \in L^2(S, \nu)$ such that $g \circ f = g$ and

$$\lim_{n\to\infty} \int_S \left| \frac{1}{n} \sum_{k=0}^{n-1} h \circ f^k - g \right|^2 d\nu = 0.$$

That is, the sequence $\left(\left(\sum_{k=0}^{n-1} h \circ f^k\right)/n\right)$ of averages converges to the invariant function g "in mean"—that is, in the space $L^2(S, \nu)$.

2. [Page 341] Paul Lévy (1886–1971), French mathematician especially noted for his work in probability theory.

Bibliography

[1] R. G. Bartle, *Elements of Integration*, Wiley, 1966, New York.

[2] S. K. Berberian, *Measure and Integration*, Macmillan, New York 1965.

[3] A. Berger, *Chaos and Chance*, de Gruyter, New York 2001.

[4] A. Berger and T. P. Hill, *Newton's method obeys Benford's Law*, Amer. Math. Monthly, **114** (2007), 588–601.

[5] G. D. Birkhoff, *Proof of the ergodic theorem*, Proc. Nat. Acad. Sci. USA, **17** (1931), 656–660.

[6] J. R. Blum and J. I. Rosenblatt, *On the moments of recurrence time*, Journ. Math. Sci. (Delhi), **2** (1967), 1–6.

[7] J. Boyle, *An application of Fourier series to the most significant digit problem*, Amer. Math. Monthly, **102** (1994), 879–886.

[8] D. M. Bressoud, *A Radical Approach to Lebesgue's Theory of Integration*, Cambridge University Press and The Mathematical Association of America, Washington DC, 2008.

[9] J. R. Brown, *Ergodic Theory and Topological Dynamics*, Academic Press, New York 1976.

[10] C. Carathéodory, *Über das lineare Mass von Punktmengen: eine Verallgemeinerung des Längenbegriffs*, Nachr. Gesell. Wiss. Göttingen, (1914), 404–426.

[11] K. Dajani and C. Kraaikamp, *Ergodic Theory of Numbers*, Carus Mathematical Monographs 29, The Mathematical Association of America, Washington DC, 2002.

[12] P. Diaconis, *The distribution of leading digits and uniform distribution* mod 1, Proc. Amer. Phil. Soc., **78** (1978), 551–572.

[13] P. and T. Ehrenfest, *Enclykopädie de mathematicschen Wissenschaften*, volume IV 2 II, No.6, Teubner, Leibzig 1912; translated as *The Conceptual Foundations of the Statistical Approach in Mechanics*, by M. J. Moravcsik, Cornell University Press, New York 1959.

[14] R. M. Fewster, *A simple explanation of Benford's Law*, The American Statistician, **63** (2009), 26–32.

[15] H. Furstenburg, *Recurrence in Ergodic Theory and Combinatorial Number Theory*, Princeton University Press, Princeton 1981.

[16] A. M. Garsia, *A simple proof of E. Hopf's maximal ergodic theorem*, Journ. Math. and Mech., **14** (1965), 381–382.

[17] P. R. Halmos, *Measure Theory*, van Nostrand, Princeton 1950.

[18] P. R. Halmos, *Lectures on Ergodic Theory,* Chelsea, New York 1956.

[19] G. H. Hardy and E. M. Wright, *An Introduction to the Theory of Numbers*, 5th Edition, Oxford University Press, Oxford, 2000.

[20] E. Hewitt and K. Stromberg, *Real and Abstract Analysis*, Springer Verlag, New York, 1965.

[21] T. P. Hill, *The significant-digit phenomenon*, Amer. Math. Monthly, **102** (1995), 322–327.

[22] M. Kac, *On the notion of recurrence in discrete stochastic processes*, Bull. Amer. Math. Soc., **53** (1947), 1002–1010.

[23] S. Kakutani, *Induced measure-preserving transformations*, Proc. Acad. Tokyo, **16** (1943), 635–641.

[24] S. Kakutani, *Examples of ergodic measure-preserving transformations which are weakly mixing but not strongly mixing*, in Recent Advances in Topological Dynamics, A. Beck (Ed.), Springer, New York, 1973.

[25] P. W. Kasteleyn, *Variations on a theme by Marc Kac*, Journ. Stat. Phys., **46**(1987), 811–827.

[26] A. Katok and B. Hasselblatt, *Modern Theory of Dynamical Systems*, Encyclopaedia of Mathematics and its Applications 54, Cambridge University Press, Cambridge 1995.

[27] A. Koeller, R. Nillsen and G. Williams, *Weakly mixing transformations and the Carathéodory definition of measurable sets*, Colloquium Mathematicum, **108** (2007), 317–328.

[28] C. S. Kubrusly, *Measure Theory: a first course*, Elsevier/Academic Press, Amsterdam 2007.

[29] J. Lüroth, *Ueber eine einedeutige entwickelung von zahlen in eine unenliche reihe*, Math. Annalen, **21**(1883), 411–423.

[30] S. Newcomb, *Note on the frequency of use of the different digits in natural numbers*, Amer. Journ. Math., **4** (1881), 39–40.

Bibliography 349

[31] R. Nillsen, *Irrational rotations motivate measurable sets*, Elem. der Mathematik, **56** (2001) 1–17.

[32] R. Nillsen, *Statistical behaviour in dynamical systems*, Lecture notes, University of Wollongong, 2006.

[33] K. Petersen, *Ergodic Theory*, Cambridge Studies in Advanced Mathematics 2, Cambridge University Press, 1983.

[34] R. Raimi, *The first digit problem*, Amer. Math. Monthly, **83** (1976), 521–538.

[35] J. J. Sánchez-Gabites, *A brief survey of discrete dynamical systems*, Invited address at the conference on Topological Groups: Introduction to Dynamical Systems, Faculty of Mathematics, Universidad Complutense de Madrid, 3rd–5th April, 2008.

[36] Ya G. Sinai, *Introduction to Ergodic Theory*, translated by V. Scheffer from the original book in Russian published by Erevan State University, USSR 1973, Princeton University Press, Princeton 1977.

[37] K. R. Stromberg, *An Introduction to Classical Real Analysis*, Wadsworth, Belmont, 1981.

[38] J. von Neumann, *Proof of the quasi-ergodic hypothesis*, Proc. Nat. Acad. Sci. USA, **18** (1932), 70–82.

[39] P. Walters, *An Introduction to Ergodic Theory*, Springer-Verlag, Berlin, 1982.

Index of Subjects

additive set function, 116–117, 185, 298
    invariant, 185, 210, 230
algebra of sets, 293, 297, 298
algorithmic complexity, 229
almost all, 122, 249, 251, 299
averaging, 10, 144, 164, 290, 323
base, 92–93, 94, 95, 203
    expansion to a base, 95, 98, 100
basic sets, 107, 109–110
    length of, 114
Benford dynamical systems, 202, 205,
Benford's Law, 92, 184–185
    Benford's data, 184
    Benford's formulas, 183, 185, 198
    relative to a base, 203
    uniqueness of, 211, 218–219
binary function or transformation, 237, 241
    is length preserving, 247
Birkhoff's Ergodic Theorem, 175, 293, 319, 320–323, 342
    as a refinement of recurrence, 292–293
Borel sets, 308, 315, 321
Borel system, 241, 267
Borel's Theorem, 144, 153–154, 175, 228
Borel-Cantelli Lemma, 164
Cantor set, 121, 122, 221, 299
Carathéodory definition, 304–305, 306
Chaitin constant, 230

chaos, 47, 70, 73, 74, 75
characteristic function, 106, 290
coin tossing, 88–89, 92–93, 143–144, 163–164, 236–237, 269
continued fractions, 293, 336, 341, 342
countable additivity property, 298
countable set, 11, 119, 120
dense set, 26–27, 40, 46, 51, 72
Diaconis' Theorem, 195
digit(s), 92, 93–94, 143
    leading significant digit(s), 181, 185, 188–189, 195, 198
dynamical system(s), 1–2, 41, 71
    and Borel's Theorem, 148–149, 152, 153–154
    and information loss and gain 8, 9
    and Kronecker's Theorem, 42, 45–46
    and Weyl's Theorem, 74, 75
    associated with Benford's Law, 202
    chaotic, 73
    conjugate, 156–157
    random, 9–10, 27, 70, 164, 179,
    replication in, 4–5, 7
    sensitive to initial conditions, 71, 72, 73
    stationary, 240, 307
    transitive, 71, 72
ergodic system, 313–314, 319

351

events, 87, 106–107, 126
  independent, 122–124
  probability of, 88, 110
expansion to a base b, 92, 93, 95
  binary expansion, 95
  eventually periodic expansion, 105–106
falsification, 22
fixed point, 3, 8, 252, 254, 294
fractional part, 28, 29
function(s), 1
  composition of, 1
  copy or replica of, 5
  domain and codomain of, 1, 2
  inverse function, 3
  one-to-one, onto and range of, 2–3
  piecewise linear, 243–244
  step function, 11–12
Gelfand's problem, 186
group, 226–227, 230
Heine-Borel theorem, 11, 259
independent events, 122–124
information gain and loss, 8–9
initial state, 2
integer part, 28
invariant set, 300, 301, 302, 303, 313
inverse image as a set of "causes", 300
irrational numbers, 25–26
iteration, 2
Kac's formula, 238, 257, 261, 264–265, 266
Kakutani's example, 331, 332
Kolmogorov complexity, 229
Kronecker system, 42, 45–46
Kronecker's Theorem, 34-35, 36–37, 54
law, statistical, 209
  and invariance, 210, 211, 217–218
leading significant digit(s), 185, 188–189, 195, 198
  formulas for, 183, 184

length 10, 53, 114
  uniqueness of, 213
  extension to outer measure, 306
Lévy's Theorem, 341
logarithmic law, 92, 182, 185, 209, 211
  and uniform distribution, 182, 185, 186
  uniqueness of, 218–219
logarithmic length, 116, 117, 185, 202, 203, 207
  uniqueness of, 217–218
Lüroth transformation(s), 332, 335, 342
mantissa, 182, 199, 205, 207, 208
  formula for, 200
mathematics
  Artin's view of, xvi
  characteristics of, 12–15
  counterexamples in, 15
  moral function of, 16–17
  status of, 13
  Weyl's view of, 13
measure, 298, 303, 304, 305
  Lebesgue, 306, 329
measurable set(s), 291, 303, 305–306
  form a σ-algebra, 307
  on the definition of, 304–305
measure zero, 117–118, 119, 122, 305
music
  of J. S. Bach, 49
  of Steve Reich, 46–47
Newcomb's observations, 180–182, 185, 186, 191, 198
normal numbers, 155
  simply normal numbers, 144, 155
  to base b, 155
  to base two, 144
Normal Numbers Theorem, 159, 164, 169, 172, 178, 179
numbers, 10
rational numbers, 10,
  countability of rationals, 11

Index of Subjects 353

open covering, 259
orbit(s), 2, 3, 90–91, 290, 292, 294
   and recurrence, 249, 252, 255
   in Benford systems, 206
   in conjugate systems, 156, 206
   in Kronecker systems, 27, 41, 46
   in the Borel system, 153–154
outer measure, 295, 296, 297–298, 303
partition, 10, 33
periodic expansion, 102, 105–106
periodic sequence, 32, 101, 104
periodic point, 223, 254
Pigeon-hole Principle, 33–34, 36
probability, 87, 88, 90
   and independence, 122–124
   and Kolmogorov, 173
   and von Mises, 173–174
   and sets of measure zero, 117
   frequency interpretation of, 87, 88, 89-90, 154, 174–175
   notions of, 110, 111, 154, 172, 173–175
proof, 12–13, 16–17, 20–21
Rademacher functions, 136–137, 138, 153, 155, 164
   and self replication, 139
randomness, 10, 92, 126, 143, 148, 164
   and independence of the past, 239
   and levels of refinement, 178, 179
   and Weyl's Theorem, 53
recurrence, 89–90, 147, 154, 156, 248–249
   and Poincaré, 249–250, 253, 255
   Recurrence Theorem, 130–131, 152, 172
recurrent points 254, 255, 256
recurrence times, 256–257, 258, 262
   average of, see Kac's formula
   standard deviation of, 271, 275, 276, 278–281
replication and self replication 4–6, 8, 139, 153
rotation(s), 42, 43, 45, 59, 60, 212, 305
   ergodicity of, 317
   equivalent to translation, 42, 43
   irrational rotations, 315, 317
sample space, 87, 88, 89, 91, 123
scale, change of, 96, 139, 148, 149, 198, 202, 208–209, 211
scale invariance, 185
sensitive to initial conditions, 72–73, 75
sequence(s), 11
   binary, 92
   eventually periodic 101, 104
   periodic, 27, 32, 101
   summable, 11
sets, 10-11
   and sample spaces, 87
   considered as events, 87–89, 110–111, 117
   countable, 11
   independence of, 122–124
   invariant, see invariant set
   of measure zero, see measure zero
σ-algebra of sets, 307
space average, 318, 319
standard deviation, 265, 270, 271
   graphs of, 280, 281
step function, 11–12
time average, 318, 319
transformation(s), 1
   ergodicity of, 313
   iteration of, 2
   length preserving, 239, 240, 241–242, 245–246, 247, 249, 307
   measure preserving, 308, 309, 315
   mixing properties of, 328–329
   uniquely ergodic, 230
translation(s) 41–42, 45, 205, 211, 212
   and sensitivity, 73–75

and weak mixing, 331
composition of, 42–43, 45
correspond to rotations, 42–43, 59
ergodicity of, 315–316, 320
invariant sets of, 305
irrational, 291, 305
length preserving property, 248
motivate Carathéodory's definition, 305

transitivity 71–72, 74

Turing machine, 228, 229

uniform distribution, 27, 52, 70, 182, 186, 205, 320

uniquely ergodic, 230

Walsh functions, 155, 158, 159–160

Weyl's Theorem, 27, 53, 54, 57, 69, 70, 323
connection with Benford's Law, 182, 185, 191, 208
ideas in the proof of, 57, 64

Index of Symbols

$x \in X$	x belongs to the set X, 1
$A \cup B$	union of sets A, B, 10
$A \cap B$	intersection of sets A, B, 10
$A \subseteq B$	denotes A is a subset of B, 10
$A \times B$	Cartesian product of sets A, B, 10
$\emptyset$	the empty set, 10
A^c	complement of set A, 10
$|A|$	number of elements in the finite set A, 10
$\mathbb{N}$	the set of positive integers or natural numbers, 10
$\mathbb{Z}$	the set of integers, 10
$\mathbb{Z}_+$	the set of non-negative integers, 10
$\mathbb{Q}$	the set of rational numbers, 10
$\mathbb{R}$	the set of real numbers, 10
$\mathbb{C}$	the set of complex numbers, 10
$\mathbb{T}$	the circle group of complex numbers of modulus 1, 30
$f : A \longrightarrow B$	function mapping domain A into codomain B, 1, 2
$f\|A$	restriction of function f to A, 3
$g \circ f$	composition of function g with function f, 1
f^n	nth iterate of transformation f, 2
$f(A)$	direct image of the set A under function f, 3
f^{-1}	inverse of function f, 3
$f^{-1}(A)$	inverse image of the set A under function f, 3
$\square$	end of a proof, 12
$\Longrightarrow$	implication, 14
$\Longleftrightarrow$	equivalence of statements, 14
$\sim P$	negation of statement P, 15
$\bigcup_{n=1}^{\infty} A_n$	union of the sets $A_1, A_2, \ldots$, 118
$\bigcap_{n=1}^{\infty} A_n$	intersection of the sets $A_1, A_2, \ldots$, 130

Index of Symbols

$\|x\|$	absolute value or modulus of x, 11
$\lim_{n\to\infty} x_n$	limit of the sequence (x_n), 11
$\liminf_{n\to\infty} x_n$	lower limit or liminf of sequence (x_n), 321
$\limsup_{n\to\infty} x_n$	upper limit or limsup of sequence (x_n), 321
$\sum_{n=1}^{\infty} x_n$	sum of the summable sequence (x_n), 11
$\mathrm{frac}(x)$	fractional part of x, 28
$\mathrm{int}(x)$	integer part of x, 28
τ_α	translation by α, 41
ρ_α	rotation anti-clockwise by $2\pi\alpha$, 43
χ_A	characteristic function of the set A, 106, 290
$\mathcal{B}$	algebra of basic sets, or an algebra of sets, 107
$\mu_v(A)$	v-length of the basic set A, 114
$\mu(A)$	length of the basic set A, 10, 114
$\int_S f d\mu$	integral of the step function f over the interval S, 138
r_n	the nth Rademacher function on $[0, 1]$, 136
$d_n(x)$	the nth digit in the expansion of x to a base, 94
w_n	the nth Walsh function, 160
$\mathrm{dig}_j(x)$	jth significant digit of x, 189
$\mathrm{dig}(x)$	first or leading significant digit of x, 188–189
$\log(x)$	the natural logarithm of x, 188
$\log_b(x)$	log to the base b of x, 116, 188
T_a	transformation in a Benford system 201, 202
λ	logarithmic length to base b, 116, 203
$\theta_U(x)$	recurrence time of the point x in U, 258
v	an additive set function, 297
v_{out}	outer measure associated with the additive set function v, 2
$A \triangle B$	symmetric difference of sets A, B, 313
$\mathcal{M}$	family of measurable sets, or a σ-algebra, 306, 308

About the Author

Rodney Nillsen was born in Launceston, Tasmania, and received his undergraduate education at the University of Tasmania. His post-graduate studies were at the Flinders University of South Australia, under Igor Kluvánek. He has held positions at the Flinders University of South Australia, the Royal University of Malta, the Open University, and the then University College of Swansea in Wales, UK. Upon his return to Australia, he took up a position at the University of Wollongong, New South Wales, where he continues to teach and research. He is a member of the Australian Mathematical Society (member of the Council of the Society 2001-2003), The Mathematical Association of America, and the American Mathematical Society. His primary research area has been harmonic analysis, but he has also published in functional analysis, differential equations and measure theory. He is the author of *Difference spaces and invariant linear forms* (Lecture Notes in Mathematics 1586, Springer Verlag 1994), in which was introduced a new class of Hilbert spaces associated with the behavior of the Fourier transform near the origin. He received a Doctor of Science degree from the University Tasmania in 2000. In 2010, he received a citation from the Australian Learning and Teaching Council in recognition of his teaching. He has interests in the application of elementary mathematics to social issues, and in higher education policy. He enjoys reading, classical music and the Australian bush. He is married with two children. Further information is on his website at www.uow.edu.au/~nillsen.